Student Solutions Manual

to accompany

Calculus for Business, Economics, and the Social and Life Sciences

Seventh Edition

Laurence D. Hoffmann
Salomon Smith Barney

Gerald L. Bradley
Claremont McKenna College

Prepared by
Henri Feiner
*West Los Angeles College,
California Academy of Mathematics and Science, Long Beach City College*

Boston   Burr Ridge, IL   Dubuque, IA   Madison, WI   New York   San Francisco   St. Louis
Bangkok   Bogotá   Caracas   Lisbon   London   Madrid
Mexico City   Milan   New Delhi   Seoul   Singapore   Sydney   Taipei   Toronto

McGraw-Hill Higher Education
A Division of The McGraw-Hill Companies

Student Solutions Manual to accompany
CALCULUS FOR BUSINESS, ECONOMICS, AND THE SOCIAL AND LIFE SCIENCES,
SEVENTH EDITION

Copyright © 2000, 1996, 1992, 1989 by The McGraw-Hill Companies, Inc. All rights reserved.
Printed in the United States of America. Except as permitted under the United States Copyright Act of 1976, no part of this publication may be reproduced or distributed in any form or by any means, or stored in a database or retrieval system, without the prior written permission of the publisher.

1 2 3 4 5 6 7 8 9 0 QPD/QPD 9 0 3 2 1 0 9

ISBN 0-07-229151-6

www.mhhe.com

Table of Contents

Chapter 1   **Functions, Graphs, and Limits** 1
    Section 1.1   Functions 1
    Section 1.2   The Graph of a Function 6
    Section 1.3   Linear Functions 12
    Section 1.4   Functional Models 16
    Section 1.5   Limits 21
    Section 1.6   Continuity 22
                    Review Problems 24

Chapter 2   **Differentiation: Basic Concepts** 37
    Section 2.1   The Derivative: Slope and Rates 37
    Section 2.2   Techniques of Differentiation 41
    Section 2.3   The Product and Quotient Rules 45
    Section 2.4   Marginal Analysis: Approximation by Increments 48
    Section 2.5   The Chain Rule 52
    Section 2.6   The Second Derivative 55
    Section 2.7   Implicit Differentiation and Related Rates 58
                    Review Problems 61

Chapter 3   **Additional Applications of the Derivative** 73
    Section 3.1   Increasing and Decreasing Functions 73
    Section 3.2   Concavity 78
    Section 3.3   Limits Involving Infinity: Asymptotes 84
    Section 3.4   Optimization 90
    Section 3.5   Practical Optimization 94
                    Review Problems 100

Chapter 4   **Exponential and Logarithmic Functions** 115
    Section 4.1   Exponential Functions 115
    Section 4.2   Logarithmic Functions 118
    Section 4.3   Differentiation of Logarithmic and Exponential Functions 121
    Section 4.4   Additional Exponential Models 124
                    Review Problems 130

Chapter 5   **Integration** 141
    Section 5.1   Antidifferentiation: The Indefinite Integral 141
    Section 5.2   Integration by Substitution 144
    Section 5.3   Introduction to Differential Equations 148
    Section 5.4   Integration by Parts 151
                    Review Problems 154

Chapter 6   **Further Topics in Integration** 163
    Section 6.1   Definite Integration 163
    Section 6.2   Applications to Business and Economics 165
    Section 6.3   Additional Applications of Definite Integration 172
    Section 6.4   Improper Integrals 177
                    Review Problems 181

Chapter 7   **Calculus of Several Variables** 193
    Section 7.1   Functions of Several Variables 193
    Section 7.2   Partial Derivatives 197
    Section 7.3   Optimizing Functions of Two Variables 201
    Section 7.4   Constrained Optimization: The Method of Lagrange Multipliers ... 208
    Section 7.5   Double Integrals over Rectangular Regions 214
                    Review Problems 217

Chapter 1

Functions, Graphs, and Limits

1.1 Functions

1.
$$f(x) = 3x^2 + 5x - 2,$$
$$f(1) = 3(1^2) + 5(1) - 2 = 6,$$
$$f(0) = 3(0^2) + 5(0) - 2 = -2,$$
$$f(-2) = 3(-2)^2 + 5(-2) - 2 = 0.$$

3.
$$g(x) = x + \frac{1}{x},$$
$$g(-1) = -1 + \frac{1}{-1} = -2,$$
$$g(1) = 1 + \frac{1}{1} = 2,$$
$$g(2) = 2 + \frac{1}{2} = \frac{5}{2}.$$

5.
$$h(t) = \sqrt{t^2 + 2t + 4},$$
$$h(2) = \sqrt{2^2 + 2(2) + 4} = 2\sqrt{3} \approx 3.46,$$
$$h(0) = \sqrt{0^2 + 2(0) + 4} = 2,$$
$$h(-4) = \sqrt{(-4)^2 + 2(-4) + 4} \approx 3.46.$$

7.
$$f(t) = (2t-1)^{-3/2} = \frac{1}{(\sqrt{2t-1})^3},$$
$$f(1) = \frac{1}{[\sqrt{2(1)-1}]^3} = 1,$$
$$f(5) = \frac{1}{[\sqrt{2(5)-1}]^3} = \frac{1}{[\sqrt{9}]^3} = \frac{1}{27},$$
$$f(13) = \frac{1}{[\sqrt{2(13)-1}]^3} = \frac{1}{[\sqrt{25}]^3} = \frac{1}{125}.$$

9.
$$f(x) = x - |x-2|,$$
$$f(1) = 1 - |1-2| = 1 - |-1| = 1 - 1 = 0,$$
$$f(2) = 2 - |2-2| = 2 - |0| = 2,$$
$$f(3) = 3 - |3-2| = 3 - |1| = 3 - 1 = 2.$$

11.
$$f(t) = \begin{cases} 3 & \text{if } t < -5 \\ t+1 & \text{if } -5 \leq t \leq 5 \\ \sqrt{t} & \text{if } t > 5 \end{cases}$$
$$f(-6) = 3,$$
$$f(-5) = -5 + 1 = -4,$$
$$f(16) = \sqrt{16} = 4.$$

13.
$$g(x) = \frac{x^2 + 5}{x + 2}.$$

Since $x + 2 \neq 0$,
the domain consists of all real numbers such that $x \neq -2$.

15.
$$y = \sqrt{x - 5}.$$

Radicands cannot be negative when the index is even (in the real number system), so $x - 5 \geq 0$, or $x \geq 5$.

17.
$$g(t) = \sqrt{t^2 + 9}.$$

Since the radicand is not negative, any real number t is valid.

19.
$$f(t) = (\sqrt{t^2 - 9})^{-1/2} = \frac{1}{\sqrt{t^2 - 9}}.$$

Radicands cannot be negative, denominators must not vanish (become 0), so $t^2 > 9$, or $|t| > 3$.
The domain consists of real numbers $|t| > 3$.

21.
$$g(t) = \frac{1}{|t - 1|}.$$

Since denominators cannot become 0, $|t - 1| \neq 0$, $t - 1 \neq 0$, or $t \neq 1$.

23. $f(u) = 3u^2 + 2u - 6$ and $g(x) = x + 2$, so

$$\begin{aligned} f[g(x)] &= f(x + 2) = 3(x + 2)^2 + 2(x + 2) - 6 \\ &= 3x^2 + 14x + 10. \end{aligned}$$

25.
$$\begin{aligned} f(u) &= (u - 1)^3 + 2u^2 \\ g(x) &= x + 1 \end{aligned}$$

so $\begin{aligned} f[g(x)] &= f(x + 1) \\ &= [(x + 1) - 1]^3 + 2(x + 1)^2 \\ &= x^3 + 2x^2 + 4x + 2. \end{aligned}$

27.
$$f(u) = \frac{1}{u^2}$$

and $g(x) = x - 1$,

so $f[g(x)] = f(x - 1) = \dfrac{1}{(x - 1)^2}$.

29.
$$f(u) = \sqrt{u + 1}$$

and $g(x) = x^2 - 1$,

so
$$\begin{aligned} f[g(x)] &= f(x^2 - 1) \\ &= \sqrt{(x^2 - 1) + 1} \\ &= \sqrt{x^2} = |x|. \end{aligned}$$

31.
$$f[g(x)] = f(x^2 + x - 1) = \frac{2}{x^2 + x - 1}$$

$$g[f(x)] = g\left(\frac{2}{x}\right) = \frac{4}{x^2} + \frac{2}{x} - 1 = \frac{-x^2 + 2x + 4}{x^2}$$

To solve $\dfrac{2}{x^2 + x - 1} - \dfrac{-x^2 + 2x + 4}{x^2} = 0$ requires solving

$$x^4 - x^3 - 5x^2 - 2x + 4 = 0$$

A calculator/computer graphing utility reveals $x \approx 0.705$ and $x \approx 2.836$.

33.
$$f[g(x)] = f\left(\frac{x + 3}{x - 2}\right) = \frac{2(\frac{x+3}{x-2}) + 3}{\frac{x+3}{x-2} - 1} = x$$

$$g[f(x)] = g\left(\frac{2x + 3}{x - 1}\right) = \frac{2(\frac{x+3}{x-1}) + 3}{\frac{2x+3}{x-1} - 2} = x$$

To solve $f[g(x)] = g[f(x)]$ for all $x \neq 1$ and $x \neq 2$ for wich $f(x)$ and $g(x)$ are not defined.

35.
$$\begin{aligned} f(x) &= 2x^2 - 3x + 1, \\ f(x - 2) &= 2(x - 2)^2 - 3(x - 2) + 1 \\ &= 2x^2 - 11x + 15. \end{aligned}$$

37.
$$\begin{aligned} f(x) &= (x + 1)^5 - 3x^2, \\ f(x - 1) &= [(x - 1) + 1]^5 - 3(x - 1)^2 \\ &= x^5 - 3x^2 + 6x - 3. \end{aligned}$$

39.
$$\begin{aligned} f(x) &= \sqrt{x}, \\ f(x^2 + 3x - 1) &= \sqrt{x^2 + 3x - 1}. \end{aligned}$$

41.
$$\begin{aligned} f(x) &= \frac{x - 1}{x}, \\ f(x + 1) &= \frac{(x + 1) - 1}{x + 1} \\ &= \frac{x}{x + 1}. \end{aligned}$$

1.1. FUNCTIONS

43.
$$f(x) = (x-1)^2 + 2(x-1) + 3$$
$$g(u) = u^2 + 2u + 3, \ h(x) = x - 1$$

45.
$$f(x) = \frac{1}{x^2 + 1}$$
can be rewritten as $g[h(x)]$

with $g(u) = \dfrac{1}{u}$

and $h(x) = x^2 + 1$

47.
$$f(x) = \sqrt[3]{2-x} + \frac{4}{2-x}$$
can be rewritten as $g[h(x)]$ with
$$g(u) = \sqrt[3]{u} + \frac{4}{u}$$
and $h(x) = 2 - x$

49. $f(x) = -x^3 + 6x^2 + 15x$

(a) $f(2) = -8 + 6 \times 4 + 15 \times 2 = 46$

(b)
$$f(1) = -1 + 6 + 15 = 20$$
$$f(2) - f(1) = 46 - 20 = 26.$$

51. $P(t) = 20 - \dfrac{6}{t+1}$

(a) $P(9) = 20 - \dfrac{6}{9+1}$ or 19,400 people.

(b)
$$P(8) = 20 - \frac{6}{8+1}.$$
$$P(9) - P(8) = 20 - \frac{3}{5} - \left(20 - \frac{2}{3}\right) = \frac{1}{15}$$

This accounts for about $\dfrac{1}{15}$ of 1,000 people or 67 people.

(c) Writing exercise — Answers will vary.

53.
$$S(r) = C(R^2 - r^2)$$
$$= 1.76 \times 10^5(1.2^2 \times 10^{-4} - r^2).$$

(a) $S(0) = 1.76 \times 10^5 \times 1.44 \times 10^{-4} = 25.344$ cm/sec.

(b)
$$S(0.6 \times 10^{-2})$$
$$= 1.76 \times 10^5(1.44 \times 10^{-4} - 0.6^2 \times 10^{-4})$$
$$= 1.76 \times 10^5(1.08 \times 10^{-4})$$
$$= 19.008 \text{ cm/sec}.$$

55.
$$f(x) = \frac{150x}{200 - x}.$$

(a) $x \ne 200$.

(b) $0 \le x \le 100$.
If $x < 0$ or $x > 200$ then $f(x) < 0$ but cost is not negative. $x > 100$ means more than 100%.

(c) $f(50) = \dfrac{150 \times 50}{200 - 50} = 50$ million dollars.

(d)
$$f(100) = \frac{150 \times 100}{200 - 100} = 150$$
$$f(100) - f(50) = 100 \text{ million dollars}.$$

(e)
$$\frac{150x}{200 - x} = 37.5.$$
$$187.5x = 37.5 \times 200,$$
$$x = \frac{7,500}{187.5} = 40\%.$$

57.

(a) Let T stand for tuition,
AF for activities fee,
CF for Course fees,
BM for book fees,
S for supplies.
1975:

T		2,000.00
AF		250.00
CF	(50)(0.30)	15.00
BM	(16)(10.5)	168.00
S		30.00
	Total:	2,463.00

CEI $= \dfrac{2,463}{2,463} = 1.000$.

1976:

T		2,150.00
AF		250.00
CF	(50)(0.31)	15.50
BM	(17.5)(10.6)	185.50
S		32.00
	Total:	2,633.00

$$\text{CEI} = \frac{2,633}{2,463} = 1.069.$$

1977:

T		2,300.00
AF		275.00
CF	(50)(0.31)	15.50
BM	(19)(10.5)	199.50
S		35.00
	Total:	2,825.00

$$\text{CEI} = \frac{2,825}{2,463} = 1.147.$$

1978:

T		2,500.00
AF		300.00
CF	(60)(0.3)	18.00
BM	(20.75)(10.4)	215.80
S		38.00
	Total:	3,071.80

$$\text{CEI} = \frac{3,071.80}{2,463} = 1.247.$$

1979:

T		2,800.00
AF		300.00
CF	(60)(0.3)	18.00
BM	(23)(10.4)	239.20
S		42.00
	Total:	3,399.20

$$\text{CEI} = \frac{3,399.20}{2,463} = 1.380.$$

1980:

T		3,200.00
AF		320.00
CF	(65)(0.29)	18.85
BM	(26.5)(10.4)	275.60
S		47.00
	Total:	3,861.45

$$\text{CEI} = \frac{3,861.45}{2,463} = 1.568.$$

1981:

T		3,600.00
AF		350.00
CF	(65)(0.28)	18.20
BM	(30)(10.4)	312.00
S		52.00
	Total:	4,332.2

$$\text{CEI} = \frac{4,332.2}{2,463} = 1.759.$$

1982:

T		4,000.00
AF		360.00
CF	(70)(0.28)	19.60
BM	(34)(10.4)	353.60
S		58.00
	Total:	4,791.2

$$\text{CEI} = \frac{4,791.20}{2,463} = 1.945.$$

(b) Writing exercise — Answers will vary.

(c) Writing exercise — Answers will vary.

59.
(a)
$$\begin{aligned} C(q) &= q^2 + q + 900 \\ \text{and } q(t) &= 25t, \\ \text{thus } C[q(t)] &= C(25t) \\ &= (25t)^2 + 25t + 900 \\ &= 625t^2 + 25t + 900. \end{aligned}$$

(b) For $t = 3$,
$$\begin{aligned} C[q(3)] &= 625(3)^2 + 25(3) + 900 \\ &= \$6,600. \end{aligned}$$

1.1. FUNCTIONS

(c)
$$625t^2 + 25t + 900 = 11,000,$$
$$625t^2 + 25t - 10,100 = 0.$$

Divide by 25 to get smaller numbers, then
$$25t^2 + t - 404 = (25t + 101)(t - 4)$$
$$= 0$$

or $t = 4$ hours.
Disregard $t = -\frac{101}{25}$.

61. For the HP48G:
$P(x) = 8x^4 + 8x^3 - x$.
To enter the function:
Turn on and (blue) CLEAR
' α P (blue) () X
α to turn off letter entry
right (blue) = 8 x α X
(do not mix upper and lower case variables names)
y^x 4 + 8 x α X y^x 3 - α X
ENTER
(blue) DEF VAR
Note P on the lower display bar.
Just as a quick test, enter 0 ENTER and press the white button under P.
$P(0) = 8 * 0^4 + 8 * 0^3 - 0 = 0$
so 0 should appear on stack level 1.
Similarly $P(1) = 8 * 1^4 + 8 * 1^3 - 1 = 15$, so enter 1 ENTER and 15 should appear on stack level 1,
and $P(-1) = 8 * (-1)^4 + 8 * (-1)^3 - (-1) = 1$, so enter 1 ± ENTER and −1 should appear on stack level 1.
This is merely a check to verify (in your own mind) that the function was entered properly. Now proceed with the problem at hand.

x	$P(x)$	x	$P(x)$
−4.0	1,540	−3.5	861
−3.0	435	−2.0	66
−1.0	1	1.0	15
1.5	66	2.0	190

For the TI-85:
Press 2nd CALC, then F1 for EVALF
Fill in using x from x-var after "evalf("
8x tiny $\wedge$ 4 + 8x $\wedge$ 3 − x, x, −4) Enter

Press 2nd ENTRY, then backspace over the 4, replace it with -3.5, Enter.
Note: the negative sign is from the (−) key just to the left of ENTER, not the subtractionkey to the right of 6.
Repeat the above procedure for the remaining values of x.

63. For the HP48G:
$$y = \frac{4x^2 - 3}{2x^2 + x - 3}.$$
Let's use the HP48G to graph y.
Turn it ON and CLEAR.
(green) plot.
'(blue) () 4 x α X y^x 2 - 3 right
÷ (blue) () 2 x αX $y \wedge x$2 + αX − 3 ENTER
Note the fraction displayed as
'(4*X^2−3)/(2*X^2+X−3)
place CHK in autoscale (white buttons, lower bar)
ERASE and DRAW on lower bar (and be patient)
The graph shows a discontinuity around $x = -1.5$ and $x = 1.1$.
Let's be more precise. The denominator cannot be 0, so plot $2x^2 + x - 3$.
CLEAR. (green) PLOT '2 * $\alpha X \wedge 2 + \alpha X - 3$,
view X from −2 to 2 and Y from −1 to 1.
Press the white buttons below ERASE and DRAW.
The parabola is partially drawn. Press TRACE and (X,Y). Move the arrow keys till $x \approx -1.5$ is about 1 and $y \approx 0$. Also $y \approx 0$ when $x \approx -1.5$.
Mathematically,
$2x^2 + x - 3 = (2x + 3)(x - 1) = 0$ at $x = 1$ and $x = -1.5$.

For the TI-85:
Press GRAPH. Then F1 for $y1 =$
Then enter $(4x \wedge 2 - 3)(2x \wedge 2 + x - 3)$.
Press 2nd F5 for graphing.
Press F3 to ZOOM. Move the cross-hair to the vertical line on the x axis. This is the value of the vertical asymptote, $x \approx -1.44$.
There is another asymptote at $x \approx 1$.

65. For the HP48G:
$$f(x) = 2\sqrt{x-1},$$

$$g(x) = x^3 - 1.2.$$

Store functions $f(x)$ and $g(x)$ in the HP48G just like we stored $P(x)$ in problem **61**.
2.3 ENTER (place 2.3 on the stack at level 1)
Press G on the lower bar to get
$g(2.3) = 10.967$
(note that $g(2) = 2^3 - 1.2 = 6.8$ and
$g(3) = 3^3 - 1.2 = 25.8$, so 10.967 is a ballpark figure — just as a check.)
Press f on the lower bar to get
$f(10.967) = 6.31$
rounded to 2 decimal places. As a quick mental check, $f(10) = 2\sqrt{10-1} = 6$.

For the TI-85: Press 2nd CALC, then F1 for EVALF
Enter $x \wedge 3 - 1.2, x, 2.3)$
Enter to get 10.97
Press F1 again to evaluate
$2\sqrt{(x-1)}, x$, 2nd ANS)
Enter generates 6.31

1.2 The Graph of a Function

1.
$$f(x) = x.$$

A function of the form
$$y = f(x) = ax + b$$
is a linear function, that is its graph is a straight line. Two points are sufficient to draw that line. Note that $f(0) = 0$ and $f(1) = 1$.

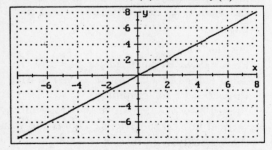

3.
$$f(x) = x^3.$$

Note that if $x > 0$ then $f(x) > 0$ and if $x < 0$, then $f(x) < 0$.

The curve will only appear in the first and third quadrants.
Since x^3 and $(-x)^3$ have the same absolute value, only their signs are opposites, the curve will be symmetric with respect to (WRT) the origin.

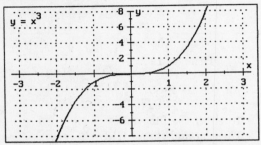

Thus only the portion of the curve in the first quadrant will need to be calculated. The portion in the third quadrant will follow automatically.
Also note that $f(x) \to \infty$ as $x \to \infty$.

x	0	1	2	3	5
$f(x)$	0	1	8	27	125

5.
$$f(x) = -x^3 + 1.$$

Note the similarities between this graph and the one in exercise 3. The y-values here are the negatives of those in 3 and the curve is translated (moved up) by 1 unit.

x	0	1	2	3	5
$f(x)$	1	0	-7	-26	-124

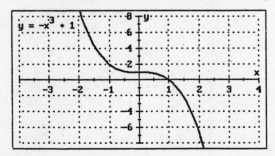

7.

1.2. THE GRAPH OF A FUNCTION

$$f(x) = 2 - 3x$$

Note that the graph is a straight line. The slope is -3. The curve falls.

x	0	$\frac{2}{3}$	2
$f(x)$	2	0	-4

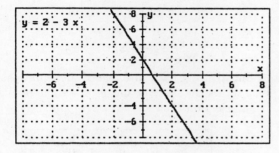

9. $f(x) = \begin{cases} x - 1 & \text{if } x \le 0 \\ x + 1 & \text{if } x > 0 \end{cases}$

Note that the graph consists of two half lines on either side of $x = 0$. $f(0) = -1$. There is no x-intercept.

x	-3	-2	-1	0	1	2	3
$f(x)$	-4	-3	-2	-1	2	3	4

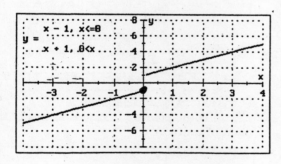

11.
$$f(x) = (x-1)(x+2).$$

The x-intercepts are $x = 1$ and $x = -2$. The y-intercept is $f(0) = -2$.

x	-3	-2	0	1	3
$f(x)$	4	0	-2	0	10

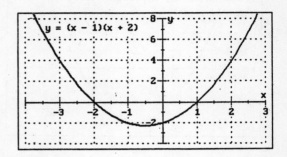

13.
$$\begin{aligned} f(x) &= -x^2 - 2x + 15 \\ &= -(x^2 + 2x - 15) \\ &= -(x+5)(x-3). \end{aligned}$$

The x-intercepts are $x = -5$ and $x = 3$. The y-intercept is $f(0) = 15$.

x	-7	-5	-3
$f(x)$	-20	0	12

x	-1	1	3	5
$f(x)$	16	12	0	-20

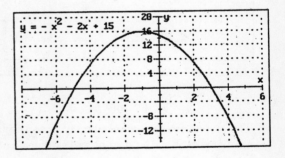

15.
$$\begin{aligned} f(x) &= 6x^2 + 13x - 5 \\ &= (3x-1)(2x+5). \end{aligned}$$

The x-intercepts are $x = -\frac{5}{2}$ and $x = \frac{1}{3}$. The y-intercept is $f(0) = -5$.

x	-3	-2	-1	0	1
$f(x)$	10	-7	-12	-5	14

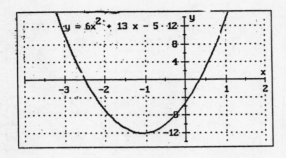

17.
$$y = 3x + 5$$
$$y = -x + 3.$$

Add 3 times the second equation to the first. Then $4y = 14$ or $y = \frac{7}{2}$. Substitute in the first, then $x = 3 - y = -\frac{1}{2}$. The point of intersection is $P\left(-\frac{1}{2}, \frac{7}{2}\right)$.

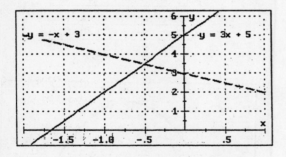

19. $y = 3x + 8$ and $y = 3x - 2$

The slopes are the same, namely 3, so the lines are parallel, but the y intercepts differ. These lines do not intersect.

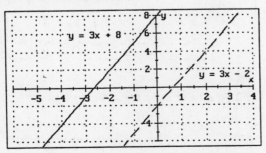

21.
$$y = x^2,$$
$$y = 6 - x.$$

The point(s) of intersection has (have) the same y-value. Thus
$$x^2 = 6 - x,$$
$$x^2 + x - 6 = 0,$$
$$(x+3)(x-2) = 0,$$
$$x_1 = -3, \qquad x_2 = 2.$$

Substituting back into one of the equations for y shows that $y_1 = 9$ and $y_2 = 4$. The points of intersection then are $P_1(-3, 9)$ and $P_2(2, 4)$.

x	-3	-1	0	2	3
$y = x^2$	9	1	0	4	9
$y = 6 - x$	9	7	6	4	3

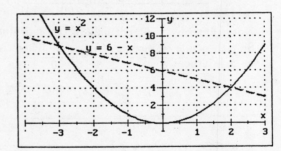

23. $y = x^2 - x$ and $y = x - 1$. Thus
$$x^2 - x = x - 1, \quad x(x-1) - (x-1) = 0$$
$(x-1)(x-1) = 0$, $x = 1$ and $y = 0$.
$P(1, 0)$ is the point of intersection (really the point of contact).

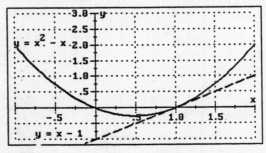

1.2. THE GRAPH OF A FUNCTION

25. The monthly profit is

$$\begin{aligned} P(x) &= \text{(number of recorders sold)} \\ &\quad \text{(price-cost)} \\ &= (120-x)(x-40) \end{aligned}$$

Thus the intercepts are at $(40, 0)$, $(120, 0)$, and $(0, -4800)$.

The graph suggests a maximum profit when $x \approx 80$, that is when 80 recorders are sold. Since

$$\begin{aligned} P(x) &= -x^2 + 160x - 4800 \\ &= -(x^2 - 160x + 80^2) + 6400 - 4800 \\ &= -(x-80)^2 + 1,600, \end{aligned}$$

This shows that $P(80) = \$1,600$.

x	0	20	40
$P(x)$	$-4,800$	$-2,000$	0

x	60	80	100	120
$P(x)$	1,200	1,600	1,200	0

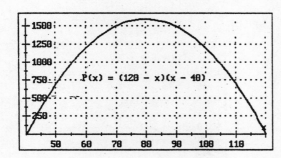

27.

(a)
$$\begin{aligned} D(p) &= -200p + 12,000 \\ &= -200(p - 60) \end{aligned}$$

where $0 \le p \le 60$.

p	0	20	40	60
$D(p)$	12,000	8,000	4,000	0

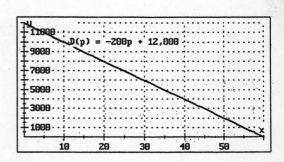

(b)
$$\begin{aligned} E(p) &= \text{(price per unit)(demand)} \\ &= -200p(p - 60) \end{aligned}$$

(c)

p	0	20	40	60
$E(p)$	0	160,000	160,000	0

(d) Writing exercise —
Answers will vary.

(e) The graph suggests a maximum expenditure when $p \approx 30$. Since

$$\begin{aligned} E(p) &= -200(p^2 - 60p) \\ &= -200(p^2 - 60p + 900) + 180,000 \\ &= -200(p - 30)^2 + 180,000 \end{aligned}$$

this shows that $E(30) = \$180,000$.

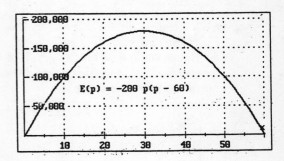

29. $C(x) = \dfrac{x^3}{6} + 2x + 5$

The average cost is

$$A(x) = \dfrac{x^2}{6} + 2 + \dfrac{5}{x}.$$

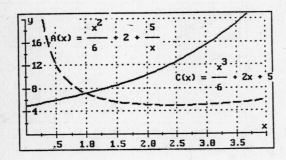

31.
$$S_1 = 4\pi r_1^2 \text{ so } r_1 = \sqrt{\frac{S_1}{4\pi}}$$
$$V_1 = \frac{4}{3}\pi r_1^3 = \frac{S_1^{3/2}}{6\sqrt{\pi}}$$

With $S_2 = \frac{4}{3}\pi r_2^2 = 2S_1$,

$$\frac{S_1}{S_2} = \frac{1}{2} = \frac{r_1^2}{r_2^2},$$

thus $r_2 = \sqrt{2}r_1$ and

$$V_2 = \frac{4}{3}\pi r_2^3 = \frac{4}{3}\pi 2^{3/2}r_1^3 = 2\sqrt{2}V_1$$

The volume is multiplied by $2\sqrt{2}$.

33.

(a) Each y–value for $y = -x^2$ is the negative of the corresponding y–value of $y = x^2$. Hence the points on the graph of $y = -x^2$ are reflections across the x–axis of the points of the graph of $y = x^2$.

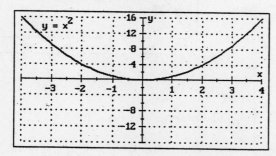

(b) If $g(x) = -f(x)$, the graph of $g(x)$ is the reflection across the x–axis of the graph of $f(x)$.

x	-2	-1	0	1	2
$y = -x^2$	-4	-1	0	-1	-4
$y = x^2$	4	1	0	1	4

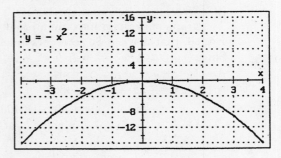

35.

(a)

x	2	5	7	10
$C(x)$	132	195	237	300

(b)
$$C(x) = 90 + 21x$$

(c)

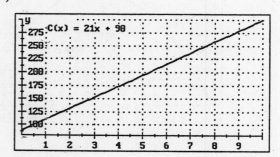

37. For the HP48G:
(green) PLOT. On the equation line type 'α X y^x 4 and ENTER. X is viewed on $[-2, 2]$ and y is viewed on $[-2, 5]$. Press ERASE and DRAW.
CANCEL. (blue) EDIT. Add $-\alpha$ X to the equation, then ENTER. DRAW (but don't press ERASE.)

1.2. THE GRAPH OF A FUNCTION

CANCEL. (blue) EDIT. Modify $-\alpha$ X to -2α X, then ENTER. DRAW (but don't press ERASE.)

CANCEL. (blue) EDIT. Modify -2α X to -3α X, then ENTER. DRAW (but don't press ERASE.)

Note: See problem 1.4.49 for plotting curves simultaneously in a more efficient manner.

$y = x^4$ is symmetric WRT the y axis. The minimum is at $(0,0)$.

When $y = -x$ is added, the curve looses symmetry and the low point shifts down and to the right. This shift becomes more pronounced for $y = -2x$ and $y = -3x$.

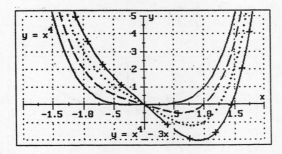

This should not be too surprising. The lines $y = -x$, $y = -2x$, and $y = -3x$ have positive y values (ordinates) which are added to x^4 over the negative x axis. Similarly negative y values are added over the positive y axis.

x	-2	-1	0	1	2
$-x$	2	1	0	-1	-2
$-2x$	4	2	0	-2	-4
$-3x$	6	3	0	-3	-6
x^4	16	1	0	1	16
$x^4 - x$	18	2	0	0	14
$x^4 - 2x$	20	3	0	-1	12
$x^4 - 3x$	22	4	0	-2	10

This is even more pronounced for $y = x^4 - x^3$, $y = x^4 - 2x^3$, and $y = x^4 - 3x^3$. Adjust the viewing window to $x:[-1,4]$ and $y:[-10,5]$.

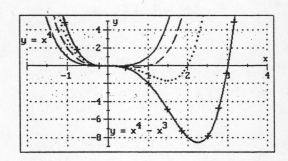

x	-2	-1	0	1	2
$-x^3$	8	1	0	-1	-8
$-2x^3$	16	2	0	-2	-16
$-3x^3$	24	3	0	-3	-24
x^4	16	1	0	1	16
$x^4 - x^3$	24	2	0	0	8
$x^4 - 2x^3$	32	3	0	-1	0
$x^4 - 3x^3$	40	4	0	-2	-8

For the TI-85:
Press GRAPH, then F1 for $y1 =$
Enter $x \wedge 4$ ENTER
Then $y2 = x \wedge 4 - x$ ENTER
$y3 = x \wedge 4 - 2x$ ENTER
$y4 = x \wedge 4 - 3x$ ENTER
2nd M2 for RANGE
Change XMIN to -2, XMAX to 2, and YMIN to -2 (-3 might be better).
2nd F5 to GRAPH.

39. Let $P(x_1, y_1)$ and $Q(x_2, y_2)$.
The point $R(x_2, y_1)$ is at the intersection of the horizontal line $y = y_1$ and the vertical line $x = x_2$.

$$PR = x_2 - x_1$$

and $QR = y_2 - y_1$

By the Pythagorean theorem,

$$d = \sqrt{(x_2 - x_1)^2 + (y_2 - y_1)^2}$$

(a) $\quad d = \sqrt{(5-2)^2 + (-1-3)^2} = 5$

(b) $\quad d = \sqrt{(2-2)^2 + (6+1)^2} = 7$

41.
$$y = Ax^2 + Bx + C$$
$$= A\left[x^2 + \frac{B}{A}x + \left(\frac{B}{2A}\right)^2\right] + C - \frac{B^2}{4A}$$
$$= A\left[\left(x + \frac{B}{2A}\right)^2 + \frac{C}{A} - \frac{B^2}{4A^2}\right].$$

If $x = -\frac{B}{2A}$

then $y = C - \frac{B^2}{4A}$,

a minimum if $A > 0$ or a maximum if $A < 0$.

43.
$$f(x) = \frac{8x^2 + 9x + 3}{x^2 + x - 1}$$

$f(x)$ is defined for all x except those where $x^2 + x - 1 = 0$ or approximately 0.6 and -1.6.

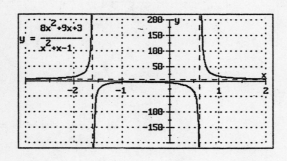

1.3 Linear Functions

1. For $P_1(2, -3)$ and $P_2(0, 4)$ the slope is
$$m = \frac{4 - (-3)}{0 - 2} = -\frac{7}{2}$$

3. For $P_1(2, 0)$ and $P_2(0, 2)$ the slope is
$$m = \frac{2 - 0}{0 - 2} = -1$$

5. For $P_1(2, 6)$ and $P_2(2, -4)$ the slope is
$$m = \frac{6 - (-4)}{2 - 2},$$
that is not defined, since the denominator is 0. The line through the given points is vertical.

7. $y = 3x$, $m = 3$,
y-intercept $b = 0$.

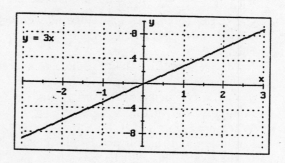

9. $y = 3x - 6$, $m = 3$,
y-intercept $b = -6$.

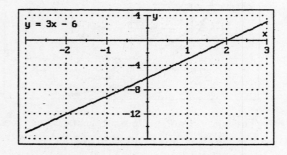

11. $3x + 2y = 6$ or $y = -\frac{3}{2}x + 3$
$m = -\frac{3}{2}$ and $b = 3$

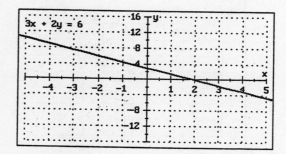

13. $5y - 3x = 4$ or $y = \frac{3}{5}x + \frac{4}{5}$
$m = \frac{3}{5}$ and $b = \frac{4}{5}$

1.3. LINEAR FUNCTIONS

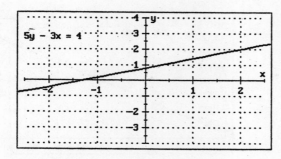

15. $\dfrac{x}{2} + \dfrac{y}{5} = 1$

or $y = -\dfrac{5}{2}x + 5$

$m = -\dfrac{5}{2}$ and $b = 5$

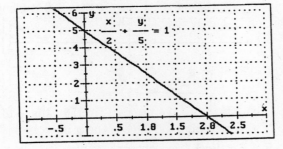

17. $x = -3$ The slope is not defined and there is no y intercept.

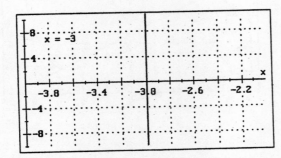

19. $m = 1$ and $P(2,0)$, so

$$y - 0 = (1)(x - 2) \text{ or } y = x - 2$$

21. $m = -\dfrac{1}{2}$ and $P(5, -2)$, so

$$y - (-2) = -\dfrac{1}{2}(x - 5) \text{ or } y = -\dfrac{1}{2}x + \dfrac{1}{2}$$

23. Since the line is parallel to the x–axis, it is horizontal and its slope is 0. For $P(2,5)$, the line is

$$y - 5 = 0(x - 2) \text{ or } y = 5$$

25. $$m = \dfrac{1 - 0}{0 - 1}$$

and for $P(1,0)$ the equation of the line is

$$y - 0 = -1(x - 1) \text{ or } y = -x + 1$$

The equation would be the same if the point $(0,1)$ had been used.

27. $$m = -\dfrac{1 - (1/4)}{-(1/5) - (2/3)} = -\dfrac{45}{52}$$

and $P\left(-\dfrac{1}{5}, 1\right)$, so

$$y - 1 = -\dfrac{45}{52}\left(x + \dfrac{1}{5}\right) \text{ or } 45x + 52y - 43 = 0$$

29. The slope is 0 because the y–values are identical. Thus

$$y = 5$$

31. The equation of the desired line is

$$2x + y = C$$

because the desired line and the given line have the same slope.
Since $(4, 1)$ belongs to the line,

$$2(4) + 1 = C = 9$$

and the desired equation is

$$2x + y = 9$$

33. The slope of the given line is $m = -1$. The perpendicular line has $m = 1$ as its slope. The equation of the desired line is

$$y = x + C$$

Since $(3, 5)$ belongs to the line,

$$5 = 3 + C$$

and the desired equation is

$$y = x + 2$$

35. Let x be the number of units manufactured. Then $60x$ is the cost of producing x units, to which the fixed cost must be added.

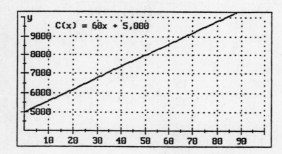

$$y = 60x + 5,000$$

37. (a) Let x be the number of hours spent registering students in person. During the first 4 hours $4 \times 35 = 140$ students were registered. Thus

$$360 - 140 = 220$$

students had pre-registered. Let y be the total number of students who register. Then

$$y = 35x + 220$$

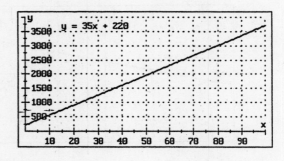

(b)

$$y = 3 \times 35 + 220 = 325$$

(c) From part (a), we see that 220 students had pre-registered.

39. The slope is
$$m = \frac{1,500 - 0}{0 - 10} = -150$$
Originally (when time $x = 0$) the value y of the books is 1500 (this is the y intercept).
$$y = -150x + 1,500$$

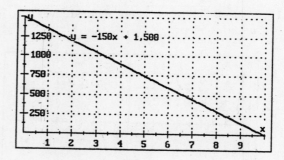

41. (a) Let x be the number of days. The slope is
$$m = \frac{200 - 164}{12 - 21} = -4$$
Thus $\dfrac{y - 200}{x - 12} = -4$
or $y = -4x + 248$

(b) $y = 248 - 4 \times 8 = 216$ million gallons.

1.3. LINEAR FUNCTIONS

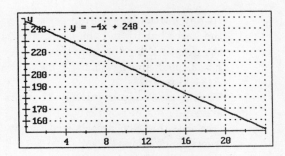

43. (a) Let x be the temperature in degrees Celsius and y the temperature in degrees Farenheit. The slope is
$$m = \frac{212 - 32}{100 - 0} = \frac{9}{5}$$
Thus $\frac{y - 32}{x - 0} = \frac{9}{5}$ or $y = \frac{9}{5}x + 32$

(b) $9 \times 15 + 160 = 5y$, $y = 59$ degrees F.

(c) $9x = 5 \times 68 - 160$, $x = 20$ degrees C.

45. (a) The original value of the book is $100 and the value doubles every 10 years. Thus

1900	1920	1940	1960	1980	2000
100	400	1,600	6,400	25,600	102,400

At the end of 90 years, in 1990, the book was worth $800.
At the end of 30 years, in 1930, the book was worth $51,200.
At the end of 100 years, in 2000, the book will be worth $102,400.

(b) Writing exercise — Answers will vary.

47. (a) Define symbols **for all the quantities** you are using!
Let x be the number of ounces of Food I, and y the number of ounces of Food II. Then $3x$ will be the number of gms of carbohydrate from the first food, and $5y$ the number of gms of protein from the second food.
Similarly $2x$ and $3y$ will be the number of gms of protein from the two foods.

$$3x + 5y$$

is the total number of gms of carbohydrate, which must equal 73, while

$$2x + 3y$$

is the total number of gms of protein, which must equal 46.

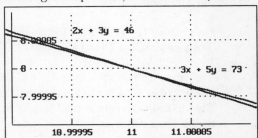

(b) Writing exercise — Answers will vary.

49. For the HP48G:
(green) PLOT ENTER
Move cursor to EQ: field
Press CHOOS on the lower bar
Press NEW on the lower bar
enter' 25 ÷ 7 * α X + 13 ÷ 2 ENTER
(the cursor is now in the name field) .
αα O N E ENTER
press OK on the lower bar
(ONE: '25/7*... now appears on top of the FUNCS IN {HOME} screen)
Similarly enter TWO: 144 * X ÷ 45 + 630 ÷ 229
Highlight fields ONE and TWO. Place a CHK from the lower bar in these fields. Press OK from the lower bar
Autoscale is not checked (turn it off in case it has a check mark)
Adjust the viewing rectangle
ERASE DRAW. The lines look like they could to be parallel. But they are not.
$m_1 = 3.5714$ while $m_2 = 3.2$.

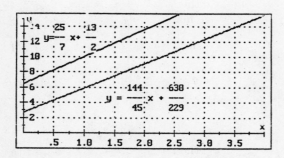

For the TI-85: Press GRAPH, then F1 for $y1 =$
Enter $25x/2 + 13/2$ ENTER
Then enter $y2 = 141x/45 + 630/229$ ENTER
Press 2nd M1 to graph.
The lines do not appear to be parallel. To check, press EXIT, then enter $25/7 - 144/45$ ENTER. The result is 0.37, not 0 (the slopes of the lines differ by 0.37.)

51. (a)

x	2	5	10	x
$R(x)$	70	85	110	$60 + 5x$

(b)
$$R(x) = 5x + 60$$

(c)

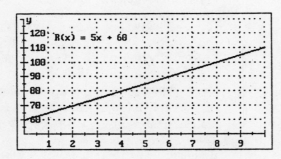

(d) For the HP48G:
Plot '60+5*x'
Let the viewing rectangle be $30 \leq x \leq 33$ and $200 \leq y \leq 230$.
Then move the cross-hair along the line using the left, down, right, and/or up keys.

At any particular position press (x, y) on the lower bar to find the coordinates of the point. The approximation $x = 3.11E1$ and $y = 2.16E2$ mean that $R(31.1) \approx 216$.
Thus it takes approximately 31 hours to accumulate a bill of \$216.
(Assume that you are allowed to pay for fractions of hours.)
For the TI-85:
Press GRAPH then F1 to get $y1 =$
Enter $5x + 60$ ENTER
Press 2nd M2 to get the range window
Change: $XMIN = 30$, $XMAX = 33$, $XSCL = 5$, $YMIN = 200$, $YMAX = 230$, $YSCL = 5$
Press F5 to graph. Press F4 to TRACE.
The coordinates of the cross-hair are $(31, 217)$. Move the left arrow key till $x \approx 216.25$. When $x = 31.26$ then $y = 216.31$.

53. Writing exercise —
Answers will vary.

55. The slope of L_1 is $m_1 = \dfrac{b}{a}$ and that of L_2 is $m_2 = \dfrac{c}{a}$.
By hypothesis, $L_1 \perp L_2$.
$$OA = \sqrt{a^2 + b^2} \text{ and}$$
$$OB = \sqrt{a^2 + c^2}$$
Since $AB = b - c$ and by the Pythagorean theorem,
$$(a^2 + b^2) + (a^2 + c^2) = (b - c)^2$$
from which $2a^2 = -2bc$ or
$$\left(\frac{b}{a}\right)\left(\frac{c}{a}\right) = m_1 m_2 = -1$$
or $m_1 = -\dfrac{1}{m_2}$.

1.4 Functional Models

1. This problem has two possible forms of the solution.

1.4. FUNCTIONAL MODELS

(a) Assume the stream is along the length, say x.
Then y is the width and
$$x + 2y = 1,000 \text{ or } x = 1,000 - 2y$$
The area is
$$A = xy = 2y(500 - y) \text{ square feet}$$

(b) Now suppose the stream flows along the width, y, then
$$2x + y = 1,000 \text{ or } x = \frac{1,000 - y}{2}$$
Thus $A = xy = \dfrac{y(1,000 - y)}{2}$ square feet

3. Let x and y be the smaller and larger numbers, respectively.
$$\text{Then } x + y = 18 \text{ or } y = 18 - x$$
$$\text{The product is } P = xy = x(18 - x)$$

5. Revenue = number of units × price per unit
$$= x(35x + 15)$$

7. Let x be the length and y the width of the rectangle. Then
$$2x + 2y = 320 \text{ or } y = 160 - x$$
The area is length × width or
$$A(x) = x(160 - x)$$

The length is estimated to be 80 meters from the graph below, which also happens to be the width.
Thus the maximum area seems to correspond to that of a square.

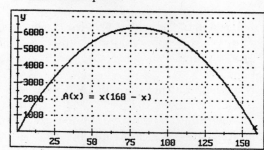

9. Let x be the length of the square base and y the height of the box.
The surface area is $2x^2 + 4xy = 4,000$
Thus $y = \dfrac{2,000 - x^2}{2x}$ and the volume is
$$V = x^2 y = x\left(1,000 - \frac{x^2}{2}\right)$$

11. Let x be the radius and y the height of the cylinder.
The surface area of the closed cylinder is
$$S = 120\pi = 2\pi x^2 + 2\pi xy \text{ or } y = \frac{60 - x^2}{x}$$
Thus $V(x) = \pi x^2 y = \pi x(60 - x^2)$

13. Let x be the radius and y the height of the cylinder.
Since the volume is
$$V = \pi x^2 y = 4\pi, \ y = \frac{4}{x^2}$$
The cost of the top or bottom is
$$C_t = C_b = 2(0.02)\pi x^2,$$
while the cost of the side is
$$2\pi xy(0.02) = \frac{0.16\pi}{x}$$
The total cost is
$$C(x) = 0.08\pi x^2 + \frac{0.16\pi}{x}$$

15. Let R denote the rate of population growth and p the population size.
Since R is directly proportional to p,
$$R(p) = kp,$$
where k is the constant of proportionality.

17. Let R denote the rate at which temperature changes, M the temperature of the medium, and T the temperature of the object.
Then $T - M$ is the difference in the temperature between the object and the medium.

Since the rate of change is directly proportional to the difference,
$$R(T) = k(T - M),$$
where k is the constant of proportionality.

19. Let R denote the rate at which people are implicated, x the number of people implicated, and n the total number of people involved. Then $n - x$ is the number of people involved but not implicated.
Since the rate of change is jointly proportional to those implicated and those not implicated,
$$R(x) = kx(n - x),$$
where k is the constant of proportionality.

21. Let s be the speed of the truck.
The cost due to wages is $\dfrac{k_1}{s}$, where k_1 is a constant of proportionality, and the cost due to gasoline is $k_2 s$, where k_2 is another constant of proportionality.
If $C(s)$ is the total cost,
$$C(s) = \frac{k_1}{s} + k_2 s$$

23.
   (a) Let x denote the taxable income and $f(x)$ the corresponding income tax.
   Thus $f(x) =$
   $$\begin{cases} 0 + .15x & \text{if } x \leq 22,100 \\ 3,315 + .28(x - 22,100) & \text{if } 22,100 < x \leq 53,500 \\ 12,107 + .31(x - 53,500) & \text{if } 53,500 < x \leq 115,000 \end{cases}$$
   or $f(x) =$
   $$\begin{cases} 0 + 0.15x & \text{if } x \leq 22,100 \\ 0.28x - 2,873 & \text{if } 22,100 < x \leq 53,500 \\ 0.31x - 4,478 & \text{if } 53,500 < x \leq 115,000 \end{cases}$$

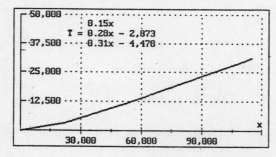

   (b) Writing exercise —
   Answers will vary.

25. Let x denote the length of a side of the square base and y the height of the box. With C the cost,
$$\begin{aligned} C &= \text{(cost per m}^2\text{ of base and top)} \\ & \quad \text{(area of base and top)} \\ & \quad +\text{(cost per m}^2\text{ of sides)(area of sides)} \end{aligned}$$
Thus $C = 2(2x^2) + 1(4xy)$. Since the volume is 250,
$$x^2 y = 250 \text{ or } y = \frac{250}{x^2}$$
it follows that $C(x) = 4x^2 + \dfrac{1,000}{x}$

27. Let x denote the length of the side of one of the removed squares and $V(x)$ the volume of the resulting box.
Then $V(x) =$(area of base)(height), or
$$V(x) = (18 - 2x)(18 - 2x)x = 4x(9 - x)^2$$
From the graph, the value of x producing a box with greatest volume is estimated to be 3 in.

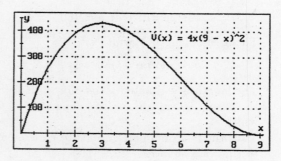

29. Let x denote the selling price of the book and $P(x)$ the corresponding profit function. Then $P(x) =$(number of books sold)(profit per book).
Now the profit per book is $x - 3$, since the cost of each book is $3.
To determine the number of books sold, first notice that $15 - x$ is the number of $1 reductions, since the initial price of the book is $15.

1.4. FUNCTIONAL MODELS

For each of these $1 reductions, 20 more books are sold, so the total number of books sold is $200 + 20(15 - x)$ Putting it all together,

$$\begin{aligned} P(x) &= [200 + 20(15 - x)](x - 3) \\ &= (500 - 20x)(x - 3) \\ &= 20(25 - x)(x - 3) \\ &= -20(x^2 - 28x + 75) \end{aligned}$$

The graph of $P(x)$ suggests that the profit is maximal when $x = 14$, that is, when the books are sold for $14 apiece. By completing the square we see that
$P(x) = -20(x - 14)^2 + 2,420$ which shows that the maximum profit of $2,420 corresponds to a selling price of $14.

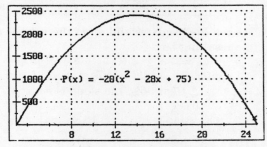

31. The speed of the car C is 60 MPH and that of the truck T 30 MPH.
At some time $t > 0$ (in hours) the car has traveled $60t$ miles and the truck $30t$ miles. Let $D(t)$ denote the distance between T and C at time t.
C is $60t$ miles from the intersection and T is $300 - 30t$ miles from it.
By the Pythagorean theorem,

$$\begin{aligned} D(t) &= \sqrt{(60t)^2 + (300 - 30t)^2} \\ &= 30\sqrt{5t^2 - 20t + 100} \end{aligned}$$

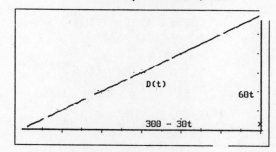

33. Let x be the number of additional trees planted per acre.
The number of oranges per tree will be $400 - 4x$ and the number of trees per acre $60 + x$.
The yield per acre is

$$\begin{aligned} y(x) &= \frac{\text{\# of oranges}}{\text{tree}} \frac{\text{\# of trees}}{\text{acre}} \\ &= 4(100 - x)(60 + x) \end{aligned}$$

The number of trees for optimal yield appears to be $60 + 20 = 80$ trees.

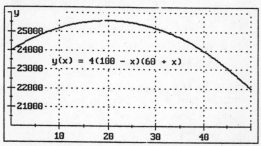

35. The equilibrium price is found when $S(p) = D(p)$ that is
$4p + 200 = -3p + 480$ or $p = 40$.
At this price the number of units supplied, as well as the number of unit demanded, is
$D(40) = 480 - 3(40) = 360$ units

37. (a) The equilibrium price is found when

$$S(p) = D(p),$$

that is $p - 10 = \dfrac{5,600}{p}$,

$$p^2 - 10p - 5,600 = (p - 80)(p + 70) = 0,$$

or $p = 80$

Only the positive number $p = 80$ is meaningful in the present exercise. At this price the number of units supplied, as well as the number of units demanded, is $D(80) = 70$ units.

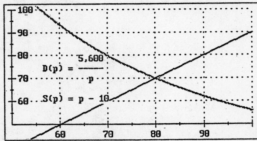

(b) An appropriate graph is shown above.

(c) Writing exercise —
Answers will vary.

39. Let x be the number of hours since the second plane took off and y the distance traveled in flight.

The first plane is ahead by $\frac{1}{2}$ hour or $0.5 \times 550 = 275$ miles.
(Distance = velocity × time, just watch the units.)
The equations for the distances are

$$y = 550x + 275$$

and $y = 650x$

for the first and second planes, respectively. They will have covered the same distance when $650x = 550x + 275$, $100x = 275$, $x = 2.75$ hrs, which is 2 hrs and 0.75×60 minutes or 2 hrs and 45 min. after the second plane took off.

41. Let x be the number of additional days beyond 80 before the club takes all its glass to the recycling center.
Let's assume that the same quantity of glass is collected daily, namely $\frac{24,000}{80} = 300$ lbs.
The daily revenue for the first 80 days would be 300 cents.
The reduction in daily revenue for x days beyond 80 is $3x$ or 1 cent per 100 lbs per day.

The club's revenue on day $80 + x$ would be

$$(300 - 3x)(80 + x) = 3(80 + x)(100 - x)$$

The key to this problem is understanding that all the glass is taken to the recycling center on day $80 + x$.
x is estimated to be 10 from the graph.

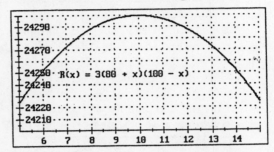

43. (a) $S(p)$ rises which implies a positive slope, so $a > 0$.
If extended below the p axis, the q intercept would be negative, so $b < 0$.
Since $D(p)$ falls, the slope is negative and $c < 0$.
The q intercept is positive, making $d > 0$.

(b)
$$ap + b = cp + d,$$
$$p = \frac{d - b}{a - c}$$

(c) If a increases the denominator will become larger and the fraction will become smaller.
The equilibrium price will decrease.

45. (a) The revenue is $R(x) = 275x$ and the cost is $C(x) = 125x + 1,500$ where x is the number of kayaks sold.
For the break even point,

$$275x = 125x + 1,500 \text{ or } x = 10$$

(b) The profit is
$$P(x) = R(x) - C(x)$$
$$= 150x - 1,500$$

For $P(x) = 1,000$,
$$15x = 250, \quad x = 17$$

47. **(a)** $$C(x) = 5.5x + 74,200$$

x	2,000	4,000	6,000	8,000
$C(x)$	85,200	96,200	107,200	118,200

(b) $$R(x) = 19.5x$$

x	2,000	4,000	6,000	8,000
$C(x)$	39,000	78,000	117,000	156,000

(c) $$y = 5.5x + 74,200$$

(d) $$y = 19.5x$$

(e)

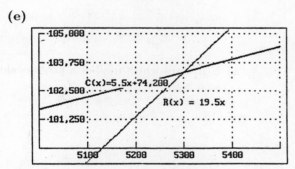

(f) The break-even point is $(5,300, 103,350)$.

(g) Approximately 4,360 books must be sold for a revenue of $85,000. The profit here is approximately $-13,160$.

1.5 Limits

1. $\lim_{x \to a} f(x) = b$, even though $f(a)$ is not defined.

3. $\lim_{x \to a} f(x) = b$ even though $f(a) = c$.

5. $\lim_{x \to a} f(x)$ does not exist since as x approaches a from the left, the function becomes unbounded.

7.
$$\lim_{x \to 2}(3x^2 - 5x + 2)$$
$$= 3\lim_{x \to 2} x^2 - 5\lim_{x \to 2} x + \lim_{x \to 2} 2$$
$$= 3(2)^2 - 5(2) + 2 = 4.$$

9.
$$\lim_{x \to 0}(x^5 - 6x^4 + 7)$$
$$= \lim_{x \to 0} x^5 - 6\lim_{x \to 0} x^4 + \lim_{x \to 0} 7 = 7.$$

11.
$$\lim_{x \to 3}(x-1)^2(x+1)$$
$$= \lim_{x \to 3}(x-1)^2 \lim_{x \to 3}(x+1)$$
$$= (3-1)^2(3+1) = 16.$$

13.
$$\lim_{x \to 2} \frac{x+1}{x+2} = \frac{\lim_{x \to 2}(x+1)}{\lim_{x \to 2}(x+2)} = \frac{3}{4}$$

15. $\lim_{x \to 5} \dfrac{x+3}{5-x}$ does not exist since the limit of the denominator is zero while the limit of the numerator is not zero.

17.
$$\lim_{x \to 1} \frac{x^2 - 1}{x - 1}$$
$$= \lim_{x \to 1} \frac{(x+1)(x-1)}{x-1} = \lim_{x \to 1}(x+1) = 2.$$

19.
$$\lim_{x \to 5} \frac{x^2 - 3x - 10}{x - 5}$$
$$= \lim_{x \to 5} \frac{(x-5)(x+2)}{x-5} = \lim_{x \to 5}(x+2) = 7.$$

21.
$$\lim_{x \to 4} \frac{(x+1)(x-4)}{(x-1)(x-4)}$$
$$= \frac{\lim_{x \to 4}(x+1)}{\lim_{x \to 4}(x-1)} = \frac{5}{3}.$$

23.
$$\lim_{x \to -2} \frac{x^2 - x - 6}{x^2 + 3x + 2}$$
$$= \lim_{x \to -2} \frac{(x-3)(x+2)}{(x+1)(x+2)}$$
$$= \frac{\lim_{x \to -2}(x-3)}{\lim_{x \to -2}(x+1)} = \frac{-5}{-1} = 5.$$

25.
$$\lim_{x \to 4} \frac{\sqrt{x} - 2}{x - 4}$$
$$= \lim_{x \to 4} \frac{\sqrt{x} - 2}{x - 4} \frac{\sqrt{x} + 2}{\sqrt{x} + 2}$$
$$= \lim_{x \to 4} \frac{x - 4}{(x - 4)(\sqrt{x} + 2)} = \frac{1}{4}.$$

27.
$$\lim_{x \to 3^+} \frac{\sqrt{x + 1} - 2}{x - 3}$$
$$= \lim_{x \to 3^+} \frac{\sqrt{x + 1} - 2}{x - 3} \frac{\sqrt{x + 1} + 2}{\sqrt{x + 1} + 2}$$
$$= \lim_{x \to 3^+} \frac{x + 1 - 4}{(x - 3)(\sqrt{x + 1} + 2)} = \frac{1}{4}.$$

29.
$$\lim_{x \to 3^-} f(x) = 2(3^2) - 3 = 15$$
$$\lim_{x \to 3^+} f(x) = 3 - 3 = 0.$$

31.

x	1	0.1	0.01
$1000(1 + 0.09x)^{1/x}$	1,090	1,093.73	1,094.13

x	0.001	0.0001
$1000(1 + 0.09x)^{1/x}$	1,094.17	1,094.17

Thus $\lim_{x \to 0^+} 1000(1 + 0.09x)^{1/x} = 1,094.17$

33. (a) $R(T)$ doubles on $[10, 15]$ as well as others.
   (b) The growth rate is constant.
   (c) The growth rate begins to decrease at $T = 45$.
   $$\lim_{T \to 50^-} R(T) = 0$$
   (d) Writing exercise — Answers will vary.

35. For problem 7 for $y = 3x^2 - 5x + 2$

y	3.33	3.93	3.993	4.73	4.07	4.007
x	1.9	1.99	1.999	2.1	2.01	2.001

For problem 9 for $y = x^5 - 6x^4 + 7$

y	6.999	7.000	7.000	7.000	7.000	6.999
x	-0.1	-0.01	-0.001	0.1	0.01	0.001

For problem 11 with $y = (x - 1)^2(x + 1)$

y	14.079	15.8008	15.98	18.081	16.2	16.02
x	2.9	2.99	2.999	3.1	3.01	3.001

For problem 13

$\frac{x + 1}{x + 2}$	0.74	0.749	.7499	.76	.751	.7501
x	1.9	1.99	1.999	2.1	2.01	2.001

For problem 15

$\frac{x + 3}{5 - x}$	79	799	7,999	-81	-801	$-8,001$
x	4.9	4.99	4.999	5.1	5.01	5.001

37.
$$\lim_{x \to 0} f(x)$$
does not exist because $f(x)$ can oscillate widely and wildly between -1 and 1, regardless how close x gets to 0.

1.6 Continuity

1. If $f(x) = 5x^2 - 6x + 1$,
   then $f(2) = 9$ and $\lim_{x \to 2} f(x) = 9$,
   and so f is continuous at $x = 2$.

3.
$$f(x) = \frac{x + 2}{x + 1},$$
then $f(1) = \frac{3}{2}$ and
$$\lim_{x \to 1} f(x) = \lim_{x \to 1} \frac{x + 2}{x + 1} = \frac{\lim_{x \to 1}(x + 2)}{\lim_{x \to 1}(x + 1)} = \frac{3}{2}$$
Hence f is continuous at $x = 1$.

5. If $f(x) = \frac{x + 1}{x - 1}$,
$f(1)$ is undefined since the denominator is zero, and hence f is not continuous at $x = 1$.

1.6. CONTINUITY

7. If $f(x) = \dfrac{\sqrt{x}-2}{x-4}$,

$f(4)$ is undefined since the denominator is zero, and hence
f is not continuous at $x = 4$.

9. If $f(x) = \begin{cases} x+1 & \text{if } x \leq 2 \\ 2 & \text{if } 2 \leq x \end{cases}$

then $f(2) = 3$ and $\lim_{x \to 2} f(x)$

must be determined.
As x approaches 2 from the right,

$$\lim_{x \to 2} f(x) = \lim_{x \to 2} 2 = 2$$

and as x approaches 2 from the left,

$$\lim_{x \to 2} f(x) = \lim_{x \to 2}(x+1) = 3$$

Hence the limit does not exists (since different limits are obtained from the left and the right), and so f is not continuous at $x = 2$.

11.
$$\lim_{x \to 3^-} f(x) = 3^2 + 1 = 10$$
$$= \lim_{x \to 3^+} f(x) = 2 \times 3 + 4 = f(3),$$

thus $f(x)$ is continuous at $x = 3$.

13.
$$\lim_{x \to a} f(x) = 3 \lim_{x \to a} x^2 - 6 \lim_{x \to a} x + 9$$
$$= 3a^2 - 6a + 9 = f(a),$$

thus $f(x)$ is continuous at all x.

15.
$$f(x) = \dfrac{x+1}{x-2}$$

is not defined at $x = 2$,
so it is not continuous there.

17.
$$f(x) = \dfrac{3x+3}{x+1} = 3 \text{ if } x \neq -1$$

$f(-1)$ is not defined, so $f(x)$ is not continuous there.

19.
$$f(x) = \dfrac{3x-2}{(x+3)(x-6)}$$

is not defined at $x = -3$ and $x = 6$,
so $f(x)$ is not continuous there.

21.
$$f(x) = \dfrac{x}{x^2 - x}$$

is not defined at $x = 0$ and $x = 1$,
so $f(x)$ is not continuous there.

23.
$$\lim_{x \to 1^-} f(x) = 2 + 3 = 5$$
$$= \lim_{x \to 1^+} f(x) = 6 - 1 = f(1),$$

thus $f(x)$ is continuous for all x.
(There are no fractions or square roots.)

25. (a) If $v = 20$ then

$$W(v) = 91.4 + (91.4 - t)(0.0203v - 0.304\sqrt{v} - 0.474)$$

i.e. when $t = 30$, $W(20) = 3.75$.
If $v = 50$ and $t = 30$ we have
$W(v) = 1.6t - 55$ or
$v(50) = 1.6(30) - 55 = -7$.

(b) For $t = 30$ we have

$$91.4 + (91.4 - 30)(0.0203v - 0.304\sqrt{v} - 0.474) = 0$$

$$1.25v - 18.67\sqrt{v} + 62.3 = 0$$

from which $v = 26$ mph or $v = 98$ mph.

(c) Yes, both expressions agree in the case where $v = 4$.
The same applies to $v = 45$.

27. $p(x)$ is discontinuous at $x = 1$, $x = 2$, $x = 3$, $x = 4$, $x = 5$

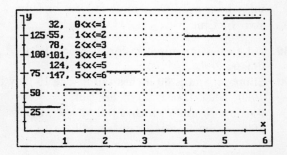

24   CHAPTER 1. FUNCTIONS, GRAPHS, AND LIMITS

29. The graph is discontinuous at $x = 6$ and $x = 12$.
What happened to cause these jumps is a writing exercise —
Answers will vary.

31.
$$\lim_{x \to 4^-} f(x) = 1 - 3 \times 4 = -11$$
$$f(4) = 16A + 8 - 3 = 16A + 5$$

To be continuous at $x = 4$ we need $16A + 5 = -11$ or $A = -1$.

33. $f(x) = x^2 - 3x$ is a polynomial in the open interval,
and thus $f(x)$ is continuous for all x in the open interval.
But at $x = 2$
$$\lim_{x \to 2^-} f(x) = 2^2 - 3 \times 2 = -2$$
$$f(2) = 4 + 2 \times 2 = 8,$$
thus $f(2) \neq \lim_{x \to 2^-} f(x)$. $f(x)$ not continuous on the closed interval.

35. Let
$$f(x) = \sqrt[3]{x} - (x^2 + 2x - 1)$$
$f(x)$ is continuous at all x and $f(0) = 1$, $f(1) = 1 - (1 + 2 - 1) = -1$.
By the root location property, there is at least one number $0 \leq c \leq 1$ such that $f(c) = 0$, and $x = c$ is a solution.

37. Your weight in pounds increased (and/or decreased) continuously from your minimum weight at birth to your present weight, like from 7 pounds to 150 pounds. Your (present) height in inches is some number, like 65. By the intermediate value property, your weight must have been 65 pounds at least once in your lifetime.

39. Let's ise some numbers for the purpose of illustration.
Suppose Nan is 60 inches tall at age 15 (say in 1980) when Dan is 30 inches tall.
Assume Nan is 70 inches tall at age 31 (in 1996) when Dan is 76 inches tall.
Draw a continuous curve (it could be a straight line) from (1980,60) to (1996,70). This represents Nan's growth curve.
Now draw a continuous curve from (1980,30) to (1996,75). This represents Dan's growth curve.
The two curves cross at one point, say in 1992 when they are both 66 inches tall. By the intermediate value property, 66 inches lies between 30 and 60 as well as between 30 and 75.

Review Problems

1. (a) The domain of the quadratic function
$$f(x) = x^2 - 2x + 6$$
consists of all real numbers x.

   (b) Since division by zero is not possible, the domain of the rational function
$$f(x) = \frac{x - 3}{x^2 + x - 2} = \frac{x - 3}{(x + 2)(x - 1)}$$
consists of all real numbers x except $x = -2$ and $x = 1$.

   (c) Since negative numbers do not have square roots, the domain of the function
$$f(x) = \sqrt{x^2 - 9} = \sqrt{(x + 3)(x - 3)}$$
consists of all values of x for which the product $(x + 3)(x - 3) \geq 0$, that is for $x \leq -3$ or $x \geq 3$.
Another way of describing this domain is $|x| \geq 3$.

2. The price x months from now is
$$P(x) = 40 + \frac{30}{x + 1} \text{ dollars. Hence:}$$

   (a) The price 5 months from now is
$$P(5) = 40 + \frac{30}{5 + 1} = \$45$$

REVIEW PROBLEMS 25

(b) The change in price during the 5th month is

$$P(5) - P(4) = 40 + \frac{30}{5+1} - \left(40 + \frac{30}{4+1}\right) = -1$$

That is the price decreases by \$1.

(c) The price will be \$43 when

$$43 = 40 + \frac{30}{x+1},$$

$3(x+1) = 30$, or $x = 9$,

that is 9 months from now.

(d) Since the term $\frac{30}{x+1}$ gets very small as x gets very large, the price

$$P(x) = 40 + \frac{30}{x+1}$$

will approach \$40 in the long run.

3. (a)

If $g(u) = u^2 + 2u + 1$ and $h(x) = 1 - x$

then
$$\begin{aligned} g[h(x)] &= g(1-x) \\ &= (1-x)^2 + 2(1-x) + 1 \\ &= x^2 - 4x + 4. \end{aligned}$$

(b) If $g(u) = \frac{1}{2u+1}$ and $h(x) = x + 2$, then

$$\begin{aligned} g[h(u)] &= g(x+2) \\ &= \frac{1}{2(x+2)+1} = \frac{1}{2x+5}. \end{aligned}$$

(c) If $g(u) = \sqrt{1-u}$ and $h(x) = 2x + 4$, then

$$\begin{aligned} g[h(x)] &= g(2x+4) = \sqrt{1-(2x+4)} \\ &= \sqrt{-2x-3}. \end{aligned}$$

4. (a) If $f(x) = x^2 - x + 4$ then

$$\begin{aligned} f(x-2) &= (x-2)^2 - (x-2) + 4 \\ &= x^2 - 5x + 10. \end{aligned}$$

(b) If $f(x) = \sqrt{x} + \frac{2}{x-1}$ then

$$\begin{aligned} f(x^2+1) &= \sqrt{x^2+1} + \frac{2}{(x^2+1)-1} \\ &= \sqrt{x^2+1} + \frac{2}{x^2}. \end{aligned}$$

(c) If $f(x) = x^2$ then

$$\begin{aligned} f(x+1) - f(x) &= (x+1)^2 - x^2 \\ &= 2x + 1. \end{aligned}$$

5. (a) $\qquad f(x) = (x^2 + 3x + 4)^5$

can be written as $g[h(x)]$ where

$g(u) = u^5$ and $h(x) = x^2 + 3x + 4$

(b)
$$f(x) = (3x+1)^2 + \frac{5}{2(3x+2)^3}$$

can be written as $g[h(x)]$ where

$$g(u) = u^2 + \frac{5}{2(u+1)^3}$$

and $h(x) = 3x + 1$

6. (a) Since the smog level Q is related to the variable p by the equation

$$Q(p) = \sqrt{0.5p + 19.4}$$

and the variable p is related to the variable t by the equation

$$p(t) = 8 + 0.2t^2$$

it follows that the composite function

$$\begin{aligned} Q[p(t)] &= \sqrt{0.5(8 + 0.2t^2) + 19.4} \\ &= \sqrt{23.4 + 0.1t^2} \end{aligned}$$

expresses the smog level as a function of the variable t.

(b) The smog level 3 years from now will be

$$\begin{aligned} Q[p(3)] &= \sqrt{23.4 + 0.1(3^2)} \\ &= \sqrt{24.3} \approx 4.93 \text{ units} \end{aligned}$$

(c) Set $Q[p(t)]$ equal to 5 and solve for t to get
$$5 = \sqrt{23.4 + 0.1t^2},$$
$1.6 = 0.1t^2, \quad t^2 = \dfrac{1.6}{0.1} = 16 \text{ or } t = 4 \text{ years}$ from now.

7. If the graph of
$$y = 3x^2 - 2x + c$$
passes through the point $(2,4)$, then $y = 4$ and $x = 2$ so that
$$4 = 3(2)^2 - 2(2) + c \text{ or } c = -4$$

8. (a) Some points on the graph of
$$\begin{aligned} y &= x^2 + 2x - 8 \\ &= (x+4)(x-2) \end{aligned}$$
are shown below.

x	-6	-5	-4	-3	-1
y	16	7	0	-5	-9

x	0	1	2	3	4
y	-8	-5	0	7	16

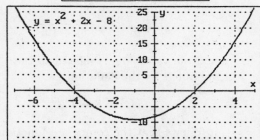

(b) Some points on the graph of $y = 3 + 4x - 2x^2$ are shown below. Note:
$$\begin{aligned} y &= -2x^2 + 4x + 3 \\ &= -2(x-1)^2 + 5 \end{aligned}$$
Thus the curve is a parabola with vertex at $(1,5)$ and opening downward.

x	-3	-2	-1
y	-27	-13	-3

x	0	1	2	3	4	5
y	3	5	3	-3	-13	-27

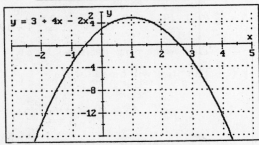

(c) Since
$$x^2 + 2x + 1 + y^2 - y + \dfrac{1}{4} = \dfrac{17}{4}$$
the equation represents a circle with center at $(-1, \tfrac{1}{2})$ and radius $\tfrac{\sqrt{17}}{2}$).

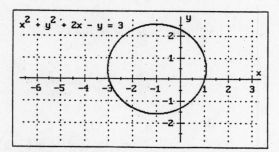

9. (a) Note: In chapter 3 we will learn how to use calculus to find the optimal price. Without calculus, we could complete the square to get
$$E(p) = -50(p-8)^2 + 3,200$$
from which it is clear that the greatest expenditure is $\$\,3,200$ and is generated when the price is $p = \$8$.

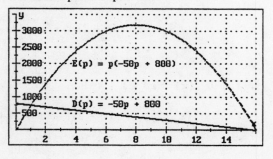

REVIEW PROBLEMS

(b) The function

$$D(p) = -50p + 800$$

is linear with slope -50 and y-intercept 800.
It represents demand for $0 \leq p \leq 16$.
The total monthly expenditure is

$$\begin{aligned} E(p) &= \text{(price per unit)(demand)} \\ &= p(-50p + 800) = -50p(p-16) \end{aligned}$$

Since the expenditure is assumed to be non-negative, the relevant interval is

$$0 \leq p \leq 16$$

(c) The graph suggests that the expenditure will be greatest if $p = 8$.

10. The number of weeks needed to reach x percent of the fund raising goal is given by

$$f(x) = \frac{10x}{150 - x}$$

(a) Since x denotes a percentage, the function
has a practical interpretation for

$$0 \leq x \leq 100$$

The corresponding portion of the graph is sketched.

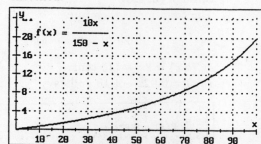

(b) The number of weeks needed to reach 50 % of the goal is

$$\frac{10(50)}{150 - 50} = 5$$

(c) The number of weeks needed to reach 100 % of the goal is

$$f(100) = \frac{10(100)}{150 - 100} = 20$$

11. (a) If $y = 3x + 2$, $m = 3$ and $b = 2$.

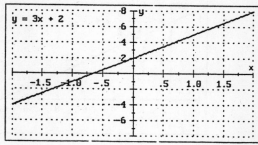

(b) If $5x - 4y = 20$ then

$$y = \frac{5}{4}x - 5$$

and so $m = \dfrac{5}{4}$, $b = -5$.

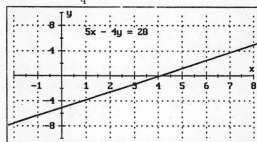

(c) If $2y + 3x = 0$ then

$$y = -\frac{3}{2}x + 0$$

and $m = -\dfrac{3}{2}$, $b = 0$.

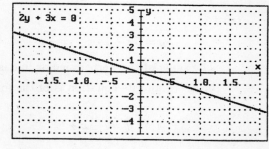

(d) If $\frac{1}{3}x + \frac{1}{2}y = 4$ then
$$y = -\frac{2}{3}x + 8$$
and $m = -\frac{2}{3}$, $b = 8$.

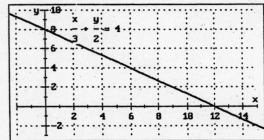

12. (a) Since $m = 5$ and $b = -4$, the equation of the line is
$$y = 5x - 4$$

(b) Since $m = -2$ and the point $(1, 3)$ is on the line, the equation of the line is
$$y - 3 = -2(x - 1) \text{ or } y = -2x + 5$$

(c) Since $\frac{x}{a} + \frac{y}{b} = 1$ is the equation of the line with $x = a \neq 0$ and $y = b \neq 0$ intercepts the equation of our line is
$$\frac{x}{3} - \frac{3y}{2} = 1$$

(d) Parallel lines have the same slope, so the desired line is of the form
$$2x + y = C$$
where $2(5) + 4 = C$. Thus
$$2x + y = 14$$

(e)
$$3y = 5x - 7 \text{ or } y = \frac{5x}{3} - \frac{7}{3}$$
The slope of the perpendicular line is $m = -\frac{3}{5}$ and its equation is of the form
$$y = -\frac{3}{5}x + C$$

Since the line contains $(-1, 3)$, $C = \frac{12}{5}$ and
$$y = -\frac{3}{5}x + \frac{12}{5}$$

13. The slope of the line joining $(-2, 3)$ and $(4, 1)$ is $m = \frac{1 - 3}{4 + 2} = -\frac{1}{3}$.
The slope of the desired line is $m = 3$ and its equation is
$$y = 3x + C \text{ where } 2 = 3 + C. \text{ Thus}$$
$$y = 3x - 1$$

14. $D(v) = 0.065v^2 + 0.148v$

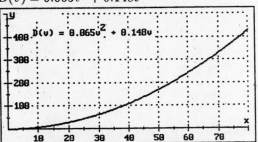

15. (a) Let x denote the time in months since the beginning of the year and $P(x)$ the corresponding price (in cents) of gasoline. Since the price increases at a constant rate of 2 cents per gallon per month, P is a linear function of x with slope $m = 2$. Since the price on June first (when $x = 5$) is 120 cents, the graph passes through $(5, 120)$.
The equation is therefore
$$P - 120 = 2(x - 5)$$
or $P(x) = 2x + 110$ cents.
$P(x) = 0.02x + 1.10$ dollars

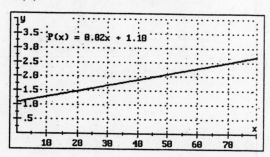

REVIEW PROBLEMS

(b) The price at the beginning of the year was $P(0) = 110$ cents per gallon.

(c) On October first, $x = 9$ and $P(9) = \$1.28$.

16. (a) Let x denote the number of months (measured from 3 months ago) and $C(x)$ the corresponding circulation. Since the circulation is increasing at a constant rate, $C(x)$ is a linear function. Three months ago (when x was 0), the circulation was 3,200, and today (when $x = 3$), the circulation is 4,400. From the points $(0, 3200)$ and $(3, 4400)$, the slope is
$$m = \frac{4,400 - 3,200}{3 - 0} = 400$$
and the y-intercept is 3,200.
Hence the equation of the line is
$$C(x) = 400x + 3,200$$

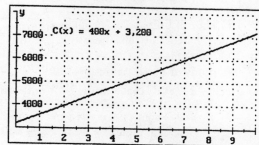

(b) Two months from now (that is, when $x = 5$) the circulation will be
$$C(5) = 400(5) + 3,200 = 5,200$$

17. (a) The graphs of
$$y = -3x + 5 \text{ and } y = 2x - 10$$
seem to intersect in the fourth quadrant. Setting the two expressions for y equal to each other yields
$$-3x + 5 = 2x - 10 \text{ or } x = 3$$
When $x = 3$, $y = -3(3) + 5 = -4$.

Hence the point of intersection is $(3, -4)$.

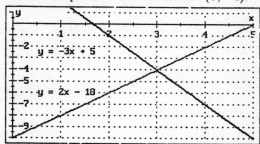

(b) The graphs of
$$y = x + 7 \text{ and } y = -2 + x$$
are parallel lines with slope 1.
Hence there are no points of intersection.

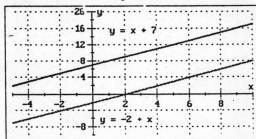

(c) The graphs of
$$y = x^2 - 1 \text{ and } y = 1 - x^2$$
seem to intersect at $(-1, 0)$ and $(1, 0)$.
Setting the two expressions for y equal to each other yields
$x^2 - 1 = 1 - x^2$, $x^2 = 1$, or $x = \pm 1$
When $x = \pm 1$, $y = (\pm 1)^2 - 1 = 0$.
Hence the points of intersection are $(-1, 0)$ and $(1, 0)$.

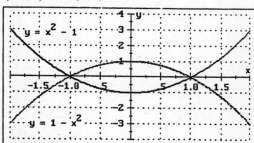

(d) The graphs of
$$y = x^2 \text{ and } y = 15 - 2x$$

seem to intersect in the first and second quadrants.
Setting the two expressions for y equal to each other yields
$$x^2 = 15 - 2x$$
$(x+5)(x-3) = 0$, or $x = -5$ and $x = 3$
When $x = -5$, $y = 25$,
and when $x = 3$, $y = 9$.
Hence the points of intersection are $(-5, 25)$ and $(3, 9)$.

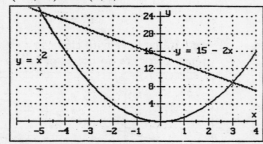

(e) the graphs of
$$y = \frac{24}{x^2} \text{ and } y = 3x$$
seem to intersect in the first quadrant.
Setting the two expressions for y equal to each other yields
$\frac{24}{x^2} = 3x$, $x^3 = 8$ or $x = 2$.
When $x = 2$, $y = 3(2) = 6$.
Hence the point of intersection is $(2, 6)$.

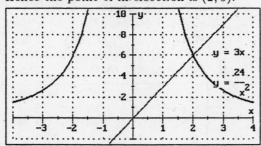

18. Let x denote the selling price of a bookcase (in dollars) and $P(x)$ the corresponding profit. Then
$$\begin{aligned} P(x) &= \text{(number of units sold)} \\ &\quad \text{(profit per unit)} \\ &= (50 - x)(x - 80) \end{aligned}$$

The graph suggests that the profit will be greatest when x is approximately \$65.
Note: In chapter 3 we will learn how to use calculus to find the optimal price exactly. Without calculus, you could complete the square to get
$P(x) = -(x - 65)^2 + 225$
from which it is clear that the maximum profit is \$225 and occurs at $x = 65$.

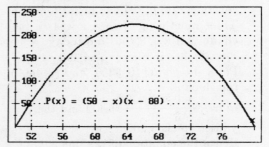

19. Let x denote the selling price (in dollars) and $P(x)$ the corresponding profit function. Then
$P(x) = $
(number of units sold)(profit per unit).
Since the cost is \$50, the profit per unit is $x - 50$.
The number of \$5 reductions is
$\frac{1}{5}(80 - x)$, and so the number of units sold is
$$40 + 10\left(\frac{1}{5}\right)(80 - x) = 200 - 2x$$

Putting it all together,
$$P(x) = -2(100 - x)(50 - x)$$

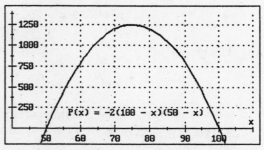

The graph suggests that the profit will be greatest when x is approximately \$75.
Note: In chapter 3, we will learn how to use calculus to find the optimal price exactly.

REVIEW PROBLEMS

Without calculus, we could complete the square to get

$$P(x) = -2(x - 75)^2 + 1,250$$

from which it is clear that the maximum profit is \$1,250 and occurs at $x = 75$.

20. Let r denote the radius, h the height, and V the volume of the can. Then

$$V = \pi r^2 h$$

To write h in terms of r, use the fact that the cost of constructing the can is to be 80 cents. That is, 80=cost of bottom + cost of side where cost of bottom =
(cost per square inch)(area)=$3\pi r^2$
and cost of side =
(cost per square inch)(area)
$= 2(2\pi rh) = 4\pi rh$. Hence

$$80 = 3\pi r^2 + 4\pi rh, \text{ or } h = \frac{20}{\pi r} - \frac{3r}{4}$$

Now substitute this expression for h into the formula for V to get

$$V(r) = \pi r^2 \left(\frac{20}{\pi r} - \frac{3r}{4}\right) = 20r - \frac{3}{4}\pi r^3$$

21. Let x denote the number of machines used and $C(x)$ the corresponding cost function. Then,

$$\begin{aligned} C(x) &= \text{(set up cost)+(operating cost)} \\ &= 80(\text{number of machines}) \\ &\quad + 5.76(\text{number of hours}). \end{aligned}$$

Since 400,000 medals are to be produced and each of the x machines can produce 200 medals per hour,

$$\text{number of hours} = \frac{400,000}{200x} = \frac{2,000}{x}$$

Putting it all together,

$$\begin{aligned} C(x) &= 80x + 5.76\left(\frac{2,000}{x}\right) \\ &= 80x + \frac{11,520}{x}. \end{aligned}$$

The graph suggests that the cost will be smallest when x is approximately 12.
In chapter 3 you will learn how to use calculus to find the optimal number of machines exactly.

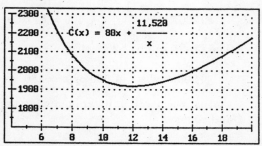

22. Assume the inventory to be maintained at the same level, continuously, over a 24-hour day period. A discontinuity occurs when the inventory drops, say, at midnight of appropriate days. Thus $x = 10$, $x = 16$, $x = 24$, $x = 25$, $x = 26$, $x = 27$, $x = 28$, $x = 29$, and $x = 30$ are points of discontinuity.

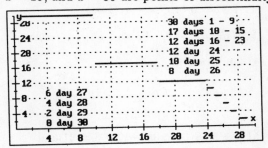

23. (a) Let x denote the number of units manufactured and sold.
$C(x)$ and $R(x)$ are the corresponding cost and revenue functions, respectively.

$$\text{Then } C(x) = 4,500 + 50x$$
$$\text{and } R(x) = 80x$$

For the manufacturer to break even, cost must be equal to revenue. That is

$$4,500 + 50x = 80x \text{ or } x = 150 \text{ units}$$

(b) Let $P(x)$ denote the profit from the manufacture and sale of x units. Then,

$$P(x) = R(x) - C(x)$$

$$= 80x - (4,500 + 50x)$$
$$= 30x - 4,500.$$

When 200 units are sold, the profit is

$$P(200) = 30(200) - 4,500 = \$1,500$$

(c) The profit will be $900 when

$$900 = 30x - 4,500 \text{ or } x = 180$$

that is, when 180 units are manufactured and sold.

24.
$$S(p) = p^2 + Ap - 3 \text{ and } D(p) = Cp + 32$$

with $3 \leq p \leq 8$.
For $p < 3$ we'll force $S(p) = 0$ (let's assume continuity and make $S(3) = 0$) so
$9 + 3A - 3 = 0$ or $3A = -6$, $A = -2$.
Similarly let $D(8) = 0$ so
$8C + 32 = 0$ or $C = -4$. Thus

$$S(p) = p^2 - 2p - 3 \text{ and } D(p) = -4p + 32$$

At equilibrium,

$$p^2 - 2p - 3 = -4p + 32,$$
$$(p+7)(p-5) = 0, \text{ or } p = 5$$

At $p = 6$, $S(6) = 36 - 12 - 3 = 21$ and $D(6) = -24 + 32 = 8$.
The difference is $21 - 8 = \$13$.

25. Royalties for publisher A are given by $P_a =$
$$\begin{cases} 0.12(5)x & , 0 < x \leq 30,000 \\ .12(5)(30,000) + .17(5)(x - 30,000) & , x \geq 30,000 \end{cases}$$
Similarly for publisher B,
$P_b =$
$$\begin{cases} 0 & \text{if } 0 < x \leq 4,000 \\ 0.15(6)(x - 4,000) & \text{if } x \geq 4,000 \end{cases}$$
Publisher A certainly offers the better deal if $x \leq 4,000$.
If $x = 30,000$,

$$P_a = 0.6(30,000) = 18,000$$

and

$$P_b = 0.9(26,000) = 23,400$$

For the break-even point

$$0.9(x - 4,000) = 0.6x,$$

$0.3x = 3,600$, $x = 12,000$.
Publisher A offers the better deal if $x < 1,200$ or if $30,000 < x$.
Publisher B offers the better deal if $1,200 < x \leq 30,000$.

26. Let x denote the number of relevant facts recalled,
n the total number of relevant facts in the person's memory,
and $R(x)$ the rate of recall.
Then $n - x$ is the number of relevant facts not recalled.

$$\text{Hence } R(x) = k(n - x)$$

where k is a constant of proportionality.

27. Let the power plant be at E,
the opposite point at O,
the point at which the cable reaches the opposite bank at P
and the factory at F.

$$\overline{OP} = x,$$
$$\overline{PF} = 3,000 - x,$$
$$\overline{EP} = \sqrt{900^2 + x^2}$$

The cost of the cable in the river is

$$C_r = 5\sqrt{900^2 + x^2}$$

and the cost of the cable on land is

$$C_l = 4(3,000 - x)$$

Thus the total cost is

$$C = 4(3,000 - x) + 5\sqrt{900^2 + x^2}$$

28. The fixed cost is $1,500 and the cost per unit is $2.

The cost is $C(x) = 1,500 + 2x$

REVIEW PROBLEMS

if $0 \leq x \leq 5,000$.

As to the question of continuity, the answer is both yes and no.

Yes, if (as we normally do) x is any real number.

No, if x is discrete ($x = 0, 1, 2, \cdots, 5,000$).

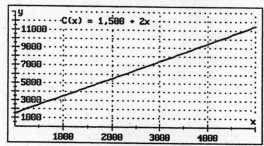

29. Let x be the time in minutes for the hour hand to move from position 3 to the position at which the hands coincide.

The hour hand moves $\frac{1}{12}$ of a tick (distance between minute tick marks on the circumference) per minute while the minute hand moves one whole tick per minute.

The minute has to cover 15 ticks before reaching position 3. Thus

$$15 + \frac{x}{12} = x,$$

$x = 16.36$, or $x = 16$ minutes and 22 seconds.

30. (a) Let T be the temperature and C the number of chirps per minute.
$m = \frac{60 - 0}{50 - 38} = 5$ and

$$C = 5(T - 38) \text{ or } T = \frac{1}{5}C + 38$$

for $38 \leq T$.

(b) If $T = 75$ $C = 5(75 - 38) = 185$ chirps per minute.

If $C = 74$, the temperature is obtained from

$74 = 5(T - 38)$, $T = 38 + \frac{74}{5} = 52.8$ degrees.

31. (a)

$$\lim_{x \to 2} \frac{x^2 - 3x}{x + 1}$$

$$= \frac{\lim_{x \to 2}(x^2 - 3x)}{\lim_{x \to 2}(x + 1)} = -\frac{2}{3}$$

(b)

$$\lim_{x \to 1} \frac{x^2 + x - 2}{x^2 - 1}$$

$$= \frac{\lim_{x \to 1}(x^2 + x - 2)}{\lim_{x \to 1}(x^2 - 1)} = \frac{0}{0}.$$

Simplifying before taking limits leads to

$$\lim_{x \to 1} \frac{x^2 + x - 2}{x^2 - 1}$$

$$= \frac{\lim_{x \to 1}(x + 2)(x - 1)}{\lim_{x \to 1}(x + 1)(x - 1)} = \frac{3}{2}.$$

(c)

$$\lim_{x \to 1} \left(\frac{1}{x^2} - \frac{1}{x} \right) = 1 - 1 = 0$$

(d)

$$\lim_{x \to 2} \frac{x^3 - 8}{2 - x} = \frac{\lim_{x \to 2}(x^3 - 8)}{\lim_{x \to 2}(2 - x)} = \frac{0}{0}$$

Simplifying before taking limits leads to

$$\lim_{x \to 2} \frac{x^3 - 8}{2 - x}$$

$$= \frac{\lim_{x \to 2}(x - 2)(x^2 + 2x + 4)}{\lim_{x \to 2} -(x - 2)}$$

$$= -\lim_{x \to 2}(x^2 + 2x + 4) = -12.$$

32. $y = \frac{1}{x}$

is the graph of a hyperbola.

The graph of $y = 3 - \frac{1}{x}$ is this hyperbola rotated 180 degrees about the x axis with 3 added to its y values.

33. (a)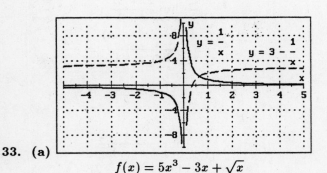
$$f(x) = 5x^3 - 3x + \sqrt{x}$$
is not continuous for $x < 0$ since square roots of negative numbers do not exist.

(b)
$$f(x) = \frac{x^2 - 1}{x + 3}$$
is not continuous at $x = -3$ since $f(-3) = \frac{10}{0}$ and division by 0 is undefined.

(c)
$$g(x) = \frac{x^3 + 5x}{(x-2)(2x+3)}$$
is not continuous at $x = 2$ and $x = -\frac{3}{2}$ since $g(2) = \frac{18}{0}$ and $g(-\frac{3}{2}) = \frac{-87/8}{0}$ (division by 0 is not defined).

(d) $h(x) = \begin{cases} x^3 + 2x - 33 & \text{if } x \le 3 \\ \dfrac{x^2 - 6x + 9}{x - 3} & \text{if } 3 < x \end{cases}$

is not continuous at $x = 3$ but $x = 3$ is not in the domain of the fraction, so $f(x)$ is continuous for all x.

34. (a) $f(x) = \begin{cases} 2x + 3 & \text{if } x < 1 \\ Ax - 1 & \text{if } 1 \le x \end{cases}$

Then $f(x)$ is continuous everywhere except possibly at $x = 1$ since $2x + 3$ and $Ax - 1$ are polynomials. Since $f(1) = A - 1$, in order that $f(x)$ be continuous at $x = 1$, A must be chosen so that
$$\lim_{x \to 1} f(x) = A - 1$$

As x approaches 1 from the right,
$$\lim_{x \to 1} f(x) = \lim_{x \to 1}(Ax - 1) = A - 1$$
and as x approaches 1 from the left,
$$\lim_{x \to 1} f(x) = \lim_{x \to 1}(2x + 3) = 5$$
$$\lim_{x \to 1} f(x)$$
exists whenever $A - 1 = 5$ or $A = 6$ and furthermore, for $A = 6$,
$$\lim_{x \to 1} f(x) = 5,$$
$f(1) = 6 - 1 = 5$.
Thus, $f(x)$ is continuous at $x = 1$ only when $A = 6$.

(b) $f(x) = \begin{cases} \dfrac{x^2 - 1}{x + 1} & \text{if } x < -1 \\ Ax^2 + x - 3 & \text{if } -1 \le x \end{cases}$

Then $f(x)$ is continuous everywhere except possibly at $x = -1$ since
$$\frac{x^2 - 1}{x + 1}$$
is a rational function and
$$Ax^2 + x - 3$$
is a polynomial.
Since $f(-1) = A - 4$, in order that $f(x)$ be continuous at $x = -1$, A must be chosen so that
$$\lim_{x \to -1} f(x) = A - 4$$

As x approaches -1 from the right,
$$\lim_{x \to -1} f(x) = \lim_{x \to -1}(Ax^2 + x - 3) = A - 4$$
and as x approaches -1 from the left,
$$\lim_{x \to -1} f(x) = \lim_{x \to -1} \frac{x^2 - 1}{x + 1}$$
$$= \lim_{x \to -1} \frac{(x+1)(x-1)}{x+1} = -2$$

REVIEW PROBLEMS

$\lim_{x \to -1} f(x)$ exists whenever $A - 4 = -2$ or $A = 2$,
and furthermore, for $A = 2$

$$\lim_{x \to -1} f(x) = -2,$$

$f(-1) = 2 - 4 = -2$.
Thus $f(x)$ is continuous at $x = -1$ only when $A = 2$.

35. **(a)** The population decreases when $5 \leq t$.

$$-8t + 72 = 0 \text{ when } t = 9$$

(b) $f(1) = 8$ and $f(7) = -56 + 72 = 16$.
Since

$$\lim_{x \to 5} f(x) = 25 + 7 = 32,$$

and $8 < 10 < 16$,
by the intermediate value property there exists a value $1 < c < 5 < 7$ such that $f(c) = 10$.

36. $w(x) = \begin{cases} Ax & \text{if } x \leq 4,000 \\ \dfrac{B}{x^2} & \text{if } x > 4,000 \end{cases}$

For continuity $4,000 A = \dfrac{B}{(4,000)^2}$
or $B = A(4,000)^3$.

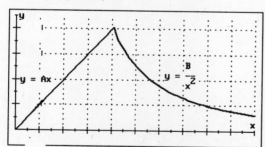

37. Use your HP48G to plot '(3*X^2 - 6*X+9)/(X^2+X-2)' with AUTOSCALE checked and trace, (x,y) on.
According to the trace, the cursor jumps near $x = 1.02$ and at $x = -2$.
Alternately use Trace on your TI-85 to estimate the x values of the vertical asymptotes. They are $x \approx -1.9$ and $x \approx 1.1$.

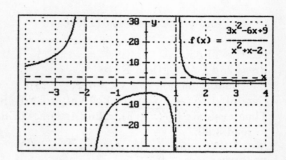

38. Use your HP48G to plot $\{$'21/9*X-84/35' '654/279*X-54/10'$\}$ with AUTOSCALE unchecked.
The lines seem to be parallel.
but they are not since
$$m_1 = \frac{21}{9} = \frac{21(31)}{9(31)} = \frac{651}{279} \neq m_2.$$

Alternately use your TI-45 to populate $y1 =$ and $y2 =$. Then graph.
Check the difference between the slopes, namely -0.01. (The lines are almost parallel.)

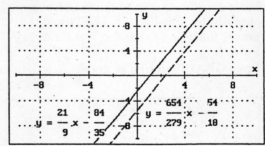

39.
$$f(x) = \sqrt{x + 3}$$

and $g(x) = 5x^2 + 4$

For the HP48G:
Define (or redefine) 'F(X)=√(X+3)' and 'G(X)=5*X^2+4'.
Set -1.28 in level 1 of the HP48G stack.
Press G on the lower bar and get
$g(-1.28) = 12.192$.
12.192 is now in level 1 on the stack.
Press F and get $f[g(-1.28)] = 3.89769162454$.
Now enter $\sqrt{}$ 2 ENTER which displays as 1.41421356237.

$$f(\sqrt{2}) = 2.10100298961$$

and $g[f(\sqrt{2})] = 26.0710678118$

For the TI-85:
Press 2nd CALC, then F1 to
EVALuateFunction $(5x \wedge 2 + 4, x, -1.28)$
Careful: Use (−). You get 12.19 after ENTER.
Then evaluate $(\sqrt{(x+3)}, x, 2ndAns)$ ENTER
to read 3.90.
Now EVALuateFunction $(\sqrt{(x+3)}, x, \sqrt{2})$
ENTER to get 2.10.
Then evaluate $(5x \wedge 2 + 4, x, 2ndAns)$ ENTER
to read 26.07. (The TI-85 was set for two
decimal place accuracy.)

40. Use your HP48G calculator with (green)
PLOT and edit to enter
<< IF 'X < 1' THEN 'X² + 1' ELSE 'X² − 2'
>>.
The viewing rectangle is $-10 < x < 10$ and
$-1 < y < 3$.
Trace and (x,y) on show a jump near $x = 1$.
For the TI-85, press GRAPH and F1 for $y1 =$
Enter $(x \wedge 2 + 1) \div (x \leq 1)$ ENTER where $\leq$ is
obtained by pressing 2nd TEST (and the
proper function key) and ÷ translates to /.
Enter $(x \wedge 2 - 1) \div (x > 1)$ ENTER.
Press GRAPH and F5 to graph.

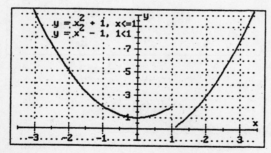

41.
$$f(x) = \frac{x^2 - 3x - 10}{1 - x} - 2$$
$$= \frac{(x-5)(x+2)}{1-x} - 2$$

$f(x) = 0$ at $x = 4$
and $x = -3$. $f(0) = -12$. The vertical
asymptote at $x = 1$ should not come as a
surprise. The oblique asymptote can be
explained as follows: Rewrite

$$\frac{-x^2 + 3x + 10}{x - 1} - 2 = -x + \frac{12}{x - 1}$$

so that the line $y = -x$ becomes an asymptote
as

$$\lim_{x \to \infty} -x + \frac{12}{x - 1} = -x$$

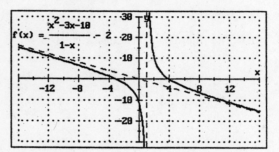

Chapter 2

Differentiation: Basic Concepts

2.1 The Derivative: Slope and Rates

1. If $f(x) = 5x - 3$ then
$$f(x+h) = 5(x+h) - 3$$
The difference quotient DQ is
$$\frac{f(x+h) - f(x)}{h}$$
$$= \frac{[5(x+h) - 3] - [5x - 3]}{h} = \frac{5h}{h} = 5.$$
Now $\lim_{h \to 0} \frac{f(x+h) - f(x)}{h} = 5$
The slope is $m = f'(2) = 5$

3. If $f(x) = 2x^2 - 3x + 5$ then
$$f(x+h) = 2(x+h)^2 - 3(x+h) + 5$$
The difference quotient DQ is
$$\frac{f(x+h) - f(x)}{h}$$
$$= \frac{[2(x+h)^2 - 3(x+h) + 5]}{h}$$
$$- \frac{[2x^2 - 3x + 5]}{h}$$
$$= \frac{4xh + 2(h)^2 - 3h}{h} = 4x + 2h - 3.$$
$$\lim_{h \to 0} \frac{f(x+h) - f(x)}{h} = 4x - 3$$
The slope is
$$m = f'(0) = -3$$

5. If $g(t) = \frac{2}{t}$ then
$$g(t+h) = \frac{2}{t+h}$$
The difference quotient DQ is
$$\frac{g(t+h) - g(t)}{h}$$
$$= \frac{\frac{2}{t+h} - \frac{2}{t}}{h}$$
$$= \frac{\frac{2}{t+h} - \frac{2}{t}}{h} \cdot \frac{t(t+h)}{t(t+h)}$$
$$= \frac{2t - 2(t+h)}{h(t)(t+h)} = \frac{-2}{t(t+h)}.$$
$$\lim_{h \to 0} \frac{g(t+h) - g(t)}{h} = -\frac{2}{t^2}$$
The slope is $m = g'\left(\frac{1}{2}\right) = -8$

7. If $f(x) = \sqrt{x}$ then
$$f(x+h) = \sqrt{x+h}$$
The difference quotient DQ is
$$\frac{f(x+h) - f(x)}{h}$$
$$= \frac{\sqrt{x+h} - \sqrt{x}}{h}$$
$$= \frac{\sqrt{x+h} - \sqrt{x}}{h} \cdot \frac{\sqrt{x+h} + \sqrt{x}}{\sqrt{x+h} + \sqrt{x}}$$
$$= \frac{x+h-x}{h(\sqrt{x+h} + \sqrt{x})}$$

$$\lim_{h\to 0} \frac{f(x+h) - f(x)}{h} = \frac{1}{2\sqrt{x}}$$

The slope is $m = f'(9) = \frac{1}{6}$

9. $f(x) = x^2 + x + 1$

$f(x+h) = x^2 + 2xh + (h)^2 + x + h + 1$

The difference quotient DQ is
$$\frac{f(x+h) - f(x)}{h}$$
$$= \frac{x^2 + 2xh + (h)^2 + x + h + 1}{h}$$
$$- \frac{(x^2 + x + 1)}{h}$$
$$= \frac{h(2x + h + 1)}{h}.$$

Thus $f'(x) = \lim_{h\to 0}(2x + h + 1) = 2x + 1$

The slope is $m = f'(2) = 5$

$f(2) = 7$. $P(2, 7)$ is a point on the curve. The equation of the tangent line is
$$\frac{y - 7}{x - 2} = 5 \text{ or } y = 5x - 3$$

11. $f(x) = \frac{3}{x^2}$

$$f(x+h) = \frac{3}{(x+h)^2}$$

$$f(x+h) - f(x) = \frac{3}{(x+h)^2} - \frac{3}{x^2},$$

which is three times that of problem 6. Thus
$$f'(x) = -\frac{6}{x^3}$$

The slope is
$$m = f'\left(\frac{1}{2}\right) = -6 \times 2^3 = -48$$

$f\left(\frac{1}{2}\right) = 12$. $P\left(\frac{1}{2}, 12\right)$ is a point on the curve.
The equation of the tangent line is
$$\frac{y - 12}{x - 1/2} = -48$$
$$\text{or } y = -48x + 36$$

13. $y = 3$ at $x = 3$ and at $x = x + h$
$$f(x+h) - f(x) = 0$$
so $\frac{dy}{dx} = 0$ for all x, including $x_0 = 2$.

15. $y = f(x) = x(1 - x) = x - x^2$

At $x + h$
$$y = f(x+h)$$
$$= x + h - x^2 - 2xh - (h)^2$$

The difference quotient DQ is
$$\frac{f(x+h) - f(x)}{h}$$
$$= \frac{x + h - x^2 - 2xh - (h)^2 - x + x^2}{h}$$
$$= \frac{h(1 - 2x - h)}{h}.$$

Thus $f'(x) = \lim_{h\to 0}(1 - 2x - h) = 1 - 2x$

$\frac{dy}{dx} = 1 - 2x$, so $y' = 1 + 2 = 3$ at $x_0 = -1$.

17. (a) If $f(x) = x^3$, then $f(1) = 1$, $f(1.1) = (1.1)^3 = 1.331$.
The slope of the secant line joining the points $(1, 1)$ and $(1.1, 1.331)$ on the graph of f is
$$m_s = \frac{\Delta y}{h} = \frac{1.331 - 1}{1.1 - 1} = 3.31$$

(b) $f(x) = x^3$

$f(x+h) = x^3 + 3x^2(h) + 3x(h)^2 + (h)^3$

The difference quotient DQ is
$$\frac{f(x+h) - f(x)}{h}$$
$$= \frac{x^3 + 3x^2(h) + 3x(h)^2 + (h)^3 - x^3}{h}$$
$$= \frac{h[3x^2 + 3xh + (h)^2]}{h}.$$

Thus $f'(x) = \lim_{h\to 0}[3x^2 + 3xh + (h)^2]$
$= 3x^2.$

2.1. THE DERIVATIVE: SLOPE AND RATES

The slope is $m_t = f'(1) = 3$

Notice that this slope was approximated by the slope of the secant in part (a).

19. $f(x) = x^3 + 3x^2 = x^2(x+3)$

$$\frac{f(x+h)-f(x)}{h}$$
$$= \frac{[(x+h)^3 + 3(x+h)^2] - [x^3 + 3x^2]}{h}$$
$$= \frac{x^3 + 3x^2h + 3x(h)^2 + (h)^3 + 3x^2}{h}$$
$$+ \frac{6xh + 3(h)^2 - x^3 - 3x^2}{h}$$

Thus $f'(x) = \lim_{h \to 0} DQ$
$$= 3x^2 + 6x = 3x(x+2)$$

The derivative is 0 when $x = 0$ or $x = -2$.
$f(0) = 0$ and $f(-2) = -8 + 12 = 4$.
$(-2, 4)$ is a maximum and $(0, 0)$ is a minimum.

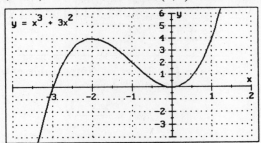

21. $P(x) = 400(15-x)(x-2) = 400(-x^2 + 17x - 30)$

(a) $P(x+h) = 400[-(x+h)^2 + 17(x+h) - 30]$

The difference quotient DQ is
$$\frac{P(x+h) - P(x)}{h}$$
$$= \frac{400[-(x+h)^2 + 17(x+h) - 30]}{h}$$
$$- \frac{400[-x^2 + 17x - 30]}{h}$$
$$= \frac{400[-2xh + (h)^2 + 17h]}{h}$$
$$= 400(-2x + h + 17)$$

$\lim_{h \to 0} \frac{f(x+h) - f(x)}{h} = 400(-2x + 17)$
$$P'(x) = 400(-2x + 17)$$

(b) $P'(x) = 0$ when $x = 8.5$. $P(8.5)$ is a maximum.

23. $y = 1 - x^2$
$$y|_{x+h} = 1 - [x^2 + 2xh + (h)^2]$$
$$y|_{x+h} - y|_x = -h(2x + h)$$
$$y' = -\lim_{h \to 0}(2x + h) = -2x$$

For a maximum (or minimum) of a function at a point for which the (first) derivative is defined, the tangent line is horizontal.
The slope is zero at this maximum. In the given problem the slope is 0 if $x = 0$. The highest point occurs when $y = 1 - 0^2 = 1$.

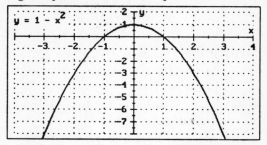

25.
$$H'(t) = \lim_{h \to 0} \frac{1}{h}[4.4(t+h)$$
$$-4.9(t+h)^2 - 4.4t + 4.9t^2]$$
$$= \lim_{h \to 0} \frac{1}{h}[4.4t + 4.4h - 4.9t^2 -$$
$$9.8th - 4.9(h)^2 - 4.4t + 4.9t^2]$$
$$= \lim_{h \to 0} \frac{1}{h}[4.4h - 9.8th - 4.9(h)^2]$$
$$= \lim_{h \to 0}[4.4 - 9.8t - 4.9h]$$
$$= 4.4 - 9.8t$$

$H'(t) = 0$ when $4.4 - 9.8t = 0$, or $t = 0.45$ sec.
$H(.45) = 4.4(.45) - 4.9(.45)^2 = 0.988$ meters

27. (a)
$$f(x) = 3x - 2$$
$$f(x+h) = 3x + 3h - 2$$

The difference quotient DQ is
$$DQ = \frac{f(x+h) - f(x)}{h}$$
$$= \frac{3x + 3h - 2 - 3x + 2}{h} = 3$$

Thus $f'(x) = 3$.

(b) $m = f'(-1) = 3$. The equation of the tangent line is

$$\frac{y - (-5)}{x + 1} = 3 \text{ or } y = 3x - 2$$

(c) The tangent to a line at a point on the line is the original line itself.

29. (a)
$$\begin{aligned} y|_{x+h} - y|_x &= x^2 + 2xh + (h)^2 + 3x \\ &\quad + 3h - x^2 - 3x \\ &= h(2x + h + 3) \\ y'_a &= 2x + 3 \end{aligned}$$

(b) If $y = x^2$ then $y'_{b1} = 2x$ (see problem 18.b). If $y = 3x$ then $y'_{b2} = 3$ (see problem 27.a).

(c) The derivative in (a) is the sum of the derivatives in (b).

(d) A possible guess is that $f(x) = g(x) + h(x)$ could lead to

$$f'(x) = g'(x) + h'(x)$$

31. The difference quotient DQ is the change in f divided by the change in h.
The derivative f' is the limit of DQ as $h \to 0$.
With $\Delta f > 0$ (and $h > 0$, as a rule) this means that the function is rising.
Similarly if $f'(x) < 0$ the graph of the function falls.

33. $$f'(x) = x^2 - x - 6$$

is such a function.

$$f'(x) = (x+2)(x-3) = x^2 - x - 6 = 0$$

at $x = -2$ and $x = 3$ qualifies for the derivative function.
Chapter 5 on antidifferentiation will show how to find a suitable function like

$$f(x) = \begin{cases} (x+2)^3 & \text{for } x \leq 1 \\ -\frac{27}{4}(x-3)^2 + 54 & \text{for } 1 < x \end{cases}$$

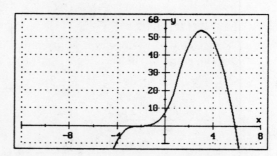

35. $$f(x) = \sqrt{x^2 - 2x - \sqrt{3x}}$$

h	$-.02$	$-.01$	$-.001$	0
$x+h$	3.83	3.84	3.849	3.85
$f(x)$	4.373	4.373	4.373	4.373
$f(x+h)$	4.352	4.363	4.372	4.373
$\frac{f(x+h)-f(x)}{h}$	1.05	1	1	

h	$.001$	$.01$	$.02$
$x+h$	3.851	3.86	3.87
$f(x)$	4.373	4.373	4.373
$f(x+h)$	4.374	4.384	4.394
$\frac{f(x+h)-f(x)}{h}$	1	1.1	1.05

Thus $m = f'(3.85) \approx 1.1$ Note: Table entries were limited to three decimal places.

37. $$f(x) = \frac{|x^2 - 1|}{x - 1}$$

is not defined at $x = 1$. There can be no point of tangency.

$$\lim_{x \to 1^+} \frac{|x^2 - 1|}{x - 1} = \lim_{x \to 1^+} \frac{|(x-1)(x+1)|}{x - 1} = 2.$$

and

$$\lim_{x \to 1^-} \frac{|x^2 - 1|}{x - 1} = \lim_{x \to 1^-} \frac{|(x-1)(x+1)|}{x - 1} = -2.$$

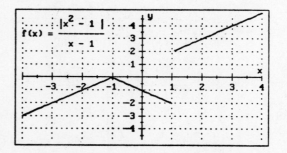

2.2 Techniques of Differentiation

1.
$$y = x^{-4}$$
$$y' = -4x^{-4-1} = -4x^{-5}$$

3.
$$y = \frac{9}{\sqrt{t}} = 9t^{-1/2}$$
$$y' = 9\left(\frac{-1}{2}\right)t^{-3/2}$$

5.
$$y = x^2 + 2x + 3,$$
$$\frac{dy}{dx} = 2x + 2$$

7.
$$f(x) = x^9 - 5x^8 + x + 12,$$
$$f'(x) = 9x^8 - 40x^7 + 1$$

9.
$$y = \frac{1}{t} + \frac{1}{t^2} - \frac{1}{\sqrt{t}}$$
$$= t^{-1} + t^{-2} - t^{-1/2},$$
$$\frac{dy}{dt} = -t^{-2} + (-2t^{-3}) - (-\frac{1}{2}t^{-3/2})$$
$$= -\frac{1}{t^2} - \frac{2}{t^3} + \frac{1}{2\sqrt{t^3}}.$$

11.
$$f(x) = \sqrt{x^3} + \frac{1}{\sqrt{x^3}} = x^{3/2} + x^{-3/2},$$
$$f'(x) = \frac{3}{2}x^{1/2} - \frac{3}{2}x^{-5/2} = \frac{3}{2}\sqrt{x} - \frac{3}{2\sqrt{x^5}}.$$

13.
$$y = -\frac{x^2}{16} + \frac{2}{x} - x^{3/2} + \frac{1}{3x^2} + \frac{x}{3}$$
$$= -\frac{1}{16}x^2 + 2x^{-1} - x^{3/2} + \frac{1}{3}x^{-2} + \frac{1}{3}x,$$
$$\frac{dy}{dx} = -\frac{1}{16}(2x) + 2(-x^{-2}) - \frac{3}{2}x^{1/2}$$
$$+ \frac{1}{3}(-2x^{-3}) + \frac{1}{3}$$
$$= -\frac{1}{8}x - \frac{2}{x^2} - \frac{3}{2}x^{1/2} - \frac{2}{3x^3} + \frac{1}{3}.$$

15.
$$y = x^2(x^3 - 6x + 7) = x^5 - 6x^3 + 7x^2$$
$$y' = 5x^4 - 6(3x^2) + 7(2x) = 5x^4 - 18x^2 + 14x$$

17.
$$y = x^5 - 3x^3 - 5x + 2$$
and $P(1, -5)$.
$$y' = 5x^4 - 9x^2 - 5$$
At $x = 1$, $y' = m = 5 - 9 - 5 = -9$.
The equation of the tangent line is
$$\frac{y+5}{x-1} = -9,$$
$$y + 5 = -9x + 9$$
or $y = -9x + 4$

19.
$$y = \sqrt{x^3} - x^2 + \frac{16}{x^2}$$
$$= x^{3/2} - x^2 + 16x^{-2} \text{ and } P(4, -7)$$
$$y' = \frac{3}{2}x^{1/2} - 2x + 16(-2)x^{-3}$$
At $x = 4$, $y' = m = 3 - 8 - \frac{1}{2} = -\frac{11}{2}$.
The equation of the tangent line is
$$\frac{y+7}{x-4} = -\frac{11}{2},$$
$$2y + 11x = 30$$

21.
$$f(x) = x^4 - 3x^3 + 2x^2 - 6$$
$$f(2) = 16 - 24 + 8 - 6 = -6$$
$f'(x) = 4x^3 - 9x^2 + 4x$,, the slope is $m = f'(2) = 4$
and the equation of the tangent line is
$$\frac{y+6}{x-2} = 4, \text{ or } y = 4x - 14$$

23.
$$f(x) = x - x^{-2}$$
$$f'(x) = 1 - (-2x^{-3}) = 1 + \frac{2}{x^3}$$
$f'(1) = 3$, $f(1) = 0$, and the equation of the tangent line is
$$y = 3x - 3$$

25.
$$f(x) = x^3 - 3x + 5$$
$$f'(x) = 3x^2 - 3 \text{ and } f'(2) = 12 - 3 = 9$$

27.
$$f(x) = x - \sqrt{x} + \frac{1}{x^2} = x - x^{-1/2} + x^{-2}$$

$$f'(x) = 1 - \frac{1}{2}x^{-1/2} - 2x^{-3}$$

$$f'(1) = 1 - \frac{1}{2} - 2 = -\frac{3}{2}$$

29.
$$f(x) = x^2 - 4x - 5 = (x-5)(x+1)$$
$$f'(x) = 2x - 4 = 0$$

when $x = 2$. At the lowest point on this curve, $f'(x) = 0$. $f(2) = -9$, so the lowest point is $(2, -9)$.

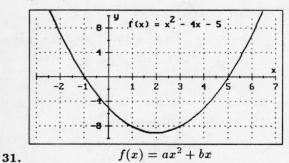

31. $f(x) = ax^2 + bx$

Since $(3, -8)$ is the lowest point, $f(3) = -8$ and $f'(3) = 0$. $f(3) = 9a + 3b = -8$ and $f'(x) = 2ax + b$, $f'(3) = 6a + b = 0$. Substitute $b = -6a$ to get $9a - 18a = -8$ or $a = \frac{8}{9}$ and $b = -\frac{16}{3}$.

33. $$f(x) = x^2 - 4x + 25$$

$f'(x) = 2x - 4$. Let (x,y) be the point of contact.
The tangent line also passes through $(0,0)$ so that the slope

$$f'(x) = \frac{y - 0}{x - 0} = \frac{y}{x}$$

Thus $2x - 4 = \frac{y}{x}$ or $2x^2 - 4x = y$.
Since $y = x^2 - 4x + 25$ also,
$2x^2 - 4x = x^2 - 4x + 25$ leads to $x = \pm 5$.
At $x = 5$ $f'(5) = 6$ and the tangent line has equation $y = 6x$.
At $x = -5$ $f'(-5) = -14$ and the tangent line has equation $y = -14x$.

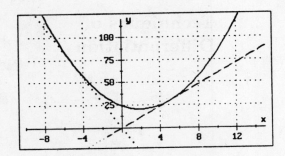

35. **(a)** With $E(p)$ denoting the total monthly expenditure,

$$E(p) = \text{(monthly demand)}$$
$$\text{(price per unit)}$$

The monthly demand
$$D(p) = -200p + 12{,}000$$

units at p dollars per unit. Thus
$$E(p) = (-200p + 12{,}000)(p) = -200p(p-60)$$

(b) Consumer expenditure will be highest where the tangent line to the graph is horizontal or

$$E'(p) = -400p + 12{,}000 = 0$$

Thus $p = 30$ and $E(30) = \$180{,}000$.

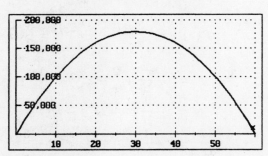

37. **(a)** Since
$$C(t) = 100t^2 + 400t + 5{,}000$$

is the circulation t years from now, the rate of change of the circulation t years from now is

$$C'(t) = 200t + 400$$

newspapers per year.

2.2. TECHNIQUES OF DIFFERENTIATION

(b) The rate of change of the circulation 5 years from now is

$$C'(5) = 200(5) + 400 = 1,400$$

newspapers per year; since this is a positive answer, circulation will be increasing at that time.

(c) The actual change in the circulation during the 6th year is $C(6) - C(5)$

$$\begin{aligned} &= [100(6^2) + 400(6) + 5,000] \\ &\quad -[100(5^2) + 400(5) + 5,000] \\ &= 1,500 \text{ newspapers} \end{aligned}$$

39. (a) Since

$$f(x) = -x^3 + 6x^2 + 15x$$

is the number of radios assembled x hours after 8:00 a.m.,
the rate at which the radios are being assembled x hours after 8:00 a.m. is

$$f'(x) = -3x^2 + 12x + 15$$

radios per hour.

(b) The rate of assembly at 9:00 a.m. (when $x = 1$) is
$f'(1) = -3(1^2) + 12(1) + 15 = 24$ radios per hour.

(c) The actual number of radios assembled between 9:00 a.m. (when $x = 1$) and 10:00 a.m. (when $x = 2$) is $f(2) - f(1)$
$= [-2^3 + 6(2^2) + 15(2)]$
$\quad - [-1^3 + 6(1^2) + 15(1)] = 26$ radios.

41. The speed of the first car C_1 is 60 kilometers per hour and the speed of the second car C_2 is 80 kilometers per hour.
After t hours, C_1 has traveled $60t$ kilometers and C_2 has traveled $80t$ kilometers, as shown in the figure.
Let $D(t)$ denote the distance between C_1 and C_2 at time t.
By the pythagorean theorem,

$$\begin{aligned} [D(t)]^2 &= (80t)^2 + (60t)^2 \\ &= 6,400t^2 + 3,600t^2 = 10,000t^2 \end{aligned}$$

and so $D(t) = \sqrt{10,000t^2} = 100t$ kilometers
The rate of change of this distance is the derivative $D'(t) = 100$ kilometers per hour and is independent of the time t.

43. (a)
$$P(x) = 2x + 4x^{3/2} + 5,000$$

is the population x months from now, the rate of population growth is

$$P'(x) = 2 + 4\left(\frac{3x^{1/2}}{2}\right) = 2 + 6x^{1/2}$$

people per month. Nine months from now, the population will be changing at the rate of

$$P'(9) = 2 + 6(9^{1/2}) = 20$$

people per month.

(b) The percentage rate at which the population will be changing 9 months from now is

$$\begin{aligned} 100 \frac{P'(9)}{P(9)} &= \frac{100(20)}{2(9) + 4(9^{3/2}) + 5,000} \\ &= \frac{2,000}{5,126} = 0.39 \text{ \% per month.} \end{aligned}$$

45. (a)
$$T(x) = 20x^2 + 40x + 600$$

dollars is the average property tax x years after 1985, the rate at which the average property tax changes is

$$T'(x) = 40x + 40$$

dollars per year. In 1991 (when $x = 6$) the rate of change is
$T'(6) = 40(6) + 40 = \$280$ dollars per year.

(b) The percentage rate at which the average property tax was increasing in 1991 was

$$\begin{aligned} 100\left(\frac{T'(6)}{T(6)}\right) &= \frac{100(280)}{20(6^2) + 40(6) + 600} \\ &= 17.95\% \end{aligned}$$

47. (a) Since your starting salary is $24,000 and you get a raise of $2,000 per year, your salary x years from now will be

$$S(x) = 24,000 + 2,000x \text{ dollars.}$$

The percentage rate of change of this salary x years from now is

$$100\left[\frac{S'(x)}{S(x)}\right] = 100\left(\frac{2,000}{24,000 + 2,000x}\right)$$
$$= \frac{100}{12 + x} \text{ percent per year.}$$

(b) The percentage rate of change after 1 year is

$$\frac{100}{13} = 7.69 \text{ percent per year.}$$

(c) In the long run, $\frac{100}{12+x}$ approaches 0. That is, the percentage rate of your salary will approach 0 (even though your salary will continue to increase at a constant rate).

49. (a) $\quad s(t) = t^2 - 2t + 6 \text{ for } 0 \leq t \leq 2$
$$v(t) = 2t - 2$$
$$a(t) = 2$$

(b) The particle is stationary when $t = 1$ since $v(1) = 0$.

51. (a)
$$s(t) = t^3 - 9t^2 + 15t + 25 \text{ for } 0 \leq t \leq 6$$
$$v(t) = 3t^2 - 18t + 15 = 3(t-1)(t-5)$$
$$a(t) = 3(2t - 6) = 6(t - 3)$$

(b) The particle is stationary when $t = 1$ and $t = 5$ since $v(1) = v(5) = 0$.

53. (a)
$$s(t) = 2t^4 + 3t^2 - 36t + 40 \text{ for } 0 \leq t \leq 3$$
$$v(t) = 8t^3 + 6t - 36$$
$$a(t) = 6(4t^2 + 1)$$

(b) The particle is stationary when 0, $t = 1.5$ and $t = 3$ since $v(1.5) = 0$.

55. (a) Since the initial speed is $S_0 = 0$ feet per second and the height of the building is $H_0 = 144$ feet,

$$H(t) = -16t^2 + S_0 t + H_0 = -16t^2 + 144$$

The stone hits the ground when $H(t) = 0$, that is, when $-16t^2 + 144 = 0$, $t^2 = 9$, or $t = 3$ seconds.

(b) The speed with which the stone hits the ground is $H'(3)$.
Since $H'(t) = -32t$, $H'(3) = -96$ feet per second.
The negative sign indicates that the direction is down.

57. Let the gravity be denoted by g.
Then the acceleration is $-g$, and the velocity is

$$v = -gt + v_0,$$

where v_0 is the initial velocity.
The rock reaches its maximum height in 2.5 seconds.
The velocity is 0 at this instant and $v_0 = \frac{5}{2}g$.
The distance is
$$s = -\frac{1}{2}gt^2 + v_0 t = \frac{1}{2}(5-t)gt.$$
$$s(2.5) = \frac{75}{2} = \frac{5}{4}\frac{5}{2}g$$

which leads to $g = 12$.
As a result our friendly spy finds himself on Mars.

59. Since y is a linear function of x, $y = mx + b$, where m and b are constants.
The rate of change is $\frac{dy}{dx} = m$, which is a constant.
Notice that this constant rate of change is the slope of the line.

61. The total manufacturing cost C is a function of q (where q is the number of units produced) and q is a function of t (where t is the number of hours during which the factory operates). Hence:

2.3. THE PRODUCT AND QUOTIENT RULES

(a) $\dfrac{dC}{dq}$ = the rate of change of cost with respect to the number of units produced in $\dfrac{\text{dollars}}{\text{unit}}$.

(b) $\dfrac{dq}{dt}$ = the rate of change of units produced with respect to time in $\dfrac{\text{units}}{\text{hour}}$.

(c) $\dfrac{dC}{dq}\dfrac{dq}{dt} = \dfrac{\text{dollars}}{\text{unit}}\dfrac{\text{units}}{\text{hour}} = \dfrac{\text{dollars}}{\text{hour}}$
= the rate of change of cost WRT time.

63. (a) $f(x+h) = (x+h)^3 = x^3 + 3hx^2 + 3h^2x + h^3$
$\dfrac{f(x+h) - f(x)}{h} = 3x^2 + 3hx + h^2$

(b) $f(x+h) = (x+h)^n = x^n + nhx^{n-1}$
$+\dfrac{n(n+1)}{2}x^{n-2}h^2 + \cdots + \dfrac{n(n-1)}{2}xh^{n-1} + h^n$
$\dfrac{f(x+h) - f(x)}{h} = nx^{n-1} + \dfrac{n(n-1)}{2}x^{n-2}h$
$+ \cdots + \dfrac{n(n-1)}{2}xh^{n-2} + h^{n-1}$

(c) $\dfrac{d}{dx}x^n$
$= \lim_{h \to 0}\left[nx^{n-1} + h\left(\dfrac{n(n+1)}{2}x^{n-2}\right)\right]$
$+ \cdots + \lim_{h \to 0}\left[\left(\dfrac{n(n+1)}{2}xh^{n-3} + h^{n-2}\right)\right]$
$= nx^{n-1}$

2.3 The Product and Quotient rules

1. $f(x) = (2x+1)(3x-2)$,
$f'(x) = (2x+1)\dfrac{d}{dx}(3x-2)$
$\qquad + (3x-2)\dfrac{d}{dx}(2x+1)$
$= 12x - 1$.

3. $y = 10(3u+1)(1-5u)$,
$\dfrac{dy}{du} = 10\dfrac{d}{du}[(3u+1)(1-5u)]$
$= 10[(3u+1)\dfrac{d}{du}(1-5u)$
$+ (1-5u)\dfrac{d}{du}(3u+1)] = -300u - 20$.

5. $f(x) = \dfrac{1}{3}(x^5 - 2x^3 + 1)$,
$f'(x) = \dfrac{1}{3}(x^5 - 2x^3 + 1)'$
$= \dfrac{1}{3}(5x^4 - 6x^2)$.

7. $y = \dfrac{x+1}{x-2}$,
$\dfrac{dy}{dx} = \dfrac{(x-2)(x+1)' - (x+1)(x-2)'}{(x-2)^2}$
$= \dfrac{-3}{(x-2)^2}$.

9. $f(t) = \dfrac{t}{t^2 - 2}$,
$f'(t) = \dfrac{(t^2-2)\dfrac{d}{dt}(t) - t\dfrac{d}{dt}(t^2-2)}{(t^2-2)^2}$
$= \dfrac{(t^2-2)(1) - (t)(2t)}{(t^2-2)^2} = \dfrac{-t^2 - 2}{(t^2-2)^2}$.

11. Method 1 (the hard way):
$y = \dfrac{3}{x+5}$,
$\dfrac{dy}{dx} = \dfrac{(x+5)\dfrac{d}{dx}(3) - 3\dfrac{d}{dx}(x+5)}{(x+5)^2}$
$= \dfrac{(x+5)(0) - 3(1)}{(x+5)^2} = \dfrac{-3}{(x+5)^2}$.

Method 2 (the easy way):

$y = \dfrac{3}{x+5} = 3(x+5)^{-1}$,

$\dfrac{dy}{dx} = -3(x+5)^{-2}$

Warning: This is not as simple as it appears.

13.
$$f(x) = \frac{x^2 - 3x + 2}{2x^2 + 5x - 1},$$

$$\begin{aligned}f'(x) &= \frac{(2x^2 + 5x - 1)\frac{d}{dx}(x^2 - 3x + 2)}{(2x^2 + 5x - 1)^2} \\ &\quad - \frac{(x^2 - 3x + 2)\frac{d}{dx}(2x^2 + 5x - 1)}{(2x^2 + 5x - 1)^2} \\ &= \frac{(2x^2 + 5x - 1)(2x - 3)}{(2x^2 + 5x - 1)^2} \\ &\quad - \frac{(x^2 - 3x + 2)(4x + 5)}{(2x^2 + 5x - 1)^2} \\ &= \frac{11x^2 - 10x - 7}{(2x^2 + 5x - 1)^2}.\end{aligned}$$

15.
$$f(x) = \frac{(2x - 1)(x + 3)}{x + 1} = \frac{u}{v}$$

Let $u = (2x - 1)(x + 3)$ then
$$u' = (2x - 1)(1) + (2)(x + 3) = 4x + 5$$

$$\begin{aligned}f'(x) &= \frac{vu' - uv'}{v^2} \\ &= \frac{(x + 1)(4x + 5) - (2x - 1)(x + 3)}{(x + 1)^2} \\ &= \frac{4x^2 + 9x + 5 - 2x^2 - 5x + 3}{(x + 1)^2} \\ &= \frac{2x^2 + 4x + 8}{(x + 1)^2}\end{aligned}$$

17.
$$y = (5x - 1)(4 + 3x)$$
$$y' = (5x - 1)(3) + (5)(4 + 3x)$$

At $x = 0$, $y' = 17$ and $y = -4$, thus
$$y + 4 = 17x$$

19.
$$\frac{x}{2x + 3}$$
$$\frac{2x + 3 - 2x}{(2x + 3)^2} = \frac{3}{(2x + 3)^2}$$

At $x = -1$ $y' = 3$ and $y = -1$, thus
$$y + 1 = 3(x + 1) \text{ or } y = 3x + 2$$

21.
$$\begin{aligned}f(x) &= (x + 2)(x + \sqrt{x}), \\ f'(x) &= (x + 2)\frac{d}{dx}(x + x^{1/2}) \\ &\quad + (x + \sqrt{x})\frac{d}{dx}(x + 2) \\ &= (x + 2)\left(1 + \frac{1}{2}x^{-1/2}\right) \\ &\quad + (x + \sqrt{x})(x), \\ f'(4) &= (4 + 2)\left[1 + \frac{1}{2}(4^{-1/2})\right] \\ &\quad + (4 + \sqrt{4})(4) \\ &= (6)\left(\frac{5}{4}\right) + (6)(4) = \frac{63}{2}\end{aligned}$$

23.
$$\begin{aligned}f(x) &= \frac{2x - 1}{3x + 5}, \\ f'(x) &= \frac{(3x + 5)(2) - (2x - 1)(3)}{(3x + 5)^2} \\ &= \frac{13}{(3x + 5)^2}, \\ f'(1) &= \frac{13}{8^2} = \frac{13}{64}.\end{aligned}$$

25.
$$y = x^2 + 3x - 5$$
$$y' = 2x + 3$$

At $x = 0$ $y' = 3$ so the slope of the perpendicular line is $m = -\frac{1}{3}$, thus $y + 5 = -\frac{1}{3}x$. The equation of the perpendicular line can be written as
$$x + 3y = -15$$

27.
$$y = (x + 3)(1 - \sqrt{x})$$
$$y' = (x + 3)\left(-\frac{1}{2\sqrt{x}}\right) + (1 - \sqrt{x})$$

At $x = 1$ $y' = -2$, so the slope of the perpendicular line is $m = \frac{1}{2}$. The equation of this line is
$$y = \frac{1}{2}(x - 1) \text{ or } x - 2y = 1$$

29.

2.3. THE PRODUCT AND QUOTIENT RULES

(a) $$y = 2x^2 - 5x - 3,$$
$$y' = 4x - 5$$

(b) $$y = (2x+1)(x-3),$$
$$y' = (2)(x-3) + (2x+1)(1) = 4x - 5$$

31. $$P(t) = \frac{24t + 10}{t^2 + 1}$$

$$P'(t) = \frac{(t^2+1)(24) - (24t+10)(2t)}{(t^2+1)^2}$$
$$= \frac{-4(6t^2 + 5t - 6)}{(t^2+1)^2}$$
$$= \frac{-4(2t+3)(3t-2)}{(t^2+1)^2}$$

Therefore $P'(t) = 0$ when $t = -\frac{3}{2}$ or $t = \frac{2}{3}$.
The population is greatest when $t = \frac{2}{3}$ hours.
The maximum population is 18 thousand.

33. $$f(x) = x(x+3)^2 = x^3 + 6x^2 + 9x$$
$$f'(x) = 3x^2 + 12x + 9$$

and the percentage rate of change is

$$\frac{300(x^2 + 4x + 3)}{x(x+3)^2}$$

When $x = 3$ the rate is 66.67 %.

35. (a) Since
$$P(t) = 20 - \frac{6}{t+1} = 20 - 6(t+1)^{-1}$$

is the population (in thousands) t years from now,
the rate at which the population is changing t years from now is

$$P'(t) = 0 - (-6)(t+1)^{-2} = \frac{6}{(t+1)^2}$$

thousand per year.

(b) One year from now, the rate of change will be
$$P'(1) = \frac{6}{(1+1)^2} = 1.5$$
or 1,500 people per year.

(c) The actual population increase during the second year is
$P(2) - P(1)$
$= 20 - \frac{6}{2+1} - \left(20 - \frac{6}{1+1}\right) = 1$
or one thousand people.

(d) Nine years from now the rate of change will be
$$P'(9) = \frac{6}{100} \text{ thousand}$$
or 60 people per year.

(e) As t increases,
$\frac{6}{(t+1)^2}$ approaches zero.
The rate of population growth will approach 0 in the long run.

37. $$N(t) = 5,175 - t^3(t-8)$$

(a) $$N'(t) = -4t^3 + 24t^2$$
$N'(3) = 108$ people per week.

(b) We need $100 \frac{N'(t)}{N(t)} \geq 25$ or

$$100 \frac{-4t^3 + 24t^2}{5,175 - t^3(t-8)} \geq 25$$

A graphics utility shows that this inequality is never satisfied on the given interval.

(c) Writing Exercise — Answers will vary.

39. (a) The ratio was greatest in 1945 and least in 1975.

(b) The ratio was increasing most rapidly in the years immediately preceding 1945. Let's estimate that in 1942 the ratio was 40% and in 1945 it was 110%, then the rate here is $\frac{110 - 40}{3} = 23.3$ % per year.

(c) In 1992 the ratio was approximately 50 and in 1993 it was 60. This rate is $\frac{60 - 50}{1} = 10$% per year.

41. (a) Let $F = fg$, then

$$\frac{d}{dx}(fgh) = \frac{d}{dx}(Fh) = \frac{dF}{dx}h + F\frac{dh}{dx}$$
$$= \left(\frac{df}{dx}g + f\frac{dg}{dx}\right)h + fg\frac{dh}{dx}$$
$$= fg\frac{dh}{dx} + fh\frac{dg}{dx} + gh\frac{df}{dx}.$$

(b)
$$y = (2x+1)(x-3)(1-4x),$$
$$\frac{dy}{dx} = (2x+1)(x-3)(-4)$$
$$+(2x+1)(1)(1-4x)$$
$$+(2)(x-3)(1-4x)$$
$$= -24x^2 + 44x + 7$$

43. $\quad \dfrac{d}{dx}(cf) = c\dfrac{d}{dx}f + (0)(f) = c\dfrac{df}{dx}$

45.
$$\frac{d}{dx}x^{-p} = \frac{d}{dx}\frac{1}{x^p}$$
$$= \frac{-px^{p-1}}{x^{2p}} = -px^{-p-1} = nx^{n-1}$$

47. $\quad f(x) = \dfrac{3x^2 - 4x + 1}{x+1}$

$f'(x) = 0$ at $x = -2.63$ and $x = 0.633$.

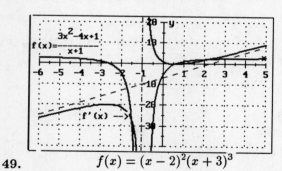

49. $\quad f(x) = (x-2)^2(x+3)^3$

$$f'(x) = 2(x-2)(x+3)^3 + 3(x+3)^2(x-2)^2$$
$$= (x-2)(x+3)^2[(2(x+3) + 3(x-2)]$$
$$= (x-2)(x+3)^2(5x)$$

$f(x)$ has a maximum at $x = 0$ and a minimum at $x = 2$. $x = -3$ will be found to be a point of inflection in chapter 3.
$f'(x)$ has x-intercepts at these same places because an intercept occurs when $f'(x) = 0$ i.e. when $f(x)$ has a horizontal tangent. In this case these extreme values of $f(x)$ occur where $f(x)$ has horizontal tangents.
[2.6in]

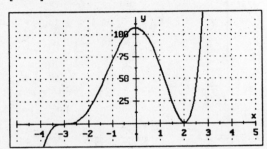

2.4 Marginal Analysis; Approximation by Increments

1. $\quad C(q) = 0.1q^3 - 0.5q^2 + 500q + 200$

(a) $\quad C'(q) = 0.3q^2 - q + 500$

The cost of the 4th unit is
$C'(3) = 0.3 \times 9 - 3 + 500 = \$499.70$.

(b)
$C(4) = 0.1 \times 64 - 0.5 \times 16 + 2,000 + 200$

$C(3) = 0.1 \times 27 - 0.5 \times 9 + 1,500 + 200$

and $C(4) - C(3) = \$500.20$.

3. (a)
$$C'(x) = \frac{2}{5}x + 4$$
$$p(x) = 9 - \frac{x}{4}$$
$$R(x) = 9x - \frac{x^2}{4}$$
$$R'(x) = 9 - \frac{x}{2}$$

(b) $\quad C'(3) = \dfrac{6}{5} + 4 = \$5.2$

(c)
$$C(4) - C(3) = \left[\frac{1}{5}4^2 + 16 + 57\right]$$
$$- \left[\frac{1}{5}3^2 + 12 + 57\right] = \$5.4$$

2.4. MARGINAL ANALYSIS; APPROXIMATION BY INCREMENTS

(d) $R'(3) = 9 - \dfrac{3}{2} = 7.5$

(e) $R(4) - R(3) = 9(4) - 4 - 9(3) + \dfrac{9}{4}$
$= 7.25$

5. (a) $C'(x) = \dfrac{2}{3}x + 2$

$p(x) = -x^2 + 4x + 10$

$R(x) = -x^3 + 4x^2 + 10x$

$R'(x) = -3x^2 + 8x + 10$

(b) $C'(3) = 2 + 2 = 4$

(c) $C(4) - C(3) = \left[\dfrac{16}{3} + 8 - 3 - 6\right] = 4.33$

(d) $R'(3) = -27 + 24 + 10 = 7$

(e) $R(4) - R(3) = -64 + 64 + 40 + 27 - 27 - 30 = 18$

7. (a) $C'(x) = \dfrac{x}{2}$

$p(x) = \dfrac{3 + 2x}{1 + x}$

$R(x) = \dfrac{3x + 2x^2}{1 + x}$

$R'(x) = \dfrac{(1 + x)(3 + 4x) - (3x + 2x^2)}{(1 + x)^2}$

$\dfrac{3 + 4x + 2x^2}{(1 + x)^2}$

(b) $C'(3) = \dfrac{3}{2} = 1.5$

(c) $C(4) - C(3) = 4 - \dfrac{9}{4} = 1.25$

(d) $R'(3) = \dfrac{3 + 12 + 18}{16} = 2.06$

(e) $R(4) - R(3) = \dfrac{12 + 32}{5} - \dfrac{9 + 18}{4} = 2.05$

9. The rate of change in the function is approximately the derivative of the function times the change in its variable, that is, $f'(x)h$

Since $f(x) = x^2 - 3x + 5$ then

$f'(x) = 2x - 3$

As x increases from 5 to 5.3, $h = 0.3$ and so

$f'(3)h = [2(5) - 3](0.3) = 2.1$

is an estimated change.

11. The percentage change in a function is

$$100\dfrac{\Delta f}{f(x)}$$

Given $f(x) = x^2 + 2x - 9$

and x increases from 4 to 4.3, then $h = 0.3$,

$\Delta f = f(4.3) - f(4)$
$= [(4.3)^2 + 2(4.3) - 9] - [4^2 + 2(4) - 9]$
$= 3.09$

and $f(4) = 15$.
Thus the percentage change is

$$f = 100\dfrac{\Delta f}{f(x)} = 100\left(\dfrac{3.09}{15}\right) = 20.6\%$$

13. Since the cost is

$$C(q) = 0.1q^3 - 0.5q^2 + 500q + 200,$$

the change in cost resulting from an increase in production from 4 units to 4.1 units ($\Delta q = 0.1$) is

$$\Delta C = C(4.1) - C(4) \approx [C'(4)](0.1)$$

Since $C'(q) = 0.3q^2 - q + 500$

and $C'(4) = 500.80$ it follows that

$$\Delta C \approx (500.80)(0.1) = 50.08$$

That is the cost will increase by approximately $50.08.

15. $$Q(L) = 300L^{2/3},$$
$L = 512 = 2^9$, and $\Delta Q = 12.5$.
$$Q'(L) = 300 \times \frac{2}{3}L^{-1/3} = 200L^{-1/3},$$
$Q'(512) = 200 \times 2^{-3} = 25$.
$\Delta Q = 12.5 = 25\Delta L$, so $\Delta L = 0.5$ worker-hours.

17. At the end of 1996 (or the beginning of 1997), t will be 3 because it will be three years after 1993.
$$T(x) = 60x^{3/2} + 40x + 1{,}200$$
$$\text{and } T'(x) = 90x^{1/2} + 40$$
$T'(3) = 90\sqrt{3} + 40 = 195.88$.
$h = 0.5$ (for $\frac{1}{2}$ year) and
$$h = 195.88 \times 0.5 = 97.94$$
Now
$$T(3) = 60 \times 3^{3/2} + 120 + 1{,}200 = 1631.77$$
The percentage rate of change is
$$\frac{100 \times 97.94}{1631.77} \approx 6.0\%.$$

19. $$f(x) = -x^3 + 6x^2 + 15x$$
$$f'(x) = -3x^2 + 12x + 15$$
$h = 0.25$ (a quarter of 60 minutes), $x = 1$ (one hour after 8:00 a.m.)
$f'(1) = 24$, $\Delta f \approx 24 \times 0.25 = 6$ radios.

21. $$Q(L) = 60{,}000L^{1/3}$$
$L = 1{,}000$, $\Delta L = 940 - 1{,}000 = -60$.
$$Q'(L) = 20{,}000L^{-2/3}$$
$$Q'(1{,}000) = \frac{20{,}000}{(\sqrt[3]{1{,}000})^2} = 200$$
$\Delta Q \approx 200 \times (-60) = -12{,}000$ units.

23. (a) $$C(q) = 3q^2 + q + 500,$$
the marginal cost is
$$C'(q) = 6q + 1,$$
$C'(40) = 241$.
(b)
$$C(41) - C(40)$$
$$= 3(41)^2 + 41 + 500 - [3(40)^2 + 40 + 500]$$
$$= 244$$

25. $$P(t) = 20 - \frac{6}{t+1}$$
$$P'(t) = (-6)\frac{(t+1)(0) - (1)(1)}{(t+1)^2} = \frac{6}{(t+1)^2}$$
$h = 0.25$, $t = 0$, $P'(0) = 6$,
$\Delta P \approx 6 \times 0.25 = 1.5$ or $1{,}500$ people (P is in units of $1{,}000$.)

27. $$A(r) = \pi r^2, \quad A(12) = 144\pi \text{ cm}^2$$
$\Delta r = 0.03 \times 12 = 0.36$ cm,
$A'(r) = 2\pi r$, $A'(12) = 24\pi$,
$$\Delta A = 24\pi \times 0.36 = 27.14 \text{ cm}^2$$
$$100\frac{\Delta A}{A} = 6\%$$

29. $$Q = 600K^{1/2}L^{1/3}$$
$$\Delta L = 0.02L$$
$$Q'(L) = 600K^{1/2}\left(\frac{1}{3}L^{-2/3}\right)$$
$$= \frac{200K^{1/2}}{L^{2/3}}$$
$$\Delta Q = \left(\frac{200K^{1/2}}{L^{2/3}}\right)(0.02L)$$
$$= 4K^{1/2}L^{1/3}$$
The percentage rate of change is
$$\frac{100 \times 4K^{1/2}L^{1/3}}{600K^{1/2}L^{1/3}} = 0.67\%.$$

31. $$\Delta S \approx 8\pi r(0.01r) = 0.08\pi r^2$$
$$\Delta V \approx 4\pi r^2(0.01r) = 0.04\pi r^3$$

2.4. MARGINAL ANALYSIS; APPROXIMATION BY INCREMENTS

33.
$$V = \frac{4\pi r^3}{3},$$
the inner radius is $\frac{17}{4}$ inches, $\Delta r = \frac{1}{8}$,
$$V'(r) = 4\pi r^2,$$

$$\Delta V \approx V'\left(\frac{17}{4}\right)\frac{1}{8}$$
$$= 4\pi\left(\frac{17}{4}\right)^2 \frac{1}{8} = 28.37 \text{ cubic inches.}$$

35. (a)
$$V = kR^4,$$
$$dV = 4kR^3 dR \text{ and } \frac{4kR^3 dR}{kR^4} = 4\frac{dR}{R} = 0.2$$
or 20 %.

(b) Writing exercise —
Answers will vary.

37. If $R(T) = kT^4$ then the percentage change in R is
$$100\,\frac{4kT^3}{kT^4}h \text{ or } \frac{400}{T}h$$
If $h = 0.02T$ then the percentage change in R is $\frac{400(0.02T)}{T} = 8\%$.

39. (a) For the HP48G:
Plot 'X³-X²-1' on you HP48G calculator. The root is
$$x = 1.46557123188$$

For the TI-85:
Graph $x^3 - x^2 - 1$.
Press ZOOM then F1 for BOX. Move the cross-hair to the left, above the root.
Press Enter.
Now move the cross-hair to the right and below the x intercept. ENTER. The box is defined and exploded.
Press M4 to TRACE. Move the cross-hair as close to the x intercept as possible by using the arrow keys. The root is about $x = 1.46$.

(b)
$$f(x) = x^3 - x^2 - 1$$
$$f'(x) = 3x^2 - 2x$$
$$x - \frac{f(x)}{f'(x)} = x - \frac{x^3 - x^2 - 1}{3x^2 - 2x}$$
$$= \frac{2x^3 - x^2 + 1}{3x^2 - 2x}.$$

Thus $x_n = \frac{2x_{n-1}^3 - x_{n-1}^2 + 1}{3x_{n-1}^2 - 2x_{n-1}}$

Pick $x_0 = 1$. Then
$$x_1 = \frac{2x_0^3 - x_0^2 + 1}{3x_0^2 - 2x_0} = 2,$$
$$x_2 = \frac{2x_1^3 - x_1^2 + 1}{3x_1^2 - 2x_1} = 1.6250,$$
$$x_3 = \frac{2x_2^3 - x_2^2 + 1}{3x_2^2 - 2x_2} = 1.4858,$$
$$x_4 = \frac{2x_3^3 - x_3^2 + 1}{3x_3^2 - 2x_3} = 1.4660,$$
$$x_5 = \frac{2x_4^3 - x_4^2 + 1}{3x_4^2 - 2x_4} = 1.4656$$

(c) The accuracy is the same since both answers were rounded to four decimal places.

41. $x = \sqrt{N}$, $f(x) = x^2 - N$, $f'(x) = 2x$
Now $x - \frac{x^2 - N}{2x} = \frac{2x^2 - x^2 + N}{2x}$
$$= \frac{1}{2}\left(x_n + \frac{N}{x_n}\right)$$

This leads to $x_{n-1} = \frac{1}{2}\left(x_n + \frac{N}{x_n}\right)$

Let's start with $x_0 = 35$ then
the root will be approximately 35.56684
because
$$x_1 = \frac{1}{2}\left(35 + \frac{1,265}{35}\right) = 35.5714$$
$$x_2 = \frac{1}{2}\left(35.5714 + \frac{1,265}{35.5714}\right) = 35.5668$$
$$x_3 = \frac{1}{2}\left(35.5668 + \frac{1,265}{35.5668}\right) = 35.566838$$

Thus $\sqrt{1,265} \approx 35.5668$.

2.5 The Chain Rule

1. $y = u^2 + 1$, $u = 3x - 2$, $\dfrac{dy}{du} = 2u$,

   $\dfrac{du}{dx} = 3$,

   $\dfrac{dy}{dx} = \dfrac{dy}{du}\dfrac{du}{dx} = (2u)(3) = 6(3x-2)$

3. $y = \sqrt{u} = u^{1/2}$, $u = x^2 + 2x - 3$,

   $\dfrac{dy}{du} = \dfrac{1}{2}u^{-1/2} = \dfrac{1}{2u^{1/2}}$,

   $\dfrac{du}{dx} = 2x + 2$,

   $\dfrac{dy}{dx} = \dfrac{dy}{du}\dfrac{du}{dx} = \dfrac{x+1}{(x^2+2x-3)^{1/2}}$.

5. $y = \dfrac{1}{u^2} = u^{-2}$, $u = x^2 + 1$,

   $\dfrac{dy}{du} = -2u^{-3}$, $\dfrac{du}{dx} = 2x$,

   $\dfrac{dy}{dx} = \dfrac{dy}{du}\dfrac{du}{dx} = \dfrac{-4x}{(x^2+1)^3}$

7. $y = \dfrac{1}{\sqrt{u}} = u^{-1/2}$, $u = x^2 - 9$,

   $\dfrac{dy}{du} = -\dfrac{1}{2}u^{-3/2} = \dfrac{-1}{2u^{3/2}}$,

   $\dfrac{du}{dx} = 2x$,

   $\dfrac{dy}{dx} = \dfrac{dy}{du}\dfrac{du}{dx} = \dfrac{-x}{(x^2-9)^{3/2}}$.

9. $y = \dfrac{1}{u-1} = (u-1)^{-1}$, $u = x^2$

   $\dfrac{dy}{du} = -(u-1)^{-2} = \dfrac{-1}{(u-1)^2}$,

   $\dfrac{du}{dx} = 2x$,

   $\dfrac{dy}{dx} = \dfrac{dy}{du}\dfrac{du}{dx}$

   $= \dfrac{-2x}{(x^2-1)^2}$.

11. $y = 3u^4 - 4u + 5$, $u = x^3 - 2x - 5$

    $\dfrac{dy}{du} = 12u^3 - 4$, $\dfrac{du}{dx} = 3x^2 - 2$,

    $\dfrac{dy}{dx} = \dfrac{dy}{du}\dfrac{du}{dx} = (12u^3 - 4)(3x^2 - 2)$.

    When $x = 2$, $u = 2^3 - 2(2) - 5 = -1$ and so

    $\dfrac{dy}{dx} = [12(-1)^3 - 4][3(2^2) - 2] = -160$

13. $y = \sqrt{u} = u^{1/2}$, $u = x^2 - 2x + 6$,

    $\dfrac{dy}{du} = \dfrac{1}{2}u^{-1/2} = \dfrac{1}{2u^{1/2}}$,

    $\dfrac{du}{dx} = 2x - 2$,

    $\dfrac{dy}{dx} = \dfrac{dy}{du}\dfrac{du}{dx}$

    $= \dfrac{x-1}{u^{1/2}}$.

    When $x = 3$, $u = 3^2 - 2(3) + 6 = 9$, and so

    $\dfrac{dy}{dx} = \dfrac{3-1}{9^{1/2}} = \dfrac{2}{3}$

15. $y = \dfrac{1}{u} = u^{-1}$, $u = 3 - \dfrac{1}{x^2} = 3 - x^{-2}$,

    $\dfrac{dy}{du} = -u^{-2} = \dfrac{-1}{u^2}$, $\dfrac{du}{dx} = \dfrac{2}{x^3}$,

    $\dfrac{dy}{dx} = \dfrac{dy}{du}\dfrac{du}{dx} = \dfrac{-1}{u^2}\dfrac{2}{x^3}$

    When $x = \dfrac{1}{2}$, $u = 3 - \dfrac{1}{(1/2)^2} = 3 - 4 = -1$,

    $\dfrac{dy}{dx} = \dfrac{-1}{(-1)^2}\dfrac{2}{(1/2)^3} = -16$.

17. $f(x) = (2x+1)^4$,

    $f'(x) = 4(2x+1)^3 \dfrac{d}{dx}(2x+1) = 8(2x+1)^3$

19. $f(x) = (x^5 - 4x^3 - 7)^8$,

    $f'(x) = 8(x^5 - 4x^3 - 7)^7(x^5 - 4x^3 - 7)'$

    $= 8x^2(x^5 - 4x^3 - 7)^7(5x^2 - 12)$

21. $f(t) = \dfrac{1}{5t^2 - 6t + 2} = (5t^2 - 6t + 2)^{-1}$,

    $f'(t) = -(5t^2 - 6t + 2)^{-2}(5t^2 - 6t + 2)'$

    $= -\dfrac{10t - 6}{(5t^2 - 6t + 2)^2}$.

2.5. THE CHAIN RULE

23. $g(x) = \dfrac{1}{\sqrt{4x^2+1}} = (4x^2+1)^{-1/2},$

$g'(x) = \dfrac{-1}{2}(4x^2+1)^{-3/2}(4x^2+1)'$

$ = \dfrac{-4x}{(4x^2+1)^{3/2}}.$

25. $f(x) = \dfrac{3}{(1-x^2)^4} = 3(1-x^2)^{-4},$

$f'(x) = -12(1-x^2)^{-5}\dfrac{d}{dx}(1-x^2)$

$ = \dfrac{24x}{(1-x^2)^5}.$

27. $h(s) = (1+\sqrt{3s})^5,$

$h'(s) = 5(1+\sqrt{3s})^4 \dfrac{d}{ds}(1+\sqrt{3s})$

$ = 5(1+\sqrt{3s})^4 \dfrac{d}{ds}(1+\sqrt{3}s^{1/2})$

$ = 5(1+\sqrt{3s})^4 (\sqrt{3})\left(\dfrac{1}{2}\right)s^{-1/2}$

$ = \dfrac{15(1+\sqrt{3s})^4}{2\sqrt{3s}}.$

29. $f(x) = (x+2)^3(2x-1)^5,$

$f'(x) = (x+2)^3 \dfrac{d}{dx}(2x-1)^5$

$ + (2x-1)^5 \dfrac{d}{dx}(x+2)^3$

$ = (x+2)^3[5(2x-1)^4(2)]$

$ + (2x-1)^5[3(x+2)^2(1)]$

$ = (x+2)^2(2x-1)^4(16x+17).$

31. $G(x) = \sqrt{\dfrac{3x+1}{2x-1}} = \left(\dfrac{3x+1}{2x-1}\right)^{1/2},$

$G'(x) = \dfrac{1}{2}\left(\dfrac{3x+1}{2x-1}\right)^{-1/2}\dfrac{d}{dx}\left(\dfrac{3x+1}{2x-1}\right)$

$ = \dfrac{1}{2}\left(\dfrac{3x+1}{2x-1}\right)^{-1/2}$

$ \dfrac{(2x-1)(3)-(3x+1)(2)}{(2x-1)^2}$

$ = \dfrac{1}{2}\dfrac{(3x+1)^{-1/2}}{(2x-1)^{-1/2}}\dfrac{-5}{(2x-1)^2}$

$ = \dfrac{-5}{2(3x+1)^{1/2}(2x-1)^{3/2}}.$

33. $f(x) = \dfrac{(x+1)^5}{(1-x)^4},$

$f'(x) = \dfrac{(1-x)^4[(x+1)^5]'}{(1-x)^8}$

$ - \dfrac{(x+1)^5 \dfrac{d}{dx}(1-x)^4}{(1-x)^8}$

$ = \dfrac{(1-x)^3(x+1)^4[5(1-x)+4(x+1)]}{(1-x)^8}$

$ = \dfrac{(x+1)^4(9-x)}{(1-x)^5}.$

35. $f(v) = \dfrac{3v+1}{\sqrt{1-4v}},$

$f'(v) = \dfrac{\sqrt{1-4v}(3v+1)' - (3v+1)[(1-4v)^{1/2}]'}{(\sqrt{1-4v})^2}$

$ = \dfrac{3\sqrt{1-4v} - \dfrac{1}{2}(3v+1)(1-4v)^{-1/2}(-4)}{1-4v}$

$ = \dfrac{3\sqrt{1-4v} + 2(3v+1)(1-4v)^{-1/2}}{1-4v}$

$ = \dfrac{3-12v+6v+2}{(1-4v)^{3/2}} = \dfrac{5-6v}{(1-4v)^{3/2}}.$

37. $f(x) = (3x^2+1)^2$

$f'(x) = 2(3x^2+1)(6x)$

$m = f'(-1) = -48.\ f(-1) = 16.$
For the tangent line

$\dfrac{y-16}{x+1} = -48 \text{ or } y = -48x-32$

39. $f(x) = (2x-1)^{-6}$

$f'(x) = -6(2x-1)^{-7}(2) = -\dfrac{12}{(2x-1)^7},$

$m = f'(1) = -12.\ f(1) = 1.$
For the tangent line

$\dfrac{y-1}{x-1} = -12 \text{ or } y = -12x+13$

41. $f(x) = (x^2+x)^2$

$f'(x) = 2(x^2+x)(2x+1) = 2x(2x^2+3x+1) = 0$

when $x = -1,\ x = 0,$ and $x = -\dfrac{1}{2}$

43. $$f(x) = \frac{x}{(3x-2)^2}$$
$$f'(x) = \frac{(3x-2)^2 - 6x(3x-2)}{(3x-2)^4}$$
$$= -\frac{3x+2}{(3x-2)^3} = 0$$

at $x = -\frac{2}{3}$

45. $$f(x) = (x^2 - 4x + 5)^{1/2}$$
$$f'(x) = \frac{x-2}{\sqrt{x^2 - 4x + 5}} = 0$$

when $x = 2$.

47. $$f(x) = (3x+5)^2$$

(a) $$f'(x) = 2(3x+5)(3) = 6(3x+5)$$

(b) $$f'(x) = (3x+5)(3) + (3)(3x+5)$$
$$= 6(3x+5).$$

49. (a) Since
$$f(t) = \sqrt{10t^2 + t + 236}$$
$$= (10t^2 + t + 236)^{1/2}$$

is the factory's gross annual earnings (in thousand-dollar units) t years after its formation in 1998, the rate at which the earnings are growing at that time is
$$f'(t) = \frac{1}{2}(10t^2 + t + 236)^{-1/2}(20t + 1)$$
$$= \frac{20t + 1}{2(10t^2 + t + 236)^{1/2}}$$

thousand dollars per year.
The rate of growth in 2003 ($t = 4$) is
$$f'(4) = \frac{20(4) + 1}{2(10(4)^2 + 4 + 236)^{1/2}} = 2.025$$

That is, in 2003 the gross annual earnings were increasing at the rate of $2,025 per year.

(b) The percentage rate of the earnings increases in 2003 was
$$100\frac{f'(4)}{f(4)}$$
$$= \frac{100(2.025)}{\sqrt{10(4^2) + 4 + 236}} = 10.125 \text{ \% per year}$$

51. $$D(p) = \frac{8,000}{p} = 8,000p^{-1}.$$
$$p(t) = 0.04t^{3/2} + 15.$$
$$p(25) = (0.04)(125) + 15 = 20.$$
$$p'(t) = \frac{3}{2}(.04t^{1/2}) = .06\sqrt{t}$$
$$D'(p) = (-8,000p^{-2}),$$

at $p = 20$ and $t = 25$
$$D' = \left(-\frac{8,000}{400}\right)(0.3) = -6$$

The demand is decreasing by 6 blenders per month.

53. $$D(p) = 40,000p^{-1}.$$
$$p(t) = 0.4t^{3/2} + 6.8.$$
$$p(4) = 3.2 + 6.8 = 10.$$
$$p'(t) = \frac{3}{2}(.4)t^{1/2} = .6\sqrt{t}$$
$$D'(p) = \left(-\frac{40,000}{p^2}\right),$$

at $p = 10$ and $t = 4$,
$$D' = -400(1.2) = -480,$$
$$D(10) = 4,000,$$

and the percentage rate of change is
$$-\frac{48,000}{4,000} = -12\%, \text{ a 12\% decrease.}$$

55. $$E = \frac{1}{v}[0.074(v-35)^2 + 32]$$
$$\frac{dE}{dV} = \frac{1}{v^2}\{v[0.074(v-35)^2 + 32]'$$
$$\quad - [0.074(v-35)^2 + 32]\}$$
$$= \frac{1}{v^2}\{0.148v^2 - 5.18v$$
$$\quad - 0.074v^2 + 5.18v - 90.65 - 32.\}$$
$$= \frac{1}{v^2}(0.074v^2 - 122.65\}$$

57. $$F = kD^2\sqrt{A-C} = kD^2(A-C)^{1/2}$$

2.6. THE SECOND DERIVATIVE

(a) $\dfrac{dF}{dC} = -\dfrac{1}{2}kD^2(A-C)^{-1/2} < 0$,

so F decreases as C increases.

(b) $\dfrac{dF}{dA} = \dfrac{1}{2}kD^2(A-C)^{-1/2}$

the percentage rate of change is

$$100\,\dfrac{dF}{F} = \dfrac{100kD^2}{2kD^2(A-C)} = \dfrac{50}{A-C}\%$$

59. $L'(x) = \dfrac{1}{x}$

(a) $f(x) = L(u),\ u = x^2,\ u' = 2x.$

$$f'(x) = L'(u)u' = \left(\dfrac{1}{u}\right)(2x)$$
$$= \dfrac{2x}{x^2} = \dfrac{2}{x}$$

(b) $f(x) = L(u),\ u = \dfrac{1}{x},\ u' = -\dfrac{1}{x^2}.$

$$f'(x) = L'(u)u' = \left(\dfrac{1}{u}\right)\left(-\dfrac{1}{x^2}\right)$$
$$= x\left(-\dfrac{1}{x^2}\right) = -\dfrac{1}{x}$$

(c) $f(x) = L(u),\ u = \dfrac{2}{3}x^{-1/2},\ u' = -\dfrac{1}{3x^{3/2}}.$

$$f'(x) = L'(u)u' = \left(\dfrac{3\sqrt{x}}{2}\right)\left(-\dfrac{1}{3x\sqrt{x}}\right)$$
$$= -\dfrac{1}{2x}$$

(d) $f(x) = L(u),\ u = \dfrac{2x+1}{1-x},\ u' =$

$\dfrac{(1-x)(2) - (2x+1)(-1)}{(1-x)^2} = \dfrac{3}{(1-x)^2}.$

$$f'(x) = L'(u)u' = \left(\dfrac{1-x}{2x+1}\right)\left(\dfrac{3}{(1-x)^2}\right)$$
$$= \dfrac{3}{(2x+1)(1-x)}$$

61. $y = h(x)[h(x)]^2.$

$$\begin{aligned}\dfrac{dy}{dx} &= h(x)[2h(x)h'(x)] + h'(x)[h(x)]^2 \\ &= 2h^2(x)h'(x) + [h(x)]^2 h'(x) \\ &= 3[h(x)]^2 h'(x)\end{aligned}$$

63. $f'(0)$ does not exist. $f(x)$ has a vertical tangent at $x = 0$ and a horizontal tangent at $x = 0.51$. $f'(4.3) = 16.626$.

2.6 The Second Derivative

1.
$$\begin{aligned} f(x) &= 5x^{10} - 6x^5 - 27x + 4 \\ f'(x) &= 50x^9 - 30x^4 - 27 \\ f''(x) &= 450x^8 - 120x^3 \end{aligned}$$

3.
$$\begin{aligned} y &= 5\sqrt{x} + \dfrac{3}{x^2} + \dfrac{1}{3\sqrt{x}} + \dfrac{1}{2} \\ &= 5x^{1/2} + 3x^{-2} + \dfrac{1}{3}x^{-1/2} + \dfrac{1}{2}. \\ \dfrac{dy}{dx} &= \dfrac{5}{2}x^{-1/2} - 6x^{-3} - \dfrac{1}{6}x^{-3/2}, \\ \dfrac{d^2y}{dx^2} &= -\dfrac{5}{4}x^{-3/2} + 18x^{-4} + \dfrac{1}{4}x^{-5/2} \\ &= -\dfrac{5}{4x^{3/2}} + \dfrac{18}{x^4} + \dfrac{1}{4x^{5/2}}. \end{aligned}$$

5. $f(x) = (3x+1)^5$

$f'(x) = 5(3x+1)^4(3) = 15(3x+1)^4,$

$f''(x) = 60(3x+1)^3(3) = 180(3x+1)^3$

7.
$$\begin{aligned} h &= (t^2+5)^8. \\ \dfrac{dh}{dt} &= 8(t^2+5)^7(2t) = 16t(t^2+5)^7, \\ \dfrac{d^2h}{dt^2} &= 16t[7(t^2+5)^6(2t)] + (t^2+5)^7(16) \\ &= 80(t^2+5)^6(3t^2+1). \end{aligned}$$

9.
$$\begin{aligned} f(x) &= \sqrt{1+x^2} = (1+x^2)^{1/2} \\ f'(x) &= \dfrac{1}{2}(1+x^2)^{-1/2}(2x) \\ &= \dfrac{x}{\sqrt{1+x^2}}, \\ f''(x) &= \dfrac{1}{1+x^2}\{(1+x^2)^{1/2}(1) \\ &\quad - x[(1/2)(1+x^2)^{-1/2}(2x)]\} \\ &= \dfrac{1}{1+x^2}(1+x^2)^{-1/2}[(1+x^2) - x^2] \\ &= \dfrac{1}{(1+x^2)^{3/2}} \end{aligned}$$

11. $z = \dfrac{2}{1+x^2} = 2(1+x^2)^{-1}.$

$\dfrac{dz}{dx} = -2(1+x^2)^{-2}(2x) = -\dfrac{4x}{(1+x^2)^2},$

$\dfrac{d^2z}{dx^2} = -\dfrac{(1+x^2)^2(4) - 4x[2(1+x^2)(2x)]}{(1+x^2)^4}$

$= -\dfrac{4(1+x^2)[(1+x^2) - 4x^2]}{(1+x^2)^4}$

$-\dfrac{4(1-3x^2)}{(1+x^2)^3} = \dfrac{4(3x^2-1)}{(1+x^2)^3}.$

13. $f(x) = x(2x+1)^4.$

$f'(x) = x(4)(2x+1)^3(2) + (2x+1)^4(1)$

$= (2x+1)^3(10x+1),$

$f''(x) = (2x+1)^3(10)$
$+ (10x+1)(3)(2x+1)^2(2)$

$= 2(2x+1)^2[5(2x+1) + 3(10x+1)]$

$= 16(2x+1)^2(5x+1)$

15. $y = \left(\dfrac{t}{t+1}\right)^2$

$\dfrac{dy}{dt} = 2\left(\dfrac{t}{t+1}\right)\dfrac{(t+1)(1) - t(1)}{(t+1)^2}$

$= \dfrac{2t}{(t+1)^3},$

$\dfrac{d^2y}{dt^2} = \dfrac{(t+1)^3(2) - 2t(3)(t+1)^2(1)}{(t+1)^6}$

$= \dfrac{(t+1)^2(2t + 2 - 6t)}{(t+1)^6} = \dfrac{2(1-2t)}{(t+1)^4}$

17. (a) $s(t) = 3t^5 - 5t^3 - 7$

$v(t) = 15t^4 - 15t^2$

$a(t) = 15(4t^3 - 2t) = 30t(2t^2 - 1)$

(b) $a(t) = 0$ at $t = 0$ and at $t = \dfrac{1}{2}\sqrt{2}$

19. (a) $s(t) = (1-t)^3 + (2t+1)^2$

$v(t) = 3(1-t)^2(-1) + 4(2t+1)$

$= -3(1-t)^2 + 8t + 4$

$a(t) = 6(1-t) + 8$

(b) $a(t) = 0$ at $3 - 3t + 4 = 0$ or $t = \dfrac{7}{3}.$

21. $Q(t) = -t^3 + 8t^2 + 15t,$

8 a.m. corresponds to $t = 0$.

(a) $Q'(t) = -3t^2 + 16t + 15$

and 9 a.m. means $t = 1$.
$Q'(1) = -3 + 16 + 15 = 28$ units per hour.

(b) $\dfrac{d}{dt}Q'(t) = -6t + 16,$

$\dfrac{d}{dt}Q'(1) = 10$ units per hour square.

(c) The approximation is $10 \times 0.25 = 2.5$ because $0.25 = 15$ minutes.

(d) $Q'(1.25) = -4.6875 + 20 + 15 = 30.3125,$
$Q'(1) = 28$, $Q'(1.25) - Q'(1) = 2.3125.$

23. (a) $P(t) = -t^3 + 9t^2 + 48t + 200,$

t years from now.

$P'(t) = -3t^2 + 18t + 48$

t years from now,
$P'(3) = -27 + 54 + 48 = 75$ that is
75,000 people per year 3 years from now.

(b) $P''(t) = -6t + 18,$

$P''(3) = 0$ 3 years from now.

(c) 4 years from now, that is at the beginning of the fourth year,

$P''(4) = -6$, $\Delta P = (-6)\left(\dfrac{1}{12}\right) = -0.5$

which corresponds to a decrease of 500 people.

(d) one month $= \dfrac{1}{12} \approx 0.08333.$
$P'(4.0833) = 71.479$, $P'(4) = 72,$
$P'(4.0833) - P'(4) = -0.521$ or a decrease of 521 people.

25. (a) $D(t) = 64t + \dfrac{10}{3}t^2 - \dfrac{2}{9}t^3$

$v(t) = 64 + \dfrac{20}{3}t - \dfrac{2}{3}t^2$

$a(t) = \dfrac{20}{3} - \dfrac{4}{3}t$

(b) $a(6) = \dfrac{20}{3} - \dfrac{24}{3} = -\dfrac{4}{3}$, which indicates that the speed is decreasing at the rate of $\dfrac{4}{3}$ km/hr^2.

2.6. THE SECOND DERIVATIVE

(c)
$$\begin{aligned}v(7) - v(6) &= 64 + \frac{20}{3}(7) - \frac{2}{3}(49) \\ &\quad - \left[64 + \frac{20}{3}(6) - \frac{2}{3}(36)\right] \\ &= -2 \text{ km/hr.}\end{aligned}$$

27. (a)
$$\begin{aligned}P(x) &= \frac{Ax}{B + x^m} \\ P'(x) &= A\frac{B + x^m - mx^m}{(B + x^m)^2} \\ &= A\frac{B + (1-m)x^m}{(B + x^m)^2}\end{aligned}$$

(b) Applying the quotient rule again we get
$$\begin{aligned}P''(x) &= \frac{A}{(B+x^m)^3} \\ &\quad \{(B+x^m)m(1-m)x^{m-1} \\ &\quad -2mx^{m-1}[B+(1-m)x^m]\} \\ &= \frac{Amx^{m-1}}{(B+x^m)^3} \\ &\quad [-B - Bm - (1-m)x^m] = 0\end{aligned}$$
when $x = \sqrt[m]{\frac{m+1}{m-1}B}$

(c) Writing exercise — Answers will vary.

29.
$$\begin{aligned}f(x) &= x^5 - 2x^4 + x^3 - 3x^2 + 5x - 6 \\ f'(x) &= 5x^4 - 8x^3 + 3x^2 - 6x + 5 \\ f''(x) &= 20x^3 - 24x^2 + 6x - 6 \\ f'''(x) &= 60x^2 - 48x + 6 \\ f^{(4)}(x) &= 120x - 48.\end{aligned}$$

Note that $f^{(4)}$ means 4th derivative while f^4 means 4th power.

31.
$$\begin{aligned}f(x) &= \frac{1}{\sqrt{3}}x^{-1/2} - 2x^{-2} + \sqrt{2} \\ f'(x) &= -\frac{1}{2\sqrt{3}}x^{-3/2} + 4x^{-3} \\ f''(x) &= \frac{3}{4\sqrt{3}}x^{-5/2} - 12x^{-4} \\ f'''(x) &= -\frac{15}{8\sqrt{3}}x^{-7/2} + 48x^{-5}.\end{aligned}$$

33. For the HP48G:
Store '(7*T-5)²*(4-T)³' in $V(T)$.
(green) Symbolic Differentiate Choos
Check V(T)
T for Var:
'14*(7*T-5)*(4-T)³+(7*T-5)²-(3*(4-T)²)'
is displayed at stack level 1.
Edit to prefix with 'A(T)=' and store in A(T)
Plot (after CHOOS) $V(T)$ and $A(T)$
Use T for var: , ERASE PLOT
For the TI-85:
Press GRAPH F1 to enter
$y1 = (7x - 5)^2(4 - x)^3$ (x is simpler than t)
ENTER.
$y2 = der1(y1, x)$ ENTER. Note: der1 is the first derivative from 2nd CALC and F3.
Change the range to $[-1,5]1$ and
$[-1000, 1000]100$.
Then GRAPH. the roots are $x \approx 0.714$, $x \approx 2$, and $x = 4$.

(a) HP48G's ROOT generates
T=0.714285714286, T=2.02857142857, and T=4
The object accelerates for
$0.7143 < t < 2.0286$.
It decelerates otherwise.

(b) HP48G's TRACE and (x,y) indicate that
$V(0) = 1,600$ is maximized.
For the TI-85, press F1 ($y(x) =$) and F5
(SELECT) $y1$, the velocity (deselect $y2$, the acceleration). 2nd M5 graph the velocity only. TRACE also shows
$V(0) = 1,600$.

(c) HP438G's FCN EXTR shows
$a(1.22343508451) = 774.277257901$.
The down arrow makes the cross-hair jump to the velocity curve to show
$v(1.23) = 271.90043$.
For the TI-85, when both functions are selected and graphed, $(1.238, 773.83)$ is the maximum traced for the acceleration.
The down atrow key jumps to
$v(1.238) = 283.25$.

(d) $a_{max} \approx 774$ and $a_{min} = -5,680$ (at $t = 0$). $\Delta a = 6454$.

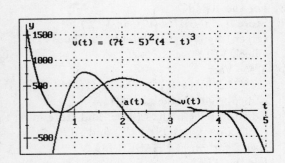

2.7 Implicit Differentiation and Related Rates

1.
$$x^2 + y^2 = 25,$$
$$2x + 2y\frac{dy}{dx} = 0,$$
$$\frac{dy}{dx} = -\frac{x}{y}.$$

3.
$$x^3 + y^3 = xy,$$
$$3x^2 + 3y^2\frac{dy}{dx} = x\frac{dy}{dx} + y,$$
$$(3y^2 - x)\frac{dy}{dx} = y - 3x^2,$$
$$\frac{dy}{dx} = \frac{y - 3x^2}{3y^2 - x}.$$

5.
$$y^2 + 2xy^2 - 3x + 1 = 0,$$
$$2y\frac{dy}{dx} + \left[2x\left(2y\frac{dy}{dx}\right) + y^2(2)\right] - 3 = 0,$$
$$(2y + 4xy)\frac{dy}{dx} = 3 - 2y^2,$$
$$\frac{dy}{dx} = \frac{3 - 2y^2}{2y(1 + 2x)}.$$

7.
$$(2x + y)^3 = x,$$
$$3(2x + y)^2\left(2 + \frac{dy}{dx}\right) = 1,$$
$$2 + \frac{dy}{dx} = \frac{1}{3(2x + y)^2},$$
$$\frac{dy}{dx} = \frac{1}{3(2x + y)^2} - 2.$$

9.
$$(x^2 + 3y^2)^5 = 2xy$$
$$5(x^2 + 3y^2)^4 \frac{d}{dx}(x^2 + 3y^2) = 2x\frac{dy}{dx} + 2y$$
$$5(x^2 + 3y^2)^4 \left(2x + 6y\frac{dy}{dx}\right) = 2x\frac{dy}{dx} + 2y$$
$$10x(x^2 + 3y^2)^4 + 30y(x^2 + 3y^2)^4\frac{dy}{dx} = 2x\frac{dy}{dx} + 2y$$
$$\frac{dy}{dx} = \frac{y - 5x(x^2 + 3y^2)^4}{15y(x^2 + 3y^2)^4 - x}$$

11. $$x^2 = y^3, \frac{dy}{dx} = \frac{2x}{3y^2}$$

When $x = 8$, the original equation gives $8^2 = y^3$, $y^3 = 64$, or $y = 4$.

Substituting in the equation for $\frac{dy}{dx}$ yields

$$\frac{dy}{dx} = \frac{2(8)}{3(4^2)} = \frac{1}{3},$$

that is, the slope of the tangent line at the point $(8, 4)$ is $\frac{1}{3}$.

13. $$xy = 2, \frac{dy}{dx} = -\frac{y}{x}$$

When $x = 2$, the original equation gives $2y = 2$ or $y = 1$.

Substituting in the equation for $\frac{dy}{dx}$ yields

$$\frac{dy}{dx} = -\frac{1}{2}$$

That is, the slope of the tangent line at the point $(2, 1)$ is $-\frac{1}{2}$.

15. $$(1 - u + v)^3 = u + 7$$
$$3(1 - u + v)^2\left[-1 + \frac{dv}{du}\right] = 1$$
$$\frac{dv}{du} = \frac{1}{3(1 - u + v)^2} + 1$$

When $u = 1$, the original equation gives $(1 - 1 + v)^3 = 1 + 7$, $v^3 = 8$, or $v = 2$.

Substituting in the equation for $\frac{dv}{du}$ yields

$$\frac{dv}{du} = \frac{1}{3(1 - 1 + 2)^2} + 1 = \frac{1}{12} + 1 = \frac{13}{12}$$

The slope of the tangent line at $(1, 2)$ is $\frac{13}{12}$.

2.7. IMPLICIT DIFFERENTIATION AND RELATED RATES

17.
$$x + y^2 = 9$$
$$1 + 2y\frac{dy}{dx} = 0$$
$\frac{dy}{dx} = -\frac{1}{2y} \neq 0$, so there is no horizontal tangent line.
$\frac{dy}{dx} = -\frac{1}{2y} = \infty$ at $(9,0)$, at which the tangent line is vertical.

19.
$$x^2 + xy + y = 3$$
$$2x + x\frac{dy}{dx} + y + \frac{dy}{dx} = 0$$
$\frac{dy}{dx} = -\frac{2x+y}{1+x} = 0$ when $y = -2x$.
Substitute into the original equation.
$x^2 + 2x + 3 = 0$ which has no real solution, so there is no horizontal tangent line.
$\frac{dy}{dx} = -\frac{2x+y}{1+x} = \infty$ at $x = -1$, at which the tangent line seems to be vertical.
Substitute $x = -1$ into the original equation you get $1 = 3$. y is not defined when $x = -1$, so there is no vertical tangent line.

21.
$$x^2 + xy + y^2 = 3$$
$$2x + x\frac{dy}{dx} + y + 2y\frac{dy}{dx} = 0$$
$\frac{dy}{dx} = -\frac{2x+y}{x+2y} = 0$ when $y = -2x$.
Substitute into the original equation. $3x^2 = 3$ at $(\pm 1, \mp 2)$ for horizontal tangent lines.
$\frac{dy}{dx} = -\frac{2x+y}{x+2y} = \infty$ at $(\pm 2, \mp 1)$, at which the tangent lines are vertical.

23.
$$xy + 2y = 5$$
(a)
$$x\frac{dy}{dx} + y + 2\frac{dy}{dx} = 0$$
$$\frac{dy}{dx} = -\frac{y}{x+2}$$
(b)
$$y = \frac{5}{x+2}$$
$$\frac{dy}{dx} = -\frac{1}{x+2}\frac{3}{x+2} = -\frac{y}{x+2}$$

25.
$$xy + 2y = x^2$$

(a)
$$x\frac{dy}{dx} + y + 2\frac{dy}{dx} = 2x$$
$$\frac{dy}{dx} = \frac{2x-y}{x+2} = \frac{x^2+4x}{(x+2)^2}$$
(b)
$$y = \frac{x^2}{x+2}$$
$$\frac{dy}{dx} = \frac{x^2+4x}{(x+2)^2}$$

27.
$$xy - x = y + 2$$
(a)
$$x\frac{dy}{dx} + y - 1 = \frac{dy}{dx}$$
$$\frac{dy}{dx} = \frac{y-1}{1-x} = \frac{-3}{(x-1)^2}$$
(b)
$$y = \frac{x+2}{x-1}$$
$$\frac{dy}{dx} = \frac{-3}{(x-1)^2}$$

29.
$$Q = 0.08x^2 + 0.12xy + 0.03y^2$$
$$0.16x + 0.12\left(x\frac{dy}{dx} + y\right) + 0.06y\frac{dy}{dx} = 0$$
$$(0.06)(2x+y)\frac{dy}{dx} = -0.16x - 0.12y$$
$$\frac{dy}{dx} = -\frac{0.16x + 0.12y}{0.06(2x+y)}$$

Note that $\frac{dy}{dx} = \lim_{h \to 0} \frac{\Delta y}{h}$
$$= -\frac{8x+6y}{3(2x+y)} \text{ and that}$$
$$\Delta y \approx \left(\frac{dy}{dx}\right)h$$
$$= -\frac{8 \times 80 + 6 \times 200}{3(2 \times 80 + 200)}(1)$$
$$= -\frac{1,840}{1,080} = -1.70$$

31.
$$3p^2 - x^2 = 12$$
$$6p\frac{dp}{dt} - 2x\frac{dx}{dt} = 0$$

When $p = 4$ we have $48 - x^2 = 12$ or $x^2 = 36$, $x = 6$.

Therefore $6(4)(0.87) - 2(6)\dfrac{dx}{dt} = 0$

$$20.88 - 12\dfrac{dx}{dt} = 0$$

$$\dfrac{dx}{dt} = \dfrac{20.88}{12} = 1.74 \text{ or } 174 \text{ units/month}$$

33.
$$V = \dfrac{4}{3}\pi r^3$$

$$\dfrac{dV}{dt} = 4\pi r^2 \dfrac{dr}{dt}$$

$$0.002\pi = 4\pi(0.005)^2 \dfrac{dr}{dt}$$

or $\dfrac{dr}{dt} = \dfrac{0.002}{4(0.005)^2} = 20$ mm/min.

35.
$$75x^2 + 17p^2 = 5,300$$

$$150x\dfrac{dx}{dt} + 34p\dfrac{dp}{dt} = 0$$

With $p = 7$, $75x^2 = 5,300 - 833$ or $x = 7.72$.

$$150(7.72)\dfrac{dx}{dt} + 34(7)(-0.75) = 0$$

so $\dfrac{dx}{dt} = \dfrac{178.5}{1158} = 0.154$

37. Let t be the time in hours and s the distance between the car and the truck.

$$s = \sqrt{3600t^2 + 2025t^2} = 75t$$

$$\dfrac{ds}{dt} = 75 \text{ MPH}$$

39.
$$Q = 60K^{1/3}L^{2/3}$$

$$\dfrac{dQ}{dt} = 20\left(2K^{1/3}L^{-1/3}\dfrac{dL}{dt} + K^{-2/3}L^{2/3}\dfrac{dK}{dt}\right)$$

With $K = 8$, $L = 1,000$, $\dfrac{dL}{dt} = 25$

$\dfrac{dQ}{dt} = 0 = 20\left(\dfrac{4(25)}{10} + \dfrac{100}{4}\dfrac{dK}{dt}\right)$ leading to

$\dfrac{dK}{dt} = -0.4$ or decreasing by \$400 per week.

41. Let x be the distance between the man and the base of the street light and L the length of the shadow.
Because of similar triangles,

$$\dfrac{L}{6} = \dfrac{x+L}{12} \text{ or } L = x$$

Thus $\dfrac{dL}{dt} = \dfrac{dx}{dt} = 4$ ft./sec

43. Let $C = \dfrac{K}{L}$, then $v = C(R^2 - r^2)$

With $\dfrac{dR}{dt} = -0.0012$ mm/min, $R = 0.007$ mm, $\dfrac{dr}{dt} = -0.0012$,

$$\dfrac{dv}{dt} = 2C\left(R\dfrac{dR}{dt} - r\dfrac{dr}{dt}\right)$$

$$= -2C(.007)(.0012) + 2C(0.0035)(0.0012)$$

$$= -8.4 \times 10^{-6}\dfrac{K}{L} \text{ mm/min}$$

45. If the output is to remain unchanged, the equation relating inputs x and y can be written as

$$Q_0 = 2x^3 + 3x^2y^2 + (1+y)^3$$

where Q_0 is a constant representing the current output.
By implicit differentiation of this equation,

$$0 = 6x^2 + 6x^2y\dfrac{dy}{dx} + 6xy^2 + 3(1+y)^2\dfrac{dy}{dx}$$

$$-[6x^2y + 3(1+y)^2]\dfrac{dy}{dx} = 6x^2 + 6xy^2$$

$$\dfrac{dy}{dx} = -\dfrac{6x^2 + 6xy^2}{6x^2y + 3(1+y)^2}$$

Use the approximation formula

$$\Delta y \approx \dfrac{dy}{dx}h$$

from section 4, with $x = 30$, $y = 20$, and $h = -0.8$ to get

$$\Delta y \approx -\dfrac{6(30^2) + 6(30)(20^2)}{6(30^2)(20) + 3(21^2)}](-0.8)$$

$$\approx 0.57$$

That is, to maintain the current level of output, input y should be decreased by approximately 0.57 unit to offset a decrease in input of 0.8 unit.

47.
$$\frac{x^2}{a^2} + \frac{y^2}{b^2} = 1$$
$$\frac{x}{a^2} + \frac{yy'}{b^2} = 0$$
$$y' = -\frac{b^2 x}{a^2 y}$$

At $P(x_0, y_0)$ $m = -\dfrac{b^2 x_0}{a^2 y_0}$,
so the equation of the tangent line is

$$y - y_0 = -\frac{b^2 x_0}{a^2 y_0}(x - x_0)$$

or
$$\frac{x_0 x}{a^2} + \frac{y_0 y}{b^2}$$
$$= \frac{x_0^2}{a^2} + \frac{y_0^2}{b^2} = 1$$

since $P(x_0, y_0)$ lies on the curve and thus satisfies the equation of the curve.

49.
$$v = \frac{dx}{dt} = ktx$$

where the straight line along which the object is propelled is the x-axis and the point P is taken as the origin.
k is the constant of proportionality.
Since $v = 4$ when $t = 5$ and $x = 20$,
$4 = k(5)(20)$ or $k = 25^{-1}$.

$$a = \frac{dv}{dt} = 25^{-1}\left(t\frac{dx}{dt} + x\right) = \frac{40}{25} = \frac{8}{5} \text{ ft/sec}^2$$

51.
$$y = x^{r/s} \text{ or } y^s = x^r$$
$sy^{s-1} dy = rx^{r-1} dx$ becomes
$$\frac{dy}{dx} = \frac{r}{s}\frac{x^{r-1}}{y^{s-1}} = \frac{r}{s} x^{r-1-r+r/s} = \frac{r}{s} x^{r/s-1}$$

Thus if $y = x^{r/s} = x^n$, $\dfrac{dy}{dx} = nx^{n-1}$.

53.
$$11x^2 + 4xy + 14y^2 = 21$$
$$22x + 4x\frac{dy}{dx} + 4y + 28y\frac{dy}{dx} = 0$$
$$\frac{dy}{dx} = -\frac{22x + 4y}{4x + 28y} = 0 \text{ when } y = -\frac{11x}{2}$$
$$11x^2 + 4x\left(-\frac{11x}{2}\right) + 14\left(-\frac{11x}{2}\right)^2 = 21$$

Solving yields $x = \pm 0.226$ and $y = \pm 1.241$.
There are two horizontal tangents at $y = 1.241$ and $y = -1.241$.

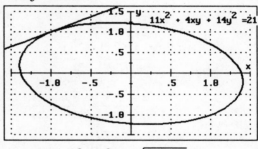

55.
$$x^2 + y^2 = \sqrt{x^2 + y^2} + x$$
$y = 1.916$, $y = -1.916$, and $y = 0$.

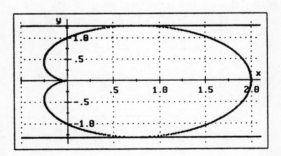

Review Problems

1. (a)
$$f(x) = x^2 - 3x + 1$$
$$\frac{1}{h}[f(x+h) - f(x)]$$
$$= \frac{1}{h}\{[(x+h)^2 - 3(x+h) + 1]$$
$$- [x^2 - 3x + 1]\}$$

$$= \frac{1}{h}[x^2 + 2xh + (h)^2 - 3x$$
$$-3h + 1 - x^2 + 3x - 1]$$
$$= 2x + h - 3.$$

As $h \to 0$, this difference quotient approaches $2x - 3$, so
$$f'(x) = 2x - 3$$

(b) $$f(x) = \frac{1}{x-2}$$

$$\frac{f(x+h) - f(x)}{h}$$
$$= \frac{\frac{1}{x+h-2} - \frac{1}{x-2}}{h}$$
$$= \frac{(x-2) - (x+h-2)}{h(x+h-2)(x-2)}$$
$$= \frac{-1}{(x+h-2)(x-2)}.$$

As $h \to 0$ this difference quotient approaches $\frac{-1}{(x-2)^2}$, so
$$f'(x) = \frac{-1}{(x-2)^2}$$

2. (a) $$f(x) = 6x^4 - 7x^3 + 2x + \sqrt{2}$$
$$f'(x) = 24x^3 - 21x^2 + 2$$

(b) $$f(x) = x^3 - \frac{1}{3x^5} + 2\sqrt{x} - \frac{3}{x} + \frac{1-2x}{x^3}$$
$$= x^3 - \frac{1}{3}x^{-5} + 2x^{1/2}$$
$$\quad -3x^{-1} + x^{-3} - 2x^{-2}.$$
$$f'(x) = 3x^2 + \frac{5}{3}x^{-6} + x^{-1/2}$$
$$\quad +3x^{-2} - 3x^{-4} + 4x^{-3}$$
$$= 3x^2 + \frac{5}{3x^6} + \frac{1}{\sqrt{x}}$$
$$\quad +\frac{3}{x^2} - \frac{3}{x^4} + \frac{4}{x^3}$$

(c) $$y = \frac{2-x^2}{3x^2+1}.$$
$$\frac{dy}{dx} = \frac{(3x^2+1)(-2x) - (2-x^2)(6x)}{(3x^2+1)^2}$$
$$= \frac{-14x}{(3x^2+1)^2}.$$

(d) $$y = (2x+5)^3(x+1)^2.$$
$$\frac{dy}{dx} = (2x+5)^3(2)(x+1)(1)$$
$$\quad +(x+1)^2(3)(2x+5)^2(2)$$
$$= 2(2x+5)^2(x+1)(5x+8).$$

(e) $$f(x) = (5x^4 - 3x^2 + 2x + 1)^{10}.$$
$$f'(x) = 10(5x^4 - 3x^2 + 2x + 1)^9$$
$$\quad (20x^3 - 6x + 2)$$

(f) $$f(x) = \sqrt{x^2+1} = (x^2+1)^{1/2}$$
$$f'(x) = \frac{1}{2}(x^2+1)^{-1/2}(2x) = \frac{x}{(x^2+1)^{1/2}}$$

(g) $$y = \left(x + \frac{1}{x}\right)^2 - \frac{5}{\sqrt{3x}}$$
$$= (x + x^{-1})^2 - \frac{5}{\sqrt{3}}x^{-1/2}$$
$$\frac{dy}{dx} = 2(x + x^{-1})\frac{d}{dx}(x + x^{-1})$$
$$\quad + \frac{5}{2\sqrt{3}}x^{-3/2}\frac{d}{dx}(x)$$
$$= 2\left(x + \frac{1}{x}\right)\left(1 - \frac{1}{x^2}\right) + \frac{5}{2\sqrt{3}x^{3/2}}$$

(h) $$y = \left(\frac{x+1}{1-x}\right)^2.$$
$$\frac{dy}{dx} = 2\left(\frac{x+1}{1-x}\right)\frac{d}{dx}\left(\frac{x+1}{1-x}\right)$$
$$= 2\frac{x+1}{1-x}\frac{(1-x) - (x+1)(-1)}{(1-x)^2}$$
$$= 2\frac{x+1}{1-x}\frac{2}{(1-x)^2}$$
$$= \frac{4(x+1)}{(1-x)^3}.$$

REVIEW PROBLEMS 63

(i) $f(x) = (3x+1)\sqrt{6x+5}$
$= (3x+1)(6x+5)^{1/2}.$

$f'(x) = (3x+1)\left(\dfrac{1}{2}\right)(6x+5)^{-1/2}(6)$
$\qquad + (6x+5)^{1/2}(3)$

$= \dfrac{3(3x+1)}{(6x+5)^{1/2}} + 3(6x+5)^{1/2}$

$= \dfrac{9(3x+2)}{\sqrt{6x+5}}.$

(j) $f(x) = \dfrac{(3x+1)^3}{(1-3x)^4}.$

$f'(x) = \dfrac{1}{(1-3x)^8}$

$\left[(1-3x)^4 \dfrac{d}{dx}(3x+1)^3 - (3x+1)^3 \dfrac{d}{dx}(1-3x)^4\right]$

$= \dfrac{1}{(1-3x)^8}\{(1-3x)^4[3(3x+1)^2(3)]$
$\qquad -(3x+1)^3[4(1-3x)^3(-3)]\}$

$= \dfrac{3(3x+1)^2(3x+7)}{(1-3x)^5}.$

(k) $y = \sqrt{\dfrac{1-2x}{3x+2}} = \left(\dfrac{1-2x}{3x+2}\right)^{1/2}.$

$\dfrac{dy}{dx} = \dfrac{1}{2}\left(\dfrac{1-2x}{3x+2}\right)^{-1/2} \dfrac{d}{dx}\left(\dfrac{1-2x}{3x+2}\right)$

$= \dfrac{1}{2}\left(\dfrac{1-2x}{3x+2}\right)^{-1/2}$

$\qquad \dfrac{(3x+2)(-2)-(1-2x)(3)}{(3x+2)^2}$

$= \dfrac{1}{2}\dfrac{(1-2x)^{-1/2}}{(3x+2)^{-1/2}}$

$\qquad \dfrac{-6x-4-3+6x}{(3x+2)^2}$

$= \dfrac{-7}{2(1-2x)^{1/2}(3x+2)^{3/2}}.$

3. (a) $f(x) = x^2 - 3x + 2$
$f'(x) = 2x - 3$

$f(1) = 0$. The slope of the tangent line at $(1,0)$ is $m = f'(1) = -1$. The equation of the tangent line is thus

$$y - 0 = -(x-1) \text{ or } y = -x + 1$$

(b) $f(x) = \dfrac{4}{x-3}$

$f'(x) = \dfrac{-4}{(x-3)^2}$

$f(1) = -2.$
The slope of the tangent line at $(1,-2)$ is $m = f'(1) = -1.$
The equation of the tangent line is

$$y - (-2) = -(x-1) \text{ or } y = -x - 1$$

(c) $f(x) = \dfrac{x}{x^2+1}.$

$f'(x) = \dfrac{(x^2+1)(1) - x(2x)}{(x^2+1)^2}$

$= \dfrac{1-x^2}{(x^2+1)^2}.$

$f(0) = 0.$
The slope of the tangent line at $(0,0)$ is $m = f'(0) = 1$. The equation of the tangent line is

$$y - 0 = x - 0 \text{ or } y = x$$

(d) $f(x) = \sqrt{x^2+5} = (x^2+5)^{1/2}$

$f'(x) = \dfrac{1}{2}(x^2+5)^{-1/2}(2x) = \dfrac{x}{\sqrt{x^2+5}}$

$f(-2) = 3$. The slope of the tangent line at $(-2,3)$

is $m = f'(-2) = -\dfrac{2}{3}.$

The equation of the tangent line is

$$y - 3 = -\dfrac{2}{3}(x+2) \text{ or } y = -\dfrac{2}{3}x + \dfrac{5}{3}$$

4. (a) The rate of change of

$$f(t) = t^3 - 4t^2 + 5t\sqrt{t} - 5$$
$$= t^3 - 4t^2 + 5t^{3/2} - 5$$
$$\text{is } f'(t) = 3t^2 - 8t + \frac{15}{2}t^{1/2}$$

at any value of $t \geq 0$ and when $t = 4$,
$f'(4) = 48 - 32 + \frac{15}{2}(2) = 31$.

(b) The rate of change of
$$f(t) = t^3(t^2 - 1) = t^5 - t^3$$
$$\text{is } f'(t) = 5t^4 - 3t^2$$

at any value of t and when $t = 0$,
$f'(0) = 0$.

5. (a) $f(t) = t^2 - 3t + \sqrt{t} = t^2 - 3t + t^{1/2}$

The percentage rate of change is
$$100\frac{f'(t)}{f(t)}$$
$$= 100\frac{2t - 3 + \frac{1}{2}t^{-1/2}}{t^2 - 3t + t^{1/2}}$$
$$= 50\frac{4t^{3/2} - 6t^{1/2} + 1}{t^{5/2} - 3t^{3/2} + t}$$

(b) $f(t) = t^2(3 - 2t)^3$

The percentage rate of change is
$$100\frac{f'(t)}{f(t)}$$
$$= 100\frac{(t^2)(3)(3-2t)^2(-2) + (3-2t)^3(2t)}{t^2(3-2t)^3}$$
$$= 100(3-2t)^2\frac{-6t^2 + 6t - 4t^2}{t^2(3-2t)^3}$$
$$= 100\frac{-10t^2 + 6t}{t^2(3-2t)} = 200\frac{(3-5t)}{t(3-2t)}$$

(c) $f(t) = \frac{1}{t+1} = (t+1)^{-1}$

The percentage rate of change is
$$100\frac{f'(t)}{f(t)} = 100\frac{-(t+1)^{-2}}{(t+1)^{-1}} = -100\frac{1}{t+1}$$

6. (a) $y = 5u^2 + u - 1, u = 3x + 1.$
$$\frac{dy}{du} = 10u + 1, \frac{du}{dx} = 3,$$
$$\frac{dy}{dx} = \frac{dy}{du}\frac{du}{dx} = (10u+1)(3)$$
$$= 3(30x + 11)$$

(b) $y = \frac{1}{u^2}, u = 2x + 3, \frac{dy}{du} = \frac{-2}{u^3}, \frac{du}{dx} = 2$
$$\frac{dy}{dx} = \frac{dy}{du}\frac{du}{dx} = \frac{-4}{(2x+3)^3}$$

7. (a) $y = u^3 - 4u^2 + 5u + 2, u = x^2 + 1.$
$$\frac{dy}{du} = 3u^2 - 8u + 5, \frac{du}{dx} = 2x,$$
$$\frac{dy}{dx} = \frac{dy}{du}\frac{du}{dx} = (3u^2 - 8u + 5)(2x)$$

When $x = 1$, $u = 2$, and so
$$\frac{dy}{dx} = [3(2^2) - 8(2) + 5][2(1)] = 2$$

(b) $y = \sqrt{u} = u^{1/2},$
$u = x^2 + 2x - 4, \frac{dy}{du} = \frac{1}{2u^{1/2}},$
$$\frac{du}{dx} = 2x + 2, \frac{dy}{dx} = \frac{dy}{du}\frac{du}{dx}$$
$$= \frac{1}{2u^{1/2}}(2x+2) = \frac{x+1}{u^{1/2}}.$$

When $x = 2$, $u = 4$, and so
$$\frac{dy}{dx} = \frac{2+1}{4^{1/2}} = \frac{3}{2}$$

8. (a) $f(x) = 6x^5 - 4x^3 + 5x^2 - 2x + \frac{1}{x},$
$$f'(x) = 30x^4 - 12x^2 + 10x - 2 - \frac{1}{x^2}$$
$$f''(x) = 120x^3 - 24x + 10 + \frac{2}{x^3}.$$

(b) $y = (3x^2 + 2)^4,$
$$\frac{dy}{dx} = 4(3x^2+2)^3(6x) = 24x(3x^2+2)^3,$$
$$\frac{d^2y}{dx^2} = 24x[3(3x^2+2)^2(6x)]$$
$$+ (3x^2+2)^3(24)$$
$$= 24(3x^2+2)^2(21x^2+2).$$

REVIEW PROBLEMS

(c) $f(x) = \dfrac{x-1}{(x+1)^2}$, then $f'(x)$

$$= \frac{(x+1)^2(1) - (x-1)(2)(x+1)(1)}{(x+1)^4}$$

$$= \frac{(x+1)[(x+1) - 2(x-1)]}{(x+1)^4} = \frac{3-x}{(x+1)^3}$$

$f''(x) = \dfrac{(x+1)^3(-1) - (3-x)(3)(x+1)^2(1)}{(x+1)^6}$

$$= \frac{(x+1)^2[-(x+1) - 3(3-x)]}{(x+1)^6}$$

$$= \frac{-x-1-9+3x}{(x+1)^4} = \frac{2(x-5)}{(x+1)^4}.$$

9. (a) $5x + 3y = 12,$

$5(1) + 3\dfrac{dy}{dx} = 0$ or $\dfrac{dy}{dx} = -\dfrac{5}{3}$

(b) $x^2 y = 1,$

$$x^2 \frac{dy}{dx} + y(2x) = 0$$

or $\dfrac{dy}{dx} = -\dfrac{2xy}{x^2} = -\dfrac{2y}{x}$

(c) $(2x + 3y)^5 = x + 1,$

$$5(2x+3y)^4 \left(2 + 3\frac{dy}{dx}\right) = 1$$

$$10(2x+3y)^4 + 15(2x+3y)^4 \frac{dy}{dx} = 1$$

$$\frac{dy}{dx} = \frac{1 - 10(2x+3y)^4}{15(2x+3y)^4}$$

(d) $(1 - 2xy^3)^5 = x + 4y$

$$5(1-2xy^3)^4 \frac{d}{dx}(1 - 2xy^3) = 1 + 4\frac{dy}{dx}$$

$$5(1-2xy^3)^4\left(-6xy^2\frac{dy}{dx} - 2y^3\right) = 1 + 4\frac{dy}{dx}$$

$$-30xy^2(1-2xy^3)^4\frac{dy}{dx} - 10y^3(1-2xy^3)^4 = 1 + 4\frac{dy}{dx}$$

$$\frac{dy}{dx} = \frac{1 + 10y^3(1-2xy^3)^4}{-4 - 30xy^2(1-2xy^3)^4}$$

10. (a) $xy^3 = 8,$

$$x\left(3y^2 \frac{dy}{dx}\right) + y^3 = 0$$

or $\dfrac{dy}{dx} = \dfrac{-y^3}{3xy^2} = -\dfrac{y}{3x}$

At $x = 1$, the original equation gives

$$y^3 = 8 \text{ or } y = 2$$

To find the slope of the tangent at the point $(1,2)$, substitute into the equation for $\dfrac{dy}{dx}$ to get the slope

$$m = \frac{dy}{dx} = -\frac{2}{3}$$

(b) $x^2 y - 2xy^3 + 6 = 2x + 2y,$

$$x^2 \frac{dy}{dx} + y(2x) - 2\left[x\left(3y^2\frac{dy}{dx}\right) + y^3(1)\right] = 2 + 2\frac{dy}{dx},$$

$$x^2 \frac{dy}{dx} + 2xy - 6xy^2\frac{dy}{dx} - 2y^3 = 2 + 2\frac{dy}{dx}$$

At $x = 0$, the original equation is

$$0^2 y - 2(0)y^3 + 6 = 2(0) + 2y,$$

$6 = 2y$ or $y = 3$.

To find the slope of the tangent line at $(0,3)$, substitute into the derivative equation and solve for $\dfrac{dy}{dx}$ to get

$$-2y^3 = 2 + 2\frac{dy}{dx}$$

$$-54 = 2 + 2\frac{dy}{dx}$$

or the slope is $m = \dfrac{dy}{dx} = -28.$

11. $3x^2 - 2y^2 = 6,$

$6x - 4y\dfrac{dy}{dx} = 0$ or $\dfrac{dy}{dx} = \dfrac{3x}{2y}$

$$\frac{d^2y}{dx^2} = \frac{2y(3) - 3x\left(2\dfrac{dy}{dx}\right)}{(2y)^2} = \frac{3y - 3x\dfrac{dy}{dx}}{2y^2}$$

Since $\dfrac{dy}{dx} = \dfrac{3x}{2y}$

$$\dfrac{d^2y}{dx^2} = \dfrac{3y - 3x\left(\dfrac{3x}{2y}\right)}{2y^2} = \dfrac{6y^2 - 9x^2}{4y^3}$$

From the original equation

$$6y^2 - 9x^2 = 3(2y^2 - 3x^2)$$
$$= -3(3x^2 - 2y^2) = -3(6) = -18$$

and so $\dfrac{d^2y}{dx^2} = \dfrac{-18}{4y^3} = \dfrac{-9}{2y^3}$

12. (a) $s(t) = -16t^2 + 160t = 0$ when $t = 0$ and $t = 10$.
    The projectile leaves the ground at $t = 0$ and returns 10 seconds later.

    (b) $\dfrac{ds}{dt} = -32t + 160$, thus $\dfrac{ds}{dt} = -160$ at $t = 10$.

    (c) $\dfrac{ds}{dt} = 0$ at $t = 5$ and
    $y_{max} = -16(25) + 160(5) = 400$ ft.

13. Since
$$C(q) = 0.1q^2 + 10q + 400$$
is the total cost of producing q units and
$$q(t) = t^2 + 50t$$
is the number of units produced during the first t hours, then

$$\dfrac{dC}{dq} = 0.2q + 10 \text{ (dollars per unit)}$$

$$\dfrac{dq}{dt} = 2t + 50 \text{ (units per hour)}$$

The rate of change of cost with respect to time is

$$\dfrac{dC}{dt} = \dfrac{dC}{dq}\dfrac{dq}{dt} = (0.2q + 10)(2t + 50)$$

dollars per hour. After 2 hours, $t = 2$, $q(2) = 104$, and so

$$\dfrac{dC}{dt} = [0.2(104) + 10][2(2) + 50] = \$1,663.20/\text{ hr}$$

14. (a) $s(t) = 2t^3 - 21t^2 + 60t - 25$
    for $1 \leq t \leq 6$.

    $v(t) = 6(t^2 - 7t + 10) = 6(t-2)(t-5)$

    The positive roots are $t = 2$, $t = 5$.
    $v(t) > 0$ for $1 < t < 2$, $5 < t < 6$,
    so the object advances.
    For $2 < t < 5$, $v(t) < 0$
    so the object retreats.

    $$a(t) = 6(2t - 7) = 0$$

    if $t = \dfrac{7}{2}$.

    $a(t) > 0$ for $\dfrac{7}{2} < t < 6$
    so the object accelerates.
    For $1 < t < \dfrac{7}{2}$ it decelerates.

    (b) $s(1) = 2 - 21 + 60 - 25 = 16$,
    $s(2) = 16 - 84 + 120 - 25 = 27$,
    $s(5) = 250 - 21(25) + 300 - 25 = 0$,
    $s(6) = 432 - 21(36) + 360 - 25 = 11$
    $\Delta s = (27 - 16) + (27 - 0) + (11 - 0) = 49$

15. (a) $s(t) = \dfrac{2t + 1}{t^2 + 12}$ for $0 \leq t \leq 3$.

    $$v(t) = \dfrac{(t^2 + 12)(2) - (2t + 1)(2t)}{(t^2 + 12)^2}$$
    $$= \dfrac{-2t^2 - 2t + 24}{(t^2 + 12)^2} = 0 \text{ if}$$
    $$t^2 + t - 12 = (t + 4)(t - 3) = 0$$

    $v(t) > 0$ for $t < 3$,
    so the object advances.

    $$a(t) = \dfrac{2(t^2 + 12)^2(-2t - 1)}{(t^2 + 12)^4}$$
    $$- \dfrac{(-t^2 - t + 12)(2)(t^2 + 12)(2t)}{(t^2 + 12)^4}$$
    $$= \dfrac{2(2t^3 + 3t^2 - 72t - 12)}{(t^2 + 12)^3}.$$

    $a(t) < 0$ for $0 < t < 3$
    so the object decelerates.

REVIEW PROBLEMS

(b) $s(0) = \frac{1}{12}$, $s(3) = \frac{1}{3}$, thus
$$\Delta s = \frac{1}{3} - \frac{1}{12} = \frac{1}{4}$$

Note: $a(t) = 0$ when
$t = 5.388$, $t = -6.722$, $t = -0.166$,
all of which are outside of the relevant interval.

16. (a) Since
$$N(x) = 6x^3 + 500x + 8,000$$
is the number of people using the system after x weeks, the rate at which use of the system is changing after x weeks is
$$N'(x) = 18x^2 + 500$$
people per week and the rate after 8 weeks is
$N'(8) = 1,652$ people per week.

(b) The actual increase in the use of the system during the 8th week is
$N(8) - N(7) = 1,514$ people.

17. (a) Since
$$Q(x) = 50x^2 + 9,000x$$
is the weekly output when x workers are employed, the marginal output is
$$Q'(x) = 100x + 9,000$$
The change in output due to an increase from 30 to 31 workers is approximately
$Q'(30) = 12,000$ units.

(b) The actual increase in output is
$Q(31) - Q(30) = 12,050$ units.

18. Since the population in t months will be
$$P(t) = 3t + 5t^{3/2} + 6,000,$$
the rate of change of the population will be
$$P'(t) = 3 + \frac{15}{2}t^{1/2},$$
and the percentage rate of change 4 months from now will be
$$100 \frac{P'(4)}{P(4)} = 100 \frac{18}{6,052} \approx 0.30\% \text{ per month}$$

19. Since daily output is
$$Q(L) = 20,000L^{1/2}$$
units when L worker-hours are used, a change in the work force from 900 worker-hours to 885 worker-hours ($\Delta L = -15$) results in a change in the output of
$$\Delta Q = Q(900) - Q(885) \approx Q'(900)(-15)$$
Since $Q'(L) = 10,000L^{-1/2}$
and $Q'(900) = 10,000(900)^{-1/2} = \frac{1,000}{3}$
it follows that
$$\Delta Q \approx \frac{1,000}{3}(-15) = -5,000,$$
that is, a decrease in output of 5,000 units.

20. The gross national product t years after 1990 is
$$N(t) = t^2 + 6t + 300 \text{ billion dollars}$$
The derivative is
$$N'(t) = 2t + 6$$
At the beginning of the second quarter of 1998, $t = 8.25$.
The change in t during this quarter is $h = 0.25$.
Hence the percentage change in N is
$$100 \frac{N'(8.25)h}{N(8.25)}$$
$$= 100 \frac{[2(8.25) + 6](0.25)}{8.25^2 + 6(8.25) + 300} \approx 1.347\%.$$

21. Let A denote the level of air pollution and p the population.
$$\text{Then } A = kp^2,$$
where k is a positive constant of proportionality.
If the population increases by 5 %, the change in population is $\Delta p = 0.05p$.

The corresponding increase in the level of air pollution is

$$\Delta A = A(p+0.05p) - A(p) \approx A'(p)(0.05p)$$
$$= 2kp(0.05p) = 0.1kp^2 = 0.1A$$

That is, an increase of 5 % in the population causes an increase of 10% in the level of pollution.

22. The output is

$$Q(L) = 600L^{2/3}$$

The derivative is

$$Q'(L) = 400L^{-1/3}$$

We are given that the percentage change in Q is 1 %, and that the goal is to find the percentage change in L, which can be represented as

$$100\frac{\Delta L}{L}$$

Apply the formula for the percentage change in

$$Q = 100\frac{Q'(L)\Delta L}{Q(L)}$$

with 1 on the left and solve for

$$\frac{\Delta L}{L} \text{ as follows:}$$

$$1 \approx 100\frac{400L^{-1/3}\Delta L}{600L^{2/3}} = 100\frac{2}{3}\frac{\Delta L}{L},$$

$$100\frac{\Delta L}{L} \approx \frac{3}{2} = 1.5\%.$$

That is, an increase in labor of approximately 1.5 % is required to increase output by 1 %.

23. By the approximation formula from section 4,

$$\Delta y \approx \frac{dy}{dx}h$$

To find $\frac{dy}{dx}$, differentiate the equation

$$Q = x^3 + 2xy^2 - 2y^3$$

implicitly WRT x, where Q is a constant representing the current level of output.

You get $0 = 3x^2 + 4xy\frac{dy}{dx} + 2y^2 + 6y^2\frac{dy}{dx}$

or $\frac{dy}{dx} = -\frac{3x^2 + 2y^2}{4xy + 6y^2}$

At $x = 10$ and $y = 20$

$$\frac{dy}{dx} = -\frac{3(10)^2 + 2(20)^2}{4(10)(20) + 6(20)^2} \approx -0.344$$

Use the approximation formula with $\frac{dy}{dx} \approx -0.344$ and $h = 0.5$
to get $\Delta y \approx -0.344(0.5) = -0.172$ unit.
That is, to maintain the current level of output, input y should be decreased by approximately 0.172 unit to offset a 0.5 unit increase in input x.

24. Let y be the length of the fence facing the river (but) not on the river bank, and x the length of the portion of fence perpendicular to the river.

Then $A = xy = 7,200$ or $y = \frac{7,200}{x}$

The length of the fence is

$$F(x) = 2x + y = 2x + \frac{7,200}{x}$$

$$F'(x) = 2 - \frac{7,200}{x^2} = 0$$

when $x = 60$.
$F''(60) > 0$ so $x = 60$ indicates a minimum.
Then $y = \frac{7,200}{60} = 120$ and the total length of the fence is $F = 2(60) + 120 = 240$.

25. The population is

$$p(t) = 10 - \frac{20}{(t+1)^2} = 10 - 20(t+1)^{-2}$$

and the carbon monoxide level is

$$c(p) = 0.8\sqrt{p^2 + p + 139}$$
$$= 0.8(p^2 + p + 139)^{1/2} \text{ units}$$

REVIEW PROBLEMS

By the chain rule, the rate of change of the carbon monoxide level with respect to time is

$$\frac{dc}{dt} = \frac{dc}{dp}\frac{dp}{dt}$$
$$= .4(p^2+p+139)^{-1/2}(2p+1)[40(t+1)^{-3}]$$
$$= \frac{0.4(2p+1)}{\sqrt{p^2+p+139}}\frac{40}{(t+1)^3}.$$

At $t=1$, $p = p(1) = 10 - \frac{20}{4} = 5$
$c = c(5) = 0.8\sqrt{169} = 10.4$.
The percentage rate of change is

$$100\frac{dc/dt}{c} = 100\frac{0.4(10+1)}{\sqrt{169}}\frac{40}{(1+1)^3}\frac{1}{10.4}$$

$\approx 16.27\ \%$ per year.

26. (a) The production function is

$$Q(t) = -t^3 + 9t^2 + 12t$$

The rate of production is the derivative

$$Q'(t) = -3t^2 + 18t + 12$$

At 9:00 a.m. $t=1$ and the rate of production is
$Q'(1) = 27$ units per hour.

(b) The rate of change of the rate of production is the second derivative

$$Q''(t) = -6t + 18$$

At 9:00 a.m., this rate is
$Q''(1) = 12$ units/hour/hour.

(c) The change in the rate of production between 9:00 a.m. and 9:06 a.m. is given by the approximation formula
$\Delta Q' \approx Q''(t)h$. At $t=1$ and $h=0.1$ hour, this gives
$Q''(1)h = 12(0.1) = 1.2$ units/hour.

(d) The actual change in the worker's rate of production between 9:00 a.m. and 9:06 a.m. is $Q'(1.1) - Q'(1) = 1.17$ units per hour.

27.
$$s(t) = 88t - 8t^2$$
$$v(t) = 88 - 16t = 0$$

at the instant t_1 the car stops.
Thus $11 - 2t_1 = 0$ or $t_1 = 5.5$ seconds.
The distance required to stop is then

$$s\left(\frac{11}{2}\right) = 8\left[(11)\left(\frac{11}{2}\right) - \left(\frac{11}{2}\right)^2\right] = 242 \text{ feet}$$

28.
$S(R) = 1.8(10^5)R^2$, $R = 1.2(10^{-2})$, $\Delta R = \pm 5(10^{-4})$

$$\Delta S = S[1.2(10^{-2}) \pm 5(10^{-4})] - S[1.2(10^{-2})]$$
$$\approx S'[1.2(10^{-2})][\pm 5(10^{-4})]$$
$$S'(R) = 3.6(10^5)R$$

$S'1.2(10^{-2}) = [3.6(10^5)][1.2(10^{-2})] = 4.32(10^3)$

$\Delta S \approx [4.3(10^3)][\pm 5(10^{-4})] = \pm 2.15$ cm/sec.

29. $A(r) = \pi r^2$, $A'(r) = 2\pi r$

$$\left|\frac{dA}{A}\right| = \left|\frac{2\pi r\, dr}{\pi r^2}\right| = 0.06 \text{ or } 6\%.$$

30.
$$V(t) = [C_1 + C_2 P(t)]\left(\frac{3t^2}{T^2} - \frac{2t^3}{T^3}\right)$$
$$\frac{dV}{dt} = [C_1 + C_2 P(t)]\left(\frac{6t}{T^2} - \frac{6t^2}{T^3}\right)$$
$$+ C_2\left[\frac{3t^2}{T^2} - \frac{2t^3}{T^3}\right]\frac{dP}{dt}.$$

31. (a) The maximum rate is the highest point on the graph, approximately 14%, which occurred around 1980.

(b) The rate increased most rapidly (the second derivative) when the tangent line to the curve is steepest.
The two-year periods 1972-74 and 1980-82 qualify.

(c) Writing Exercise —
Answers will vary.

32. Let x be the distance between the woman and the building, and L the length of the shadow. Since

$$h(t) = 150 - 16t^2,$$

the lantern will be 10 ft from the ground when

$$10 = 150 - 16t^2$$

which leads to $t = \frac{1}{4}\sqrt{140}$ seconds.

When $h = 10$ and $x = 5t = \dfrac{5\sqrt{140}}{4}$

from similar right triangles we get

$$\frac{x}{h-5} = \frac{x+L}{h}$$

$$\frac{5\sqrt{140}}{4(10-5)} = \frac{\frac{5}{4}\sqrt{140}+L}{10} \text{ or } L = \frac{5\sqrt{140}}{4}$$

$$\begin{aligned}
hx &= hx + hL - 5x - 5L \\
hL' + h'L &= 5\frac{dx}{dt} + 5L' \\
(h-5)L' &= 5\frac{dx}{dt} - h'L \\
5L' &= 5(5) + 32\left(\frac{1}{4}\right)\sqrt{140}\left(\frac{5}{4}\sqrt{140}\right) \\
L' &= 5 + 2(140) = 285\text{ft/sec}.
\end{aligned}$$

33. Let x be the distance from the player to third base and L the distance to home plate. Then

$$x^2 + 90^2 = L^2$$

from which

$$x\frac{dx}{dt} = L\frac{dL}{dt}$$

When $x = 15$, $L = \sqrt{15^2 + 90^2}$.
The distance between the runner and home base is changing at

$$\frac{dL}{dt} = \frac{15(-20)}{\sqrt{15^2 + 90^2}} = -3.2880 \text{ ft/sec}.$$

34. (a) $\quad s(t) = -16t^2 + 87t + 129$

$s(t) = 0$ at $t = \dfrac{-87 - 125.8}{-32} = 6.65$ sec.

(b) $v(t) = -32t + 87$, $v(6.65) = -125.8$ ft/sec.

(c) $v(t) = 0$ at $t = 2.72$ sec. $s_{max} = -16(2.72)^2 + 87(2.72) + 129 = 247.3$ ft.

35. For the HP48G
Enter '$(3x+5)(2x^3-5x+4)$' using the symbolic equation writer.
Store in $f(x)$.
CHOOSe this $f(x)$ to be differentiated
[(green) Symbolic OK VAR: X OK]
Edit to prefix with 'FPR(X)=' ENTER (blue) DEF VAR
(green) PLOT EDT
{ 'F(X)' 'FPR(X)' } ENTER ERASE DRAW
Note: In case of error, you may use the programming symbol << X → in the EQ: field. Edit and delete so that only quantities enclosed by '' are left.
$f'(x) = 0$ near $x = -1.78$, $x = -0.35$, $x = 0.88$
Note: $f(0) = 20$ so the upper curve is $f(x)$ while the lower one is $f'(x)$.
For the TI-85:
Press GRAPH and $y(x) =$ to enter
$y1 = (3x+5)(2x^3-5x+4)$ ENTER
Then enter $y2 = der1(y1, x)$ ENTER. Note: der1 is from 2nd CALC and y1 is lower case (2nd alpha).
Press GRAPH, set $[-2, 1]1$ and $[-50, 50]10$, and press F5 to graph
$y2 = 0$ near $X = -1.78$, $x = -35$, and $x = 0.88$.

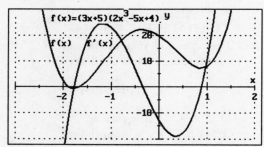

36. For the HP48G:
Enter 'F(X)=(2*X+3)/(1-3*X)'
(green) SYMBOLIC Differentiate NEW
CHOOSe F: OK
Note: you may need to strip characters outside the single quotes
Name: FPR OK VAR: X OK OK
The result is displayed at stack level 1
Enter this result in FPR
PLOT CHOOSe F(check) FPR(check) OK

REVIEW PROBLEMS

Edit to leave only quantities between single quotes (if necessary)
ERASE PLOT
Note that $f(x)$ and $f'(x)$ are discontinuous at $x = 0.33$.
$-\infty < f(x) < \infty$ while $0 < f'(x)$. This allows us to distinguish between curves.
$f'(x) \neq 0$.

For the TI-85:
Press GRAPH, F1. Enter
$y1 = (2x + 3)(1 - 3x)$ ENTER.
Enter $y2 = der1(y1, x)$ ENTER. Note: 2nd CALC for differentiation and 2nd M2 for lower case y. Select (F5) $y2$ and deselect $y1$.
The range $[-20, 20]1, [-.2, .2].05$ suggests $y2 \neq 0$

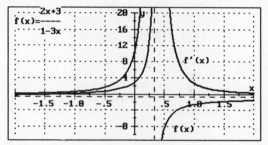

37. For the HP48G:
Enter 'Y^2*(2-X)=X^3' (strip if necessary) into EQ:
CHOOSe CONIC for function
ERASE DRAW
Note that $0 \leq x < 2$ because $x < 0$ makes the left member positive and the right member of $y^2(2-x) = x^3$ negative while the left member is negative.
Also $x < 2$ otherwise the left member would be negative while the right member is positive.
According to the HP48G the points $P(0.52, .314)$ and $Q(0.455, .276)$ are on the curve.
The graph exhibits a vertical asymptote as $x \to 2^-$.
The tangent line at $(0,0)$ is horizontal.
For the TI-85:
Since $y^2(2-x) = x^3$, $y = \pm\sqrt{\dfrac{x^3}{2-x}}$.

Press GRAPH and F1.
Enter $y1 = x^3/(2-x)$ ENTER
Enter $y2 = \sqrt{(y1)}$ ENTER where y is lower case (2nd alpha Y)
Enter $y3 = -\sqrt{(y1)}$ ENTER and F5.
The points $P(.476, .266)$ and $Q(.523, .312)$ are on the curve.
The slope of the tangent line is approximately
$\dfrac{.312 - .266}{.523 - .476} = 0.98$. The point $R(0.5, .289)$ is also on the curve, so

$$y - .289 = 0.98(x - 0.5)$$

is the approximate equation of the tangent line.

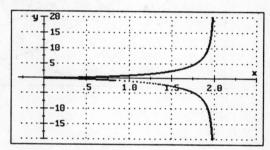

38. (a)
$$s(t) = 0.73t^{9/2} - 3.1t^{7/2} + 2.7t^{5/2},$$
$$v(t) = 3.285t^{7/2} - 10.85t^{5/2} + 6.75t^{3/2},$$
$$a(t) = 11.4975t^{5/2} - 27.125t^{3/2} + 10.125t^{1/2},$$

Define the functions above.
Enter (stripped) into EQ: of the Function graph of the HP48G or into $y1 =$, $y2 =$, $y3 =$ of the TI-85..
Leave the plot environment temporarily.
Obtain $s(2) = -3.2$, $v(2) = -5.1$, $a(2) = (0, 1, 018)$.
Thus the graph of $v(t)$ is the one that crosses $t = 2$ at the lowest point.

(b) Back to the Plot environment.
Use TRACE and (x,y) and control the graph by the up/down arrow keys.
$v(0.825) \approx 0$.

(c) TRACE (X,Y) shows a minimum near $P(1.25, -6.51)$.

Press FCN EXTR for the HP48G to get an extremum at
$Q(1.27767710557, -6.5139414007)$
According to the TRACE of the TI-85,
$Q(1.2777, -6.513941254)$ is the minimum.
$a(t)$ is minimized when $t = 1.2777$,

$s(1.2777) = -0.1276, v(1.2777) = -2.5114,$

and $a(1.2777) = -6.5139$

$s(1.25) \approx -0.06$ and $v(1.25) \approx -2.35$

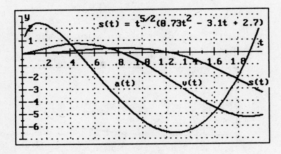

Chapter 3

Additional Applications of the Derivative

3.1 Increasing and Decreasing Functions

1.

$$\xrightarrow{\quad\searrow\quad\underset{\underset{\text{min}}{\bullet}}{-2}\nearrow\underset{\underset{\text{max}}{\bullet}}{2}\searrow\quad}$$

3.

$$\xrightarrow{\quad\nearrow\underset{\underset{\text{max}}{\bullet}}{-4}\searrow\underset{\underset{\text{none}}{\bullet}}{-2}\searrow\underset{\underset{\text{min}}{\bullet}}{0}\nearrow\underset{\underset{\text{max}}{\bullet}}{2}\searrow\quad}$$

5. $f(x) = x^2 - 4x + 5$
   $f'(x) = 2(x-2) = 0$ when $x = 2$

   If $x > 2$, $f'(x) > 0$; else $f'(x) < 0$

   $$\xrightarrow{\quad\searrow\underset{\underset{\text{min}}{\bullet}}{2}\nearrow\quad}$$

7. $f(x) = x^3 - 3x - 4$
   $f'(x) = 3(x+1)(x-1) = 0$ when $x = -1, 1$

   If $x < -1$ or $x > 1$, $f'(x) > 0$; else $f'(x) < 0$

   $$\xrightarrow{\quad\nearrow\underset{\underset{\text{max}}{\bullet}}{-1}\searrow\underset{\underset{\text{min}}{\bullet}}{1}\nearrow\quad}$$

9. $g(t) = t^5 - 5t^4 + 100$
   $g'(t) = 5t^3(t-4) = 0$ when $t = 0, 4$

   $$\xrightarrow{\quad\nearrow\underset{\underset{\text{max}}{\bullet}}{0}\searrow\underset{\underset{\text{min}}{\bullet}}{4}\nearrow\quad}$$

11. $f(x) = 3x^4 - 8x^3 + 6x^2 + 2$
    $f'(x) = 12x(x-1)^2 = 0$ when $x = 0, 1$

    $f(0) = 2$, $f(1) = 3$.
    $(0, 2)$ is a relative minimum.

    $$\xrightarrow{\quad\searrow\underset{\underset{\text{min}}{\bullet}}{0}\nearrow\underset{\underset{\text{none}}{\bullet}}{1}\nearrow\quad}$$

13. $f(t) = 2t^3 + 6t^2 + 6t + 5$
    $f'(t) = 6(t+1)^2 \geq 0$

    There are no relative extrema.
    $f'(-1) = 0$ but $(-1, 3)$ is not an extremum.

    $$\xrightarrow{\quad\nearrow\underset{\underset{\text{none}}{\bullet}}{-1}\nearrow\quad}$$

15. $g(x) = (x-1)^5$
    $g'(x) = 5(x-1)^4 = 0$ when $x = 1$

    $g(1) = 0$ but $(1, 0)$ is not a relative extremum.

    $$\xrightarrow{\quad\nearrow\underset{\underset{\text{none}}{\bullet}}{1}\nearrow\quad}$$

17. $S(t) = (t^2 - 1)^5$
    $S'(t) = 10t(t^2-1)^4 = 0$ when $t = -1, 0, 1$

$S(-1) = 0$, $S(0) = -1$, $S(1) = 0$.
$(0, -1)$ is a relative minimum. $(-1, 0)$ and $(1, 0)$ are not extrema.
Note the symmetry WRT the y-axis since t can be replaced by $-t$ in the original equation.

$$\xrightarrow{} \underset{\text{none}}{\overset{-1}{\bullet}} \xrightarrow{} \underset{\text{min}}{\overset{0}{\bullet}} \xrightarrow{} \underset{\text{none}}{\overset{1}{\bullet}} \xrightarrow{}$$

19.
$$f(x) = (x^3 - 1)^4$$
$$f'(x) = 12x^2(x^3 - 1)^3 = 0 \text{ when } x = 0, 1$$

$f(0) = 1$, $f(1) = 0$.
$(0, 1)$ is not an extremum,
$(1, 0)$ is a relative minimum.

$$\xrightarrow{} \underset{\text{none}}{\overset{0}{\bullet}} \xrightarrow{} \underset{\text{min}}{\overset{1}{\bullet}} \xrightarrow{}$$

21.
$$f(x) = x^3 - 3x^2$$
$$f'(x) = 3x^2 - 6x = 3x(x - 2) = 0$$

at $x = 0$ and $x = 2$.

$$f'(x) > 0 \quad \text{for} \quad x < 0$$
$$f'(x) < 0 \quad \text{for} \quad 0 < x < 2$$
$$f'(x) > 0 \quad \text{for} \quad 2 < x$$

$f(0) = 0$ and $f(2) = -4$. Thus $(0, 0)$ is a relative maximum and $(2, -4)$ is a relative minimum.

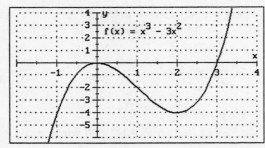

23.
$$f(x) = 3x^4 - 8x^3 + 6x + 2$$
$$f'(x) = 6(2x^3 - 4x^2 + 1) = 0$$

when $x = -0.46$, $x = 0.6$, and $x = 1.85$.

$$f'(x) < 0 \quad \text{for} \quad x < -0.45$$
$$f'(x) > 0 \quad \text{for} \quad -0.45 < x < 0.6$$
$$f'(x) < 0 \quad \text{for} \quad 0.6 < x < 1.85$$
$$f'(x) > 0 \quad \text{for} \quad 1.85 < x$$

$f(-0.46) = 0.15$, $f(0.6) = 4.3$, $f(1.85) = -2.4$.
Thus $(-0.46, 0.15)$ is a relative minimum, $(0.6, 4.3)$ is a relative maximum, and $(1.85, -2.4)$ is a relative minimum.

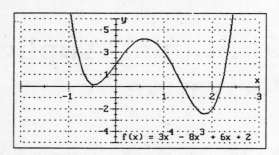

25.
$$f(t) = 2t^3 + 6t^2 + 6t + 5$$
$$f'(t) = 6t^2 + 12t + 6 = 6(t + 1)^2 = 0$$

when $t = -1$. $f'(t) \geq 0$ for all t except $t = -1$. $f(-1) = 3$ is not a relative minimum or maximum.

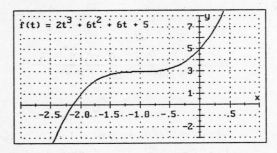

27.
$$g(t) = \frac{t}{t^2 + 3}$$
$$g'(t) = \frac{-t^2 + 3}{(t^2 + 3)^2} = 0$$

when $t = \pm\sqrt{3} \approx \pm 1.732$.

$$g'(t) < 0 \quad \text{for} \quad t < -1.732$$
$$g'(t) > 0 \quad \text{for} \quad -1.732 < t < 1.732$$
$$g'(t) < 0 \quad \text{for} \quad -1.732 < x$$

$g(\pm\sqrt{3}) = \pm 0.289$.
Thus $(-\sqrt{3}, -0.289)$ is a relative minimum, $(\sqrt{3}, 0.289)$ is a relative maximum.

3.1. INCREASING AND DECREASING FUNCTIONS

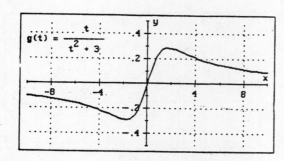

29.
$$f(x) = 3x^5 - 5x^3 + 4$$
$$f'(x) = 15x^4 - 15x^2 = 15x^2(x+1)(x-1) = 0$$

when $x = -1$, $x = 0$, and $x = 1$.

$f'(x) > 0$ for $x < -1$
$f'(x) < 0$ for $-1 < x < 0$ and $0 < x < 1$
$f'(x) > 0$ for $1 < x$

$f(-1) = 6$, $f(0) = 4$, $f(1) = 2$.
Thus $(-1, 6)$ is a relative maximum, $(1, 2)$ is a relative minimum. $(0, 4)$ is neither a relative minimum nor a relative minimum.

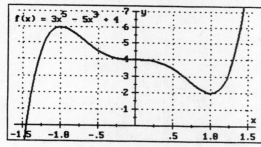

31.
$$f'(x) = x^3(2x-7)^2(x+5) = 0$$

when $x = 0$, $x = \frac{7}{2}$, and $x = -5$.

Thus $x = -5$ produces a relative maximum and $x = 0$ produces a relative minimum.

33.
$$H(T) = -0.53T^2 + 25T - 209.$$
$$H'(T) = -1.06T + 25 = 0 \text{ at } T = 23.5849$$

$(23.5849, 85.8113)$ is a relative maximum.
There are no asymptotes.
The prescribed domain is $15 \leq t \leq 30$.

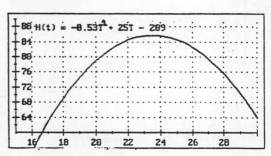

35. For the revenue function
$$R(x) = x(10 - 3x)^2.$$
$$= 100x - 60x^2 + 9x^3.$$
$$R'(x) = 100 - 120x + 27x^2 = 0$$

at $x = 1.1111$, $x = 3.3333$.
$(1.1111, 49.3827)$ is a relative maximum
$(3.3333, 0)$ is a relative minimum.

For the marginal revenue is
$$MR(x) = 100 - 120x + 27x^2.$$

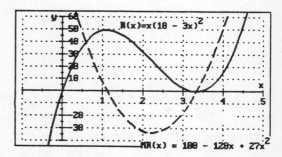

37.
$$C(t) = \frac{0.15t}{t^2 + 0.81}.$$
$$C'(t) = \frac{0.15(-t^2 + 0.81)}{(t^2 + 0.81)^2} = 0 \text{ at } t = 0.9$$

The maximum concentration appears at $(0.9, 0.08333)$
$y = 0$ is a horizontal asymptote.

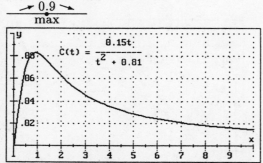

39. Integration, in chapter 5, will teach the technique of finding $f(x)$ when $f'(x)$ is known.

$$\text{If } f(x) = -3x^5 + 15x^4 - 25x^3 + 15x^2$$
$$\text{then } f'(x) = -15x(x-1)^2(x-2)$$

for which the graph is

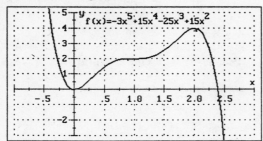

Note that
$$f'(0) = f'(1) = f'(2) = 0,$$
$$f'(x) < 0 \text{ when } x < 0 \text{ and } x > 2,$$
$$f'(x) > 0 \text{ when } 0 < x < 2.$$

41. (a) Since $f'(x) > 0$ when $x < -5$ and $x > 1$, $f(x)$ is increasing on these intervals.
    (b) Since $f'(x) < 0$ when $-5 < x < 1$, $f(x)$ is decreasing on this interval.
    (c) Since $f(-5) = 4$ and $f(1) = -1$, $f(x)$ has a relative maximum at $(-5, 4)$ and a relative minimum at $(1, -1)$.

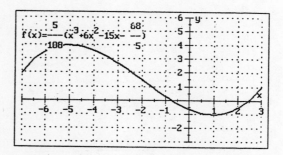

43. Since $f(x) = ax^2 + bx + c$ crosses the y-axis at $(0,3)$, $f(0) = 3$ and $a(0^2) + b(0) + c = 3$. Hence $c = 3$,
$f(x) = ax^2 + bx + 3$ has a relative maximum at $(5, 12)$ so $f'(5) = 0$.

Thus $f'(x) = 2ax + b$, $\quad 2a(5) + b = 0$ or $b = -10a$

Note that $(5, 12)$ is on the curve,
so $f(5) = 12$, $25a + 5b = 25a + 5(-10a) = 9$,
or $a = -\dfrac{9}{25}$.

Substituting for b leads to $b = \dfrac{18}{5}$.
The desired function is
$$f(x) = -\frac{9}{25}x^2 + \frac{18}{5} + 3$$

45. $f(x) = x^{2/3} > 0$ for all x.
$f'(x) = \dfrac{2}{3x^{1/3}} \neq 0$.
$f'(x) < 0$ for $x < 0$, $f(x)$ decreases.
$f'(x) > 0$ for $x > 0$, $f(x)$ increases.
$f(x)$ is symmetric WRT the y-axis.
$f(0) = 0$ but $f'(0)$ is not defined.
Note: The graph of the curve does not show up on the screen for $x < 0$. You need to analyze your results. A calculator (and computer) has its limitations.

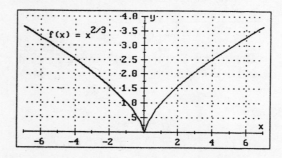

3.1. INCREASING AND DECREASING FUNCTIONS

47.
$$y = ax^2 + bx + c$$
$$y' = 2ax + b = 0 \text{ if } x = -\frac{b}{2a}$$

Assume $a > 0$.
$y' < 0$ for $x < -\frac{b}{2a}$, y decreases.
$y' > 0$ for $x > -\frac{b}{2a}$, y increases.
$x = -\frac{b}{2a}$ is the x value of a minimum.
Now assume $a < 0$.
$y' > 0$ for $x < -\frac{b}{2a}$, y increases.
$y' < 0$ for $x > -\frac{b}{2a}$, y decreases.
$x = -\frac{b}{2a}$ is the x value of a maximum.

49. $f(x) = x^4 + 3x^3 - 9x^2 + 4$.
Plot the curve. See a relative minimum in the third quadrant and another in the second.
For the HP48G Plot $f'(x) = 4x^3 + 9x^2 - 18x$.
Press ERASE.
Use TRACE to move the cross-hair close to the leftmost root of $f'(x)$.
FCN and EXTR locate $x = -3.5262$.
ZOOM BOXZ if needed to refine the graph on $-1 < x < 3$.
Move the cross-hair to the second root and press ROOT. $x = 0$.
Move the cross-hair to the third root and press ROOT. $x = 1.2762$.
For the TI-85 graph $y1 = f(x)$ and $y2 = f'(x)$.
You can deselect $y1$ if it interferes with decent viewing. TRACE and ZOOM (BOX) are used.
$f'(-3.528) = -0.096$, $f'(-0.002) = 0.029$, and $f'(1.2778) = 0.039$.

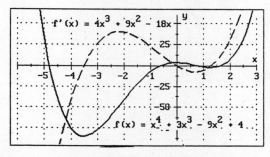

51.

$$f(x) = x^5 - 7.6x^3 + 2.1x^2 - 5$$
$$f'(x) = 5x^4 - 22.8x^2 + 4.2x$$

$f'(x) = 0$ at $x = -2.2$, $x = 0$, $x = 0.19$, and $x = 2$.

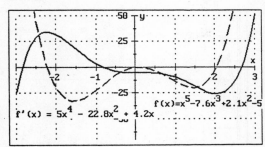

53. The curve in (a) is $f(x)$ (like $f(x) = x^4 + 3$.)
The slope of the tangent line is negative for $x < 0$ and positive for $0 < x$. It is 0 when $x = 0$.
The curve in (b) is $f'(x)$ (like $f'(x) = 4x^3$.)
The ordinates (y values) are negative for $x < 0$ and positive for $0 < x$. $y(0) = 0$.

55. Plot $f(x) = x^3 + 3x^2 - 5x + 11$ and then
plot $g(x) = (x+1)^3 + 3(x+1)^2 - 5(x+1) + 11$

Finally plot
$h(x) = (x+3)^3 + 3(x+3)^2 - 5(x+3) + 13$.
Suggestion: use a viewing window with $-10 < x < 3$ and $-5 < y < 40$.
Note: HP48G's Plot is replaced by Graph on the TI-85.

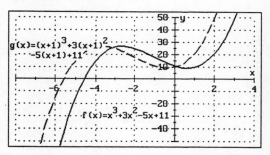

57. Plot $f(x) = x^3 - 6x^2 + 5x - 11$ and then
plot $g(x) = (2x)^3 - 6(2x)^2 + 5(2x) - 11$

Suggestion: use a viewing window with $-5 < x < 10$ and $-40 < y < 80$.

Note: HP48G's Plot is replaced by Graph on the TI-85.

The second plot seems to be a copy of the first but with the x-values drawn (scaled) out by a factor of 2.

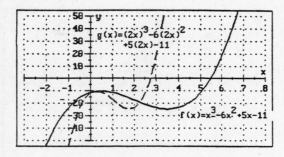

3.2 Concavity

1. The graph is
   concave downward ($f'' < 0$) for $x < 2$,
   and concave upward ($f'' > 0$) for $x > 2$.

3. $$f(x) = \frac{x^3}{3} - 9x + 2$$
    $$f'(x) = (x+3)(x-3) = 0,\ x = -3,\ x = 3$$
    $$f''(x) = 2x = 0 \text{ when } x = 0$$

    $(-3, 20)$ is a relative maximum,
    $(3, -16)$ is a relative minimum,
    $(0, 2)$ is a point of inflection.

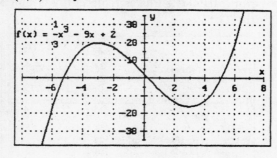

5. $$f(x) = x^4 - 4x^3 + 10$$
    $$f'(x) = 4x^2(x-3) = 0 \text{ when } x = 0,\ 3$$
    $$f''(x) = 12x(x-2) = 0 \text{ when } x = 0 \text{ and } x = 2$$

    $(3, -17)$ is a relative minimum,
    $(0, 10)$ is not an extremum, and
    $(0, 10)$ as well as $(2, -6)$ are points of inflection.

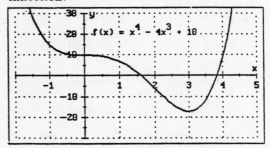

7. $$f(x) = (x-2)^3$$
    $$f'(x) = 3(x-2)^2 = 0 \text{ when } x = 2$$
    $$f''(x) = 6(x-2) = 0 \text{ when } x = 2$$

    $(2, 0)$ is a point of inflection.

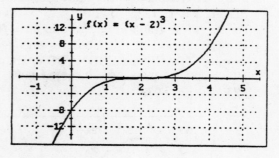

3.2. CONCAVITY

9.
$$f(x) = (x^2 - 5)^3$$
$$f'(x) = 6x(x^2 - 5)^2 = 0$$

when $x = -\sqrt{5}, x = 0, x = \sqrt{5}$.

$$f''(x) = 6x[2(x^2-5)(2x)] + (x^2-5)^2(6)$$
$$= 30(x^2-5)(x+1)(x-1) = 0$$

when $x = -\sqrt{5}, x = -1, x = 1, x = \sqrt{5}$.
$(0, -125)$ is a relative minimum.
$(-\sqrt{5}, 0), (-1, -64), (1, -64), (\sqrt{5}, 0)$ are points of inflection.

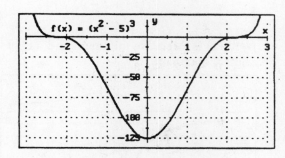

Note the symmetry with respect to the y axis. This was predictable because the points (x, y) and $(-x, y)$ satisfy the original equation $f(x) = (x^2 - 5)^3$. In other words, x can be replaced by $-x$ without changing the the equation.

```
  ↘-√5 ↘-1  ↘  0  ↗  1  ↗-√5 ↗
  neither neither  min   neither neither
   ∪_-√5 ∩_-1    ∪    _1 ∩_√5 ∪
```

11.
$$f(s) = 2s(s+4)^3.$$
$$f'(s) = 2s[3(s+4)^2(1)] + 2(s+4)^3$$
$$= 8(s+4)^2(s+1) = 0$$

when $s = -4$ and $s = -1$.

$$f''(s) = 8[(s+4)^2(1) + (s+1)(2)(s+4)]$$
$$= 24(s+4)(s+2) = 0 \text{ at } s = -4, -2$$

$(-1, -54)$ is a relative minimum,
$(-4, 0)$ and $(-2, -32)$ are inflection points.

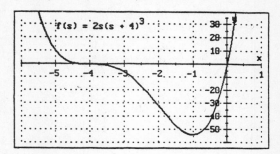

```
  ↘ -4 ↘ -2 ↘ -1 ↗
  neither neither  min
   ∪_-4 ∩_-2   ∪
```

13.
$$f(x) = (x+1)^{1/3}$$
$$f'(x) = \frac{1}{3(x+1)^{2/3}} \text{ which is never 0}$$

It is undefined when $x = -1$.

$$f''(x) = \frac{-2}{9(x+1)^{5/3}} \text{ which is never 0}$$

but undefined at $x = -1$.
$(-1, 0)$ is an inflection point.

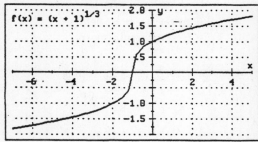

```
  ↗ -1 ↗
  neither
```

15.
$$f(x) = (x+1)^{4/3}$$
$$f'(x) = \frac{4(x+1)^{1/3}}{3} = 0 \text{ when } x = -1$$
$$f''(x) = \frac{4}{9(x+1)^{2/3}} \text{ which is never } 0$$

but undefined at $x = -1$.
$(-1, 0)$ is a relative minimum.

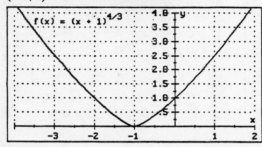

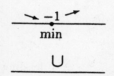

17.
$$g(x) = \sqrt{x^2+1} = (x^2+1)^{1/2}$$
$$g'(x) = \frac{x}{(x^2+1)^{1/2}} = 0 \text{ when } x = 0$$
$$g''(x) = x[-\frac{1}{2}(x^2+1)^{-3/2}(2x)] + (x^2+1)^{-1/2}(1)$$
$$= \frac{-x^2 + (x^2+1)}{(x^2+1)^{3/2}} = \frac{1}{(x^2+1)^{3/2}}$$

which is never 0 nor undefined.
$(0, 1)$ is a relative minimum.

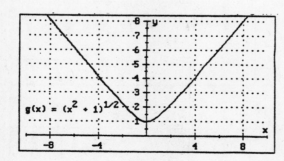

19.
$$f(x) = \frac{x}{x^2+x+1}$$
$$f'(x) = \frac{-x^2+1}{(x^2+x+1)^2} = 0$$

when $x = -1$ or $x = 1$.
$$f''(x) = \frac{2x^3 - 6x - 2}{(x^2+x+1)^3} = 0$$

when $x \approx -1.53$, $x \approx -0.35$, and $x \approx 1.88$.
$(-1, -1)$ is a relative minimum and $(1, \frac{1}{3})$ is a relative maximum.

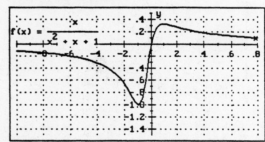

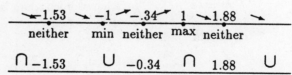

21.
$$f(x) = x^3 + 3x^2 + 1$$
$$f'(x) = 3x(x+2) = 0 \text{ at } x = -2, x = 0$$

$f(-2) = 5$ and $f(0) = 1$.
$f''(x) = 6(x+1)$.
$f''(0) = 6 > 0$ and $f''(-2) = -6 < 0$,
so $(0, 1)$ is a relative minimum,
$(-2, 5)$ is a relative maximum.

3.2. CONCAVITY

23.
$f(x) = (x^2 - 9)^2$
$f'(x) = 4x(x-3)(x+3) = 0$ at $x = -3, 0, 3$

$f(\pm 3) = 0$ and $f(0) = 81$.

$$f''(x) = 12(x^2 - 3)$$

$f''(-3) = 72 > 0$, $f''(0) = -36 < 0$,
$f''(3) = 72 > 0$,
so $(0, 81)$ is a relative maximum,
$(-3, 0), (3, 0)$ are relative minima.

25.
$f(x) = 2x + 1 + \dfrac{18}{x}$

$f'(x) = 2 - \dfrac{18}{x^2}$

$= \dfrac{2(x-3)(x+3)}{x^2} = 0$

at $x = -3$, $x = 3$. $f(-3) = -11$, $f(3) = 13$.

$$f''(x) = \dfrac{36}{x^3}$$

$f''(-3) < 0$, $f''(3) > 0$,
so $(-3, -11)$ is a relative maximum,
$(3, 13)$ is a relative minimum.

27.
$f(x) = x^2(x-5)^2 = x^4 - 10x^3 + 25x^2$
$f'(x) = 2x(x-5)(2x-5) = 4x^3 - 30x^2 + 50x$
$f''(x) = 2(6x^2 - 30x + 25)$

$f'(0) = 0$, $f'(2.5) = 0$, and $f'(5) = 0$.
$f''(0) > 0$ and $f''(5) > 0$ so
$(0, 0)$ and $(5, 0)$ are relative minima,
$f''(2.5) < 0$ so $(2.5, 39)$ is a relative maximum.

29.
$h(t) = \dfrac{2}{1+t^2} = 2(1+t^2)^{-1}$

$h'(t) = \dfrac{-4t}{(1+t^2)^2} = 0$ when $t = 0$

$h''(t) = \dfrac{-4(1+t^2)^2 - (-4t)(2)(1+t^2)(2t)}{(1+t^2)^4}$

$= \dfrac{4(3t^2 - 1)}{(1+t^2)^3}$

$h''(0) < 0$ so $(0, 2)$ is a relative maximum.

31.
$f(x) = \dfrac{(x-2)^3}{x^2}$

$f'(x) = \dfrac{x^2[3(x-2)^2(1)] - (x-2)^3(2x)}{x^4}$

$= \dfrac{(x-2)^2(x+4)}{x^3} = 0$ when $x = -4$, $x = 2$

$f''(x) = \dfrac{1}{x^6}\{x^3[(x-2)^2(1) + (x+4)(2)(x-2)]$
$\quad - (x-2)^2(x+4)(3x^2)\}$

$= \dfrac{1}{x^4}\{(x-2)[x^2 - 2x + 2x^2 + 8x$
$\quad -3x^2 - 6x + 24]\}$

$= \dfrac{24(x-2)}{x^4}$

$f''(-4) < 0$ so $\left(-4, -\dfrac{27}{2}\right)$ is a relative
maximum,
$f''(2) = 0$ so the test fails.

33. (a) $f'(x) = x^2 - 4x = x(x-4) = 0$

when $x = 0$ and $x = 4$, leading to possible extrema. The curve is risng for $x < 0$, falling for $0 < x < 4$, and rising for $4 < x$.

(b)
$$f''(x) = 2x - 4 = 0$$

when $x = 2$. The curve is concave down for $x < 2$ and concave up for $2 < x$.
A possible graph is shown in part (d) below.

(c) There is a relative maximum when $x = 0$ and a relative minimum at $x = 4$.
When $x = 2$, the function has an inflection point.

(d)

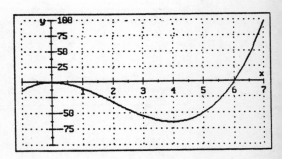

35. (a) $$f'(x) = 5 - x^2 = 0$$
    when $x = \pm\sqrt{5}$, leading to possible extrema. The curve is falling for $x < -\sqrt{5}$, rising for $-\sqrt{5} < x < \sqrt{5}$, and falling for $\sqrt{5} < x$.

    (b) $$f''(x) = -2x = 0$$
    when $x = 0$. The curve is concave up for $x < 0$ and concave down for $0 < x$, or vice-versa.

    A possible graph is shown in part (d) below.

    (c) There is a relative minimum at $x = -\sqrt{5}$ and a relative maximimum at $x = \sqrt{5}$. When $x = 0$, the function has an inflection point.

    (d)

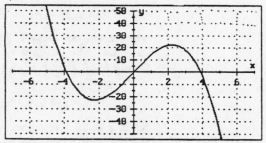

37. (a) The curve rises for $x < -1$ and $3 < x$.
    (b) It falls when $-1 < x < 3$.
    (c) The concavity is down for $x < 2$.
    (d) The concavity is up for $2 < x$.
    Here is a possible graph.

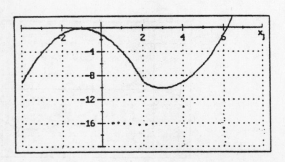

39. From the graph of $f'(x)$,
    $f'(x) < 0$ (f decreases) when $x < 2$,
    $f'(x) > 0$ (f increases) when $x > 2$.
    Hence f has a relative minimum at $x = 2$.
    The graph shows $f'(x)$ increasing for all x, which implies $f''(x) > 0$ and
    $f(x)$ is concave $\uparrow$ for all x.
    The following is an example for the desired graph

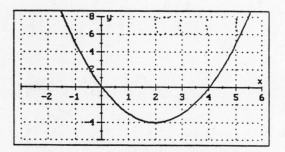

41. From the graph of $f'(x)$, we see that
    $f'(x) \leq 0$ (f decreases) when $x < 2$,
    and $f'(x) > 0$ (f increases) when $x > 2$.
    Hence f has a relative minimum at $x = 2$.
    The graph shows $f'(x)$ is increasing when
    $x < -3$ and $x > -1$, which implies $f''(x) > 0$
    and $f(x)$ is concave $\uparrow$ on these intervals.

    $f'(x)$ decreases when $-3 < x < -1$, which
    implies f has inflection points at $x = -3$ and
    $x = -1$.
    The following is an example of a possible
    graph which fits the prescribed conditions.

3.2. CONCAVITY

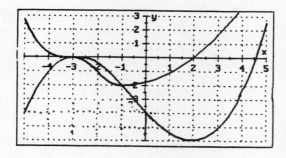

43. (a)
$$C(x) = 0.3x^3 - 5x^2 + 28x + 200$$
$$M(x) = C'(x) = 0.9x^2 - 10x + 28$$

(b)
$$M'(x) = C''(x) = 1.8x - 10 = 0$$

if $x = 5.56$.
The roots of $M'(x)$ correspond to the extremum (extrema) of $M(x)$.
Thus the x-value of the inflection point(s) of $C(x)$ correspond to the root(s) of $C''(x)$.

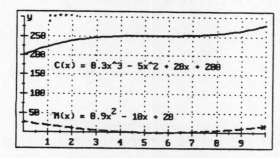

45. We want to maximize the rate of output on the interval $0 \leq t \leq 4$. Since
$$Q(t) = -t^3 + \frac{9}{2}t^2 + 15t$$
is the output, the rate of output is
$$R(t) = Q'(t) = -3t^2 + 9t + 15.$$
$$R'(t) = Q''(t) = -3(2t - 3) = 0 \text{ when } t = \frac{3}{2}$$

$R(0) = 15$, $R(1.5) = 21.75$, and $R(4) = 3$.
Thus an absolute maximum occurs at $t = 1.5$ and an absolute minimum when $t = 4$.

(a) That is, the worker is performing most efficiently at 9:30 a.m.

(b) and least efficiently at 12:00 noon.

47. The rate of population growth is the derivative
$$R(t) = P'(t) = -3t^2 + 18t + 48.$$

(a) We need to optimize $R(t)$ on the interval $0 \leq t \leq 5$.
$$R'(t) = P''(t) = -6t + 18 = 0 \text{ when } t = 3$$
Now $R(0) = 48$, $R(3) = 75$, and $R(5) = 63$.
Thus the rate of growth is greatest 3 years from now, when it is 75 thousand people per year.

(b) It is smallest now (at $t = 0$), when it is 48 thousand people per year.

(c) The rate of growth of $R(t)$, $R'(t) = -6t + 18$, is largest when $t = 0$.

49.
$$\frac{100 P'(t)}{P(t)} = A - BP(t)$$
$$P'(t) = \frac{P(t)(A - BP(t))}{100}$$
$$P''(t) = \frac{1}{100^2}[P(t)(A - BP(t))(A - 2BP(t))]$$
$P''(t) = 0$ when $P(t) = \frac{A}{B}$ or $P(t) = \frac{A}{2B}$.
These are inflection points. $P(t)$ is changing most rapidly at these points.

51.
$$y = ax^2 + bx + c$$
$$y' = 2ax + b, \text{ and } y'' = 2a$$

$y'' < 0$ when $a < 0$, the curve is concave downward and the graph has a maximum. Similarly, when $a > 0$, the graph exhibits a minimum and is concave up.

x	-4	-2	-1
$f(x)$	1,300.42	106.74	10.03
$f'(x)$	$-1,204.64$	-186.76	-33.89
$f''(x)$	835.12	241.96	78.58

x	0	1	2
$f(x)$	-0.7	-0.03	26.26
$f'(x)$	0	3.71	66.04
$f''(x)$	4	18.22	121.24

(c) The x intercepts are $x \approx 1.01$, $x \approx -0.4$, the y intercept is $y = -0.7$.

(d) According to the graphing utility, $(0, -0.7)$, $(0.5, -0.6)$ are extrema. But the utility is lying. $(0.5, -0.6)$ is an inflection point. This is another reminder to think about your calculator results.

(e) $f(x)$ is increasing on $0 < x$.

(f) $f(x)$ is decreasing on $x < 0$.

(g) $f(x)$ has a point of inflection at $(0.18, -0.66)$ and $(0.5, -0.6)$.

(h) $f(x)$ is concave upward on $x < 0.18$ and $0.5 < x$.

(i) $f(x)$ is concave downward on $0.18 < x < 0.5$.

(j) $f''(0.177) = 0.05$, $f''(0.185) = -0.06$, $f''(0.45) = -0.59$, $f''(0.55) = 0.832$.

(k) $(-4, 1,300.42)$ is the absolute maximum, $(0, -0.7)$ is the absolute minimum.

53. (a)

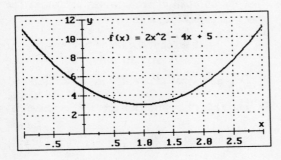
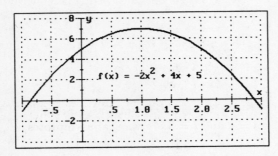

(b)

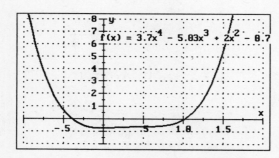

$$f(x) = 3.7x^4 - 5.03x^3 + 2x^2 - 0.7$$
$$f'(x) = 14.8x^3 - 15.09x^2 + 4x$$
$$f''(x) = 44.4x^2 - 30.18x + 4$$

3.3 Limits Involving Infinity; Asymptotes

1. $$f(x) = x^3 - 4x^2 - 4,$$
$$\lim_{x \to \infty} f(x) = \infty^3 - 4\infty^2 - 4 = \infty$$

3. $$f(x) = (1 - 2x)(x + 5)$$
$$\lim_{x \to \infty} f(x) = (-\infty)(\infty) = -\infty$$

3.3. LIMITS INVOLVING INFINITY; ASYMPTOTES

5.
$$f(x) = \frac{x^2 - 2x + 3}{2x^2 + 5x + 1}$$
$$\lim_{x \to \infty} f(x) = \lim_{x \to \infty} \frac{1 - \frac{2}{x} + \frac{3}{x^2}}{2 + \frac{5}{x} + \frac{1}{x^2}} = \frac{1}{2}$$

7.
$$f(x) = \frac{2x + 1}{3x^2 + 2x - 7},$$
$$\lim_{x \to \infty} f(x) = \lim_{x \to \infty} \frac{\frac{2}{x} + \frac{1}{x^2}}{3 + \frac{2}{x} - \frac{7}{x^2}} = 0$$

9.
$$f(x) = \frac{3x^2 - 6x + 2}{2x - 9},$$
$$\lim_{x \to \infty} f(x) = \lim_{x \to \infty} \frac{3 - \frac{6}{x} + \frac{2}{x^2}}{\frac{2}{x} - \frac{9}{x^2}} = \infty.$$

11.
$$\lim_{x \to 0} f(x) = +\infty$$
$$\text{and } \lim_{x \to \infty} f(x) = 0$$

thus $y = 0$ is a horizontal asymptote and $x = 0$ is a vertical asymptote.

13.
$$\lim_{x \to +\infty} f(x) = 1$$
$$\text{and } \lim_{x \to -\infty} f(x) = -1$$

thus $y = 1$ and $y = -1$ are horizontal asymptote.

15.
$$\lim_{x \to \infty} f(x) = 1$$
$$\lim_{x \to -2^-} f(x) = \infty$$
$$\text{and } \lim_{x \to 2^+} f(x) = \infty$$

thus $y = 1$ is a horizontal asymptote, $x = -2$ and $x = 2$ are vertical asymptotes.

17.
$$\lim_{x \to -2} \frac{3x - 1}{x + 2} = \infty$$
$$\lim_{x \to \infty} \frac{3x - 1}{x + 2} = 3$$

thus $y = 3$ is a horizontal asymptote and $x = -2$ is a vertical asymptote.

19.
$$\lim_{x \to \pm 1} \frac{x^2 + 2}{x^2 - 1} = \pm\infty$$
$$\lim_{x \to \infty} \frac{x^2 + 2}{x^2 - 1} = 1$$

thus $y = 1$ is a horizontal asymptote, $x = -1$ and $x = 1$ are vertical asymptotes.

21.
$$t^2 - 5t + 6 = (t - 3)(t - 2)$$
$$\lim_{t \to 2} \frac{t^2 + 3t - 5}{t^2 - 5t + 6} = \pm\infty$$
$$\lim_{t \to 3} \frac{t^2 + 3t - 5}{t^2 - 5t + 6} = \infty$$
$$\lim_{t \to \infty} \frac{t^2 + 3t - 5}{t^2 - 5t + 6} = 1$$

thus $y = 1$ is a horizontal asymptote, $t = 2$ and $t = 3$ are vertical asymptotes.

23.
$$\frac{1}{x} - \frac{1}{x - 1} = -\frac{1}{x(x - 1)}$$
$$\lim_{x \to 0} \frac{1}{x(x - 1)} = \pm\infty$$
$$\lim_{x \to 1} \frac{1}{x(x - 1)} = \pm\infty$$
$$\lim_{x \to \infty} \frac{1}{x(x - 1)} = 0$$

thus $y = 0$ is a horizontal asymptote, $x = 0$ and $x = 1$ are vertical asymptotes.

25.
$$f(x) = x^3 + 3x^2 - 2$$
$$f'(x) = 3x(x + 2)$$
$$f''(x) = 6(x + 1)$$

$f'(-2) = f'(0) = 0$ and $f''(-1) = 0$.

x	$-\infty$	-2		-1		0	∞
$f(x)$				0		-2	
$f'(x)$	$+$	0	$-$	$-$	$-$	0	$+$
$f''(x)$	$-$	$-$	$-$	0	$+$	$+$	$+$

86   CHAPTER 3. ADDITIONAL APPLICATIONS OF THE DERIVATIVE

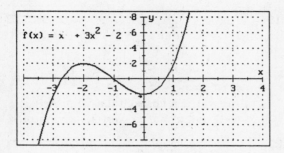

27.
$$f(x) = x^4 + 4x^3 + 4x^2$$
$$f'(x) = 4x(x+1)(x+2)$$
$$f''(x) = 4(3x^2 + 6x + 2)$$

$f'(-2) = f'(-1) = f'(0) = 0$ and
$f''(-1.577) = f''(-0.423) = 0$.

x	$-\infty$	-2		-1.577		-1
$f(x)$		0				1
$f'(x)$	$-$	0	$+$	$+$	$+$	0
$f''(x)$	$+$	$+$	$+$	0	$-$	$-$

x	-1		-0.423		0	∞
$f(x)$	1				0	
$f'(x)$	0	$-$	$-$	$-$	0	$+$
$f''(x)$	$-$	$-$	0	$+$	$+$	$+$

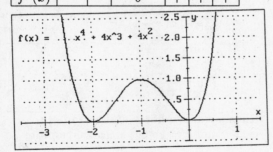

29.
$$f(x) = (2x-1)^2(x^2-9)$$
$$f'(x) = 2(2x-1)(4x^2-x-18)$$
$$f''(x) = 2(24x^2-12x-35)$$

$f'(-2) = f'(0.5) = f'(2.25) = 0$ and
$f''(-0.983) = f''(1.483) = 0$.

x	$-\infty$	-2		-0.983		0.5
$f(x)$						
$f'(x)$	$-$	0	$+$	$+$	$+$	0
$f''(x)$	$+$	$+$	$+$	0	$-$	$-$

x	0.5		1.483		2.25	∞
$f(x)$						
$f'(x)$	0	$-$	$-$	$-$	0	$+$
$f''(x)$	$-$	$-$	0	$+$	$+$	$+$

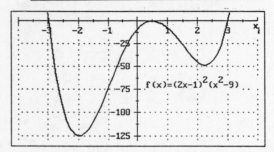

31.
$$f(x) = (2x+3)^{-1} \neq 0$$
$$f'(x) = -2(2x+3)^{-2} < 0$$
$$f''(x) = 4(2x+3)^{-3}$$

$y = 0$ is a horizontal asymptote and
$x = -1.5$ is a vertical asymptote.

x	$-\infty$		-1.5		∞
$f(x)$			∞		
$f'(x)$	$-$	$-$	∞	$-$	$-$
$f''(x)$	$-$	$-$	∞	$+$	$+$

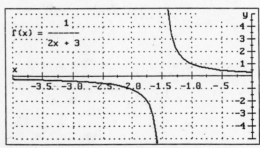

33.
$$f(x) = x - x^{-1}$$
$$f'(x) = 1 + x^{-2} \neq 0$$
$$f''(x) = -2x^{-3}$$

$y = x$ is an oblique asymptote and
$x = 0$ is a vertical asymptote.

x	$-\infty$		0		∞
$f(x)$			$\pm\infty$		
$f'(x)$	$+$	$+$	∞	$+$	$+$
$f''(x)$	$+$	$+$	$\pm\infty$	$-$	$-$

3.3. LIMITS INVOLVING INFINITY; ASYMPTOTES

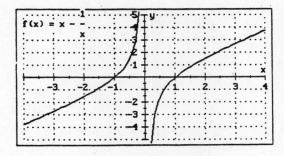
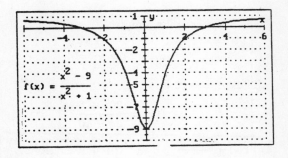

35.
$$f(x) = \frac{1}{x^2 - 9}$$
$$f'(x) = \frac{-2x}{(x^2 - 9)^2}$$
$$f''(x) = \frac{6(x^2 + 3)}{(x^2 - 9)^3}$$

$x = \pm 3$ are vertical asymptotes and $y = 0$ is a horizontal asymptote.

x	$-\infty$	-3		0		3	∞
$f(x)$	0	∞		-0.11		∞	0
$f'(x)$	$+$	∞	$+$	0	$-$	∞	$-$	$-$
$f''(x)$	$+$	∞	$-$	$-$	$-$	∞	$+$	$+$

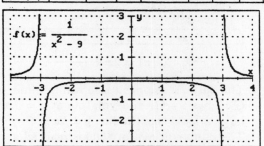

37.
$$f(x) = \frac{x^2 - 9}{x^2 + 1}$$
$$f'(x) = \frac{20x}{(x^2 + 1)^2}$$
$$f''(x) = \frac{20(-3x^2 + 1)}{(x^2 + 1)^3}$$

$y = 1$ is a horizontal asymptote.

x	$-\infty$	$-.58$		0		$.58$	∞
$f(x)$				-9			
$f'(x)$	$-$	$-$	$-$	0	$+$	$+$	$+$
$f''(x)$	$-$	0	$+$	$+$	$+$	0	$-$

39. $y = \dfrac{1}{x^2}$

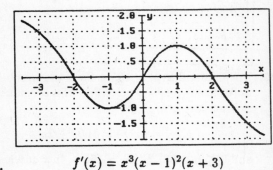

41.

43. $f'(x) = x^3(x-1)^2(x+3)$

$$\xrightarrow{} \underset{-3}{\bullet} \xleftarrow{} \underset{0}{\bullet} \xrightarrow{} \underset{1}{\bullet} \xrightarrow{}$$

(a) $f(x)$ decreases on $-3 < x < 0$ and increases elsewhere.

(b) A maximum is indicated at $x = -3$ a minimum at $x = 0$.

45. For a vertical asymptote the denominator must vanish at $x = 2$.

$$\text{Thus } B = -\frac{5}{2}$$

For a horizontal asymptote,
$$\lim_{x \to \infty} f(x) = \frac{A}{B} = 4$$
Thus $A = -10$ and
$$f(x) = \frac{20x + 6}{5x - 10}$$

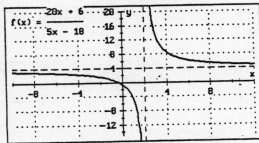

47.

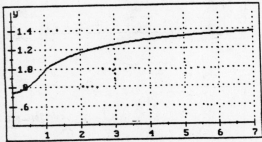

49. The demand $x(p) = 160 - 2p$ is the number of units bought at the price p. The consumer expenditure is
$$E(p) = px(p) = 160p - 2p^2$$
and $0 \leq p \leq 80$. $E'(p) = 160 - 4p = 0$ when $p = 40$. $E(0) = E(80) = 0$.
Thus the total consumer expenditure is greatest when $p = 40$ and $E(p) = 3,200$.

51. (a)
$$I(S) = \frac{aS}{S+c}$$

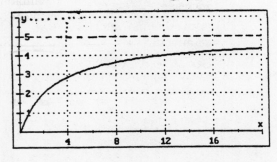

(b) Writing exercise —
Answers will vary.

53.
$$f(x) = \frac{600x}{300 - x},$$
where x represents the percent of homes.
There is a discontinuity at $x = 300$.
The practical values of x here are $0 \leq x \leq 100$.

x	-900	0	100
$f(x)$	-450	0	300

x	290	310	$1,100$
$f(x)$	$17,400$	$-18,600$	-825

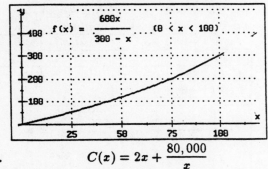

55.
$$C(x) = 2x + \frac{80,000}{x}$$
The optimal shipment size will minimize $C(x)$.
$C(x)$ has a vertical asymptote at $x = 0$ and for large x looks like the line $y = 2x$.
The minimum cost of $800 is reached at $x = 200$.

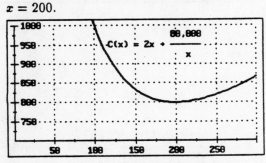

57.
$$\left(P + \frac{a}{V^2}\right)(V - b) = nRT$$

3.3. LIMITS INVOLVING INFINITY; ASYMPTOTES

(a)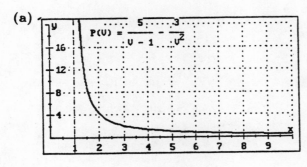

(b)
$$P(V) = \frac{nRT_c}{V-b} - \frac{a}{V^2}$$
$$P'(V) = -\frac{nRT_c}{(V-b)^2} + \frac{2a}{V^3} = 0$$
$$\frac{nRT_c}{(V-b)^2} = \frac{2a}{V^3}$$
and $b = \frac{8a}{27nRT_c}$ if $V = 3b$
$$P''(V_c) = \frac{2nRT_c}{(V_c-b)^3} - \frac{6a}{V_c^4}$$
$$= \frac{2nRT_c}{8b^3} - \frac{6a}{27b^4} = 0 \text{ if } V = 3b$$

(c)
$$P(V_c) = \frac{nRT_c}{2b} - \frac{a}{9b^2}$$
$$= \frac{1}{18b^2}\left(\frac{72ab}{27b} - 2a\right)$$
$$= \frac{a}{27b^2}$$
$$= \frac{nRT_c}{2b} - \frac{a}{9b^2}$$
or $T_c = \frac{8a}{27bnR}$

59. (a)
$$f(x) = x^{1/3}(x-4)$$
$$f'(x) = \frac{4(x-1)}{3x^{2/3}}$$
$$f''(x) = \frac{4(x+2)}{9x^{5/3}}$$

x	$-\infty$	-2	0	1	∞
$f(x)$			0	-3	
$f'(x)$	$-$	$-$	$-\infty$	0	$+$
$f''(x)$	$+$	0	$-$	$\pm\infty$	$+$	$+$

$(1, -3)$ is the only minimum.

(b) The graph is concave down for $-2 < x < 0$ and concave up elsewhere. $(0,0)$ and $(-2, 7.6)$ are inflection points.

(c) $(0,0)$ and $(4,0)$ are intercepts.

(d)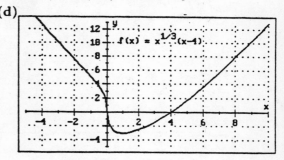

61. (a) Let $D = 25 - 1.1x - x^2$
$$f(x) = \frac{x+9.4}{D}$$
$$f'(x) = \frac{D - (x+9.4)(-2x-11)}{D^2}$$
$$= \frac{x^2 + 18.8x + 35.34}{D^2}$$
$$f''(x) = \frac{D^2(2x+18.8)}{D^4}$$
$$\quad - \frac{(x^2+18.8x+35.34)[2D(-1.1-2x)]}{D^4}$$
$$= \frac{D(2x+18.8)}{D^3}$$
$$\quad - \frac{(x^2+18.8x+35.34)[2(-1.1-2x)]}{D^3}$$
$$= \frac{2(x^3 + 28.2x^2 + 106.02x + 273.87)}{D^3}$$

According to the graphing utility, $(-16.68, 0.0309)$, $(-2.1185, 0.3188)$ are extrema.
You may also get $(10^{125}, 10^{-125})$, but ignore this point. It is just a reminder that there is a horizontal asymptote at $y = 0$.

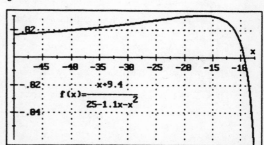

$f(x)$ is increasing on $x < -16.68$, $-2.1185 < x < 4.48$, and $4.48 < x$.
$f(x)$ is decreasing on $-16.68 < x < -5.5802$ and $-5.5802 < x < -2.1185$.

(b)

```
   ↗ -16.68 ↘ -5.58 ↘ -2.12 ↗ 4.48 ↗
      max      neither   min    neither

  ∪  -24.3  ∩   -5.58   ∪   4.48  ∩
```

(c) The x intercept is $x = -9.4$,
the y intercept is $y = 0.376$.

(d) There are vertical asymptotes at $x = -5.58$ and $x = 4.48$.
There is a horizontal asymptote at $x = 0$.

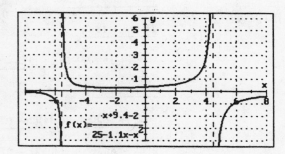

3.4 Optimization

1.
$$f(x) = x^2 + 4x + 5, \quad -3 \leq x \leq 1$$
$$f'(x) = 2x + 4 = 2(x+2) = 0$$

at $x = -2$, which is in the interval.
$f(-2) = 1$, $f(-3) = 2$, and $f(1) = 10$.
Thus
$$f(1) = 10$$
is the absolute maximum while
$$f(-2) = 1$$
is the absolute minimum.

3.
$$f(x) = \frac{x^3}{3} - 9x + 2, \quad 0 \leq x \leq 2$$
$$f'(x) = (x+3)(x-3) = 0$$

at $x = -3$ and $x = 3$, which are not in the interval.
$f(0) = 2$ and $f(2) = -\frac{40}{3}$.

Thus $f(0) = 2$
is the absolute maximum while
$$f(2) = -\frac{40}{3}$$
is the absolute minimum.

5.
$$f(t) = 3t^5 - 5t^3, \quad -2 \leq t \leq 0$$
$$f'(t) = 15t^4 - 15t^2$$
$$= 15t^2(t+1)(t-1) = 0$$

at $t = -1$, $t = 0$, and $t = 1$,
of which only $t = -1$ and $t = 0$ are in the interval.
$f(0) = 0$, $f(-1) = 2$, and $f(-2) = -56$.

Thus $f(-1) = 2$
is the absolute maximum while
$$f(-2) = -56$$
is the absolute minimum.

7.
$$f(x) = (x^2 - 4)^5, \quad -3 \leq x \leq 2$$
$$f'(x) = 10x(x-2)^4(x+2)^4 = 0$$

at $x = -2$, $x = 0$, and $x = 2$,
all of which are in the interval.
$f(0) = -1,024$, $f(-2) = 0$, $f(2) = 0$, and $f(-3) = 3,125$.

Thus $f(-3) = 3,125$
is the absolute maximum while
$$f(0) = -1,024$$
is the absolute minimum.

3.4. OPTIMIZATION

9.
$$g(x) = x + \frac{1}{x}, \quad \frac{1}{2} \le x \le 3$$
$$g'(x) = 1 - \frac{1}{x^2} = \frac{(x-1)(x+1)}{x^2} = 0$$

at $x = -1$ and $x = 1$, of which only $x = 1$ is in the interval.
Note that $g'(x)$ is undefined when $x = 0$, which is not in the domain of g and hence not a critical point.
$g(1) = 2$, $g\left(\frac{1}{2}\right) = \frac{5}{2}$, and $g(3) = \frac{10}{3}$.

$$\text{Thus } g(3) = \frac{10}{3}$$

is the absolute maximum while

$$g(1) = 2$$

is the absolute minimum.

11.
$$f(u) = u + \frac{1}{u}, \quad 0 < u$$
$$f'(u) = 1 - \frac{1}{u^2} = \frac{(u-1)(u+1)}{u^2} = 0$$

at $u = -1$ and $u = 1$, of which only $u = 1$ is in the interval.
Note that $f'(u)$ is undefined when $u = 0$, which is not in the domain of f and hence not a critical point.
$f(1) = 2$ and there are no other endpoints.
$f'(u) < 0$ when $0 < u < 1$ (i.e. f is decreasing) and $f'(u) > 0$ when $u > 1$ (i.e. f is increasing),
it follows that there is no absolute maximum

$$f(1) = 2$$

is the absolute minimum.

13.
$$f(x) = \frac{1}{x}, \quad 0 < x$$
$$f'(x) = -\frac{1}{x^2} \ne 0$$

There are no endpoints and no critical points. Hence there are no absolute extrema.

15.
$$f(x) = \frac{1}{x+1} = (x+1)^{-1}, \quad 0 \le x$$
$$f'(x) = -\frac{1}{(x+1)^2} < 0 \text{ for all } x$$

Hence the graph of f begins at $f(0) = 1$ and decreases for all $x > 0$.

$$f(0) = 1$$

is the absolute maximum
and there is no absolute minimum.

17. $p(q) = 49 - q$ and $C(q) = \frac{1}{8}q^2 + 4q + 200$.

(a)
$$R(q) = q\,p(q) = 49q - q^2$$
$$MR = R'(q) = 49 - 2q$$
$$MC = C'(q) = \frac{1}{4}q + 4$$

The profit
$$P(q) = R(q) - C(q)$$
$$= -\frac{9}{8}q^2 + 45q - 200$$
$$P'(q) = -\frac{9}{4}q + 45 = 0 \text{ at } q = 20$$

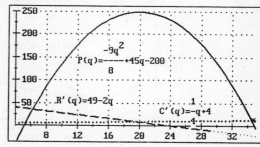

(b)
$$A(q) = \frac{C(q)}{q} = \frac{1}{8}q + 4 + \frac{200}{q}$$
$$A'(q) = \frac{1}{8} - \frac{200}{q^2} = 0 \text{ at } q = 40$$

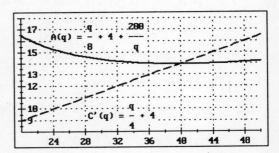

19. $p(q) = 180 - 2q$ and $C(q) = q^3 + 5q + 162$.

(a)
$$R(q) = q\,p(q) = 180q - 2q^2$$
$$MR = R'(q) = 180 - 4q$$
$$MC = C'(q) = 3q^2 + 5$$

The profit $P(q) = R(q) - C(q)$
$$= -q^3 - 2q^2 + 175q - 162$$
$$P'(q) = -3q^2 - 4q + 175 = 0 \text{ at } q = 7 > 0$$

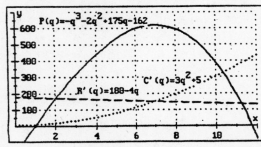

(b)
$$A(q) = \frac{C(q)}{q} = q^2 + 5 + \frac{162}{q}$$
$$A'(q) = 2q - \frac{162}{q^2} = 0 \text{ at } q = 4.327$$

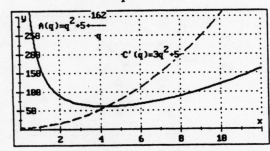

21. $p(q) = 1.0625 - 0.0025q$ and $C(q) = \dfrac{q^2 + 1}{q + 3}$.

(a)
$$R(q) = q\,p(q) = 1.0625q - 0.0025q^2$$
$$MR = R'(q) = 1.0625 - 0.005q$$
$$MC = C'(q) = \frac{q^2 + 6q - 1}{(q+3)^2}$$

The profit $P(q)$
$$= R(q) - C(q)$$
$$= \frac{1}{q+3}[-0.0025q^3 + 0.055q^2 + 3.1875q - 1]$$

$$P'(q) = \frac{1}{(q+3)^2}[(q+3)(-.0075q^2 + .11q + 3.1875) + .0025q^3 - .055q^2 - 3.1875q + 1]$$
$$= \frac{1}{(q+3)^2}[-.005q^3 + .0325q^2 + .33q + 10.5625] = 0$$

at $q = 17.3361$ (calculator root).

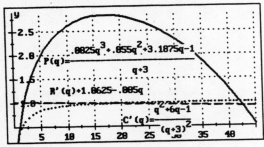

(b)
$$A(q) = \frac{C(q)}{q} = \frac{q^2 + 1}{q(q+3)}$$
$$A'(q) = \frac{1}{(q^2 + 3q)^2}[2q(q^2 + 3q) - (q^2 + 1)(2q + 3)]$$
$$= \frac{3q^2 - 2q - 3}{(q^2 + 3q)^2} = 0$$

at $q = 1.3874 > 0$.

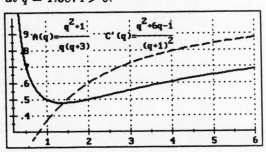

23. (a) If the total revenue is
$$R(q) = -2q^2 + 68q - 128,$$
the average revenue per unit is
$$A(q) = \frac{R(q)}{q} = -2q + 68 - \frac{128}{q}$$

3.4. OPTIMIZATION

The marginal revenue is
$$R'(q) = -4q + 68$$
and is equal to the average revenue when
$$-4q + 68 = -2q + 68 - \frac{128}{q},$$
or $2q = \frac{128}{q}$, $q^2 = 64$, or $q = 8$.

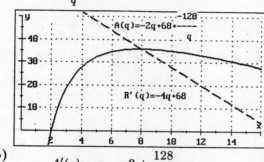

(b)
$$\begin{aligned} A'(q) &= -2 + \frac{128}{q^2} \\ &= \frac{-2(q+8)(q-8)}{q^2}. \end{aligned}$$

If $0 < q < 8$, $A'(q) > 0$, and $A(q)$ is increasing.
If $8 < q$, $A'(q) < 0$, and $A(q)$ is decreasing.

(c) The average and marginal revenue functions are sketched on the same graph. Notice that the average revenue has a maximum at $q = 8$ and q–intercepts are $q = 2$ and $q = 32$.

25.
$$f(x) = 2x^2 - \frac{1}{3}x^3$$

The slope of the tangent is
$$f'(x) = 4x - x^2$$

To maximize $f'(x)$ on $-1 \leq x \leq 4$ we compute
$$f''(x) = 4 - 2x = 0 \text{ at } x = 2$$

Compare $f'(-1) = -5$, $f'(2) = 4$, $f'(4) = 0$.
$f(x)$ is steepest at $x = -1$.
The slope is greatest at $x = 2$, however.

27. (a) The membership of the association x years after 1978 is given by the function
$$f(x) = 100(2x^3 - 45x^2 + 264x)$$

The period of time between 1978 and 1992 corresponds to the interval $0 \leq x \leq 14$.
$$\begin{aligned} f'(x) &= 100(6x^2 - 90x + 264) \\ &= 600(x^2 - 15x + 44) \\ &= 600(x - 4)(x - 11) = 0 \end{aligned}$$

when $x = 4$ and $x = 11$. $f(11) = 12,100$, $f(4) = 46,400$, $f(0) = 0$, and $f(14) = 36,400$. $f(4) = 46,400$ is the absolute maximum.
Hence the membership was greatest in 1982, four years after the founding of the association, when there were 46,400 members.

(b) The period of time between 1978 and 1992 corresponds to the interval $0 \leq x \leq 14$. $f(0) = 0$, $f(11) = 12,100$, $f(4) = 46,400$, $f(14) = 36,400$.
Thus the membership was smallest in 1989, eleven years after the founding of the association, when there were 12,100 members.

29.
$$\begin{aligned} P &= \frac{w^2}{2\rho S v} + \frac{\rho A v^3}{2} \\ P' &= -\frac{w^2}{2\rho S v^2} + \frac{3Av^2\rho}{2} = 0 \\ &\quad \text{when } v = \left(\frac{w^2}{3A\rho^2 S}\right)^{1/4} \\ P'' &= \frac{w^2}{\rho S v^3} + 3Av\rho > 0 \end{aligned}$$

so P is a relative minimum.

31. Let S denote the speed of the blood, R the radius of the artery, and r the distance from the central axis.
Poiseuille's law states that
$$S(r) = c(R^2 - r^2)$$

where c is a positive constant. The relevant interval is $0 \leq r \leq R$.
$$S'(r) = -2cr = 0 \text{ at } r = 0$$
(the left-hand endpoint of the interval). With $S(0) = cR^2$ and $S(r) = 0$, we see that the speed of the blood is greatest when $r = 0$, that is, at the central axis.

33.
$$E(v) = \frac{1}{v}[0.074(v-35)^2 + 32]$$
$$E'(v) = -\frac{1}{v^2}[0.074(v-35)^2 + 32]$$
$$+ \frac{0.148}{v}(v-35) = 0$$
at $v = 40.7$. This reflects a minimum since $E(40.7) > 0$.

35.
$$R(D) = D^2\left(\frac{C}{2} - \frac{D}{3}\right) = \frac{CD^2}{2} - \frac{D^3}{3},$$
$R'(D) = D(C-D) = 0$ when $D = 0$ or $D = C$. $R''(C) < 0$ so $D = C$ represents a maximum for sensitivity.

37.
$$I = \frac{E}{r+R}, \quad P(r) = I^2R = \frac{E^2R}{(r+R)^2}$$
$$P'(R) = \frac{E^2[(r+R)^2 - 2R(r+R)]}{(r+R)^4}$$
$$= \frac{E^2}{(r+R)^3}(r-R) = 0 \text{ if } R_1 = r$$
This is a maximum since $P(0) = 0$, $P'(0) > 0$, and R_1 is the only critical point.

39. $V(t) = -6.8 \times 10^{-8}T^3 + 8.5 \times 10^{-6}T^2$
$\quad\quad\quad -6.4 \times 10^{-5}T + 1$

(a) $V(T)$ is minimized at $T = 3.95$, $V(3.94) = 0.999875$

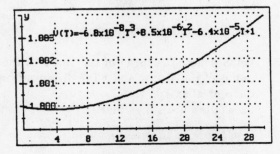

(b) Writing exercise—
Answers will vary.

41. $A(r) = \dfrac{1}{(1-r^2)^2 + kr^2} = [(1-r^2)^2 + kr^2]^{-1}$

$A'(r) = -2r[(1-r^2)^2 + r^2]^{-2}[-2(1-r^2) + k] = 0$

when $2r^2 = 2 - k$ or $2r^2 = 2 - k$,

$r_1 = \sqrt{\dfrac{2-k}{2}}$, $(k \leq 2)$.

$$A''(r_1) = \frac{64(k-2)}{k^2(k-4)^2} < 0,$$

$A(r_1)$ is the absolute maximum.

43. (a) If the rate of cost increases more rapidly as q increases,
$$\frac{\Delta C(q)}{\Delta q}$$
increases as q, which means that the graph of the cost curve is concave upward or $C''(q) > 0$.

(b)
$$A(q) = \frac{C(q)}{q}$$
$$A'(q) = \frac{qC' - C}{q^2}; \quad A'(q_c) = 0$$
$$A''(q) = \frac{q^2(qC' - C)' - 2q(qC' - C)}{q^4}$$
so $A''(q_c) = C''(q_c)$ and $A''(q_c) > 0$ if and only if $C''(q_c) > 0$.
Thus, q_c is a minimum by the second derivative test.

3.5 Practical Optimization

1. Let x denote the number that exceeds its square, x^2, by the largest amount.
$$\text{Then } f(x) = x - x^2$$
is the function to be maximized.

3.5. PRACTICAL OPTIMIZATION

There is no restriction on the domain of $f(x)$.

$$f'(x) = 1 - 2x = 0 \text{ when } x = \frac{1}{2}$$

$f''(x) = -2 < 0$ so that $\left(\frac{1}{2}, f\left(\frac{1}{2}\right)\right) = \left(\frac{1}{2}, \frac{1}{4}\right)$ is a relative maximum.

Since the graph of $f(x)$ is a parabola opening down, the relative maximum is an absolute maximum. Thus, $x = \frac{1}{2}$ is the number that maximizes $f(x) = x - x^2$.

3. Let x be the first number and y the second with $0 \leq x \leq 50$.

Then xy is the product and $y = 50 - x$.

To maximize $P(x) = x(50 - x) = 50x - x^2$ find

$$P'(x) = 50 - 2x = 0 \text{ or } x = 25.$$

$P(25) = 625$, $P(0) = 0$, $P(50) = 0$.

Thus $x = 25$ and $y = 25$ maximize the product.

5. Let x be the $1.00 price increments above $40.00.

Then $40 + x$ will be the price per computer game,

$50 - 3x$ will be the number of units sold per month, and the profit will be

$$P(x) = (50 - 3x)(40 + x - 25) = 750 + 5x - 3x^2$$

$$P'(x) = 5 - 6x = 0 \text{ when } x = \frac{5}{6} \approx 1.$$

The selling price for maximum profit is $41.00

7. Let x be the number of additional trees planted per acre.
The number of oranges per tree will be $400 - 4x$ and the number of trees per acre $60 + x$
The yield per acre is

$$y(x) = \frac{\# \text{ of oranges}}{\text{tree}} \cdot \frac{\# \text{ of trees}}{\text{acre}}$$
$$= 4(100 - x)(60 + x)$$

The number of trees for optimal yield appears to be $60 + 20 = 80$ trees.

9. Let x be the length of the field and y the width. The area is

$$A = xy = 3,600 \text{ square meters}$$

Thus $y = \dfrac{3,600}{x}$ and the perimeter

$$P = 2x + 2y = 2x + \frac{7,200}{x}$$
$$P' = 2\left(1 - \frac{3,600}{x^2}\right) = 0$$

at $x = 60$. $P''(60) > 0$, so the least amount of fencing will enclose a rectangular playground of dimensions 60×60 m^2.

11. Let x be the length of the rectangle and y the width.
The area is $A = xy$ square units.
The perimeter is $2(x + y) = 2p$. Thus $y = p - x$ and the area

$$A = px - x^2$$
$$A' = p - 2x = 0$$

at $x = \dfrac{p}{2}$. $A'' < 0$, so the maximum area corresponds to $x = \dfrac{p}{2}$ and $y = p - x = \dfrac{p}{2}$, namely a square.

13. Let x be the length of the rectangle and $5 - y$ the width. By the similarity of right triangles,

$$\frac{12 - x}{5 - y} = \frac{12}{5}$$

or $5x = 12y$. The area of the rectangle is

$$A = x(5 - y)$$
$$A(x) = 5\left(x - \frac{x^2}{12}\right)$$
$$A'(x) = 5\left(1 - \frac{x}{6}\right) = 0$$

if $x = 6$ and $y = \dfrac{5}{2}$. This is a maximum since $A'' < 0$ and $A(0) = A(12) = 0$.

15. Let the side of the rectangular base be x and the height of the box y.

The cost of the sides is $3 per square meter or $3(4xy)$ dollars.

The cost of the bottom is $4 per square meter, and since the area of the bottom is x^2, the cost of the bottom is $4x^2$ dollars.

The fact that the total cost is to be $48 implies $12xy + 4x^2 = 48$ or $y = \dfrac{4}{x} - \dfrac{x}{3}$. Since

$$V = x^2 y = 4x - \frac{x^3}{3}$$
$$V' = 4 - x^2 = 0$$

at $x = 2$. This is an absolute maximum since $V'' < 0$ for $x > 0$. Now

$$y = \frac{4}{2} - \frac{2}{3} = \frac{4}{3}$$

Hence the box of greatest volume has dimensions 2 by 2 by $\dfrac{4}{3}$ meters.

17. Let x be the number of hours after the reference time, which we take as 0.
The distance and the square of the distance both increase and decrease simultaneously.
The distance traveled by the car going north is $60x$ miles, while the distance of the truck is $300 - 30x = 30(10 - x)$.
These distances correspond to the legs of a right triangle.
The square of the hypotenuse is the square of the distance between the vehicles, namely

$$\begin{aligned} D &= d^2 = (60x)^2 + (30^2)(10-x)^2 \\ &= 900(5x^2 - 20x + 100) \\ D' &= 900(10x - 20) = 0 \end{aligned}$$

when $x = 2$. $D'' > 0$ indicates that the distance between the vehicles will be minimized 2 hours from the reference time.

19. Let x be the distance along the river bank that corresponds to the portion of the cable underwater.
Then $2,000 - x$ meters of cable are above ground.
The distance underwater is

$$D_u = \sqrt{x^2 + 1{,}200^2}$$

and the cost

$$C_u = 25\sqrt{x^2 + 1{,}200^2}$$

The cost above ground is

$$C_o = 20(2{,}000 - x)$$

The total cost

$$C = 25\sqrt{x^2 + 1{,}440{,}000} + 20(2{,}000 - x)$$

has a critical point when

$$C' = \frac{(25)(2x)}{2\sqrt{x^2 + 1{,}440{,}000}} - 20 = 0$$

or $25x = 20\sqrt{x^2 + 1{,}440{,}000}$.
Dividing by 5 and squaring both members of the equation leads to $x = 1{,}600$ meters.
Note that $C' < 0$ if $x < 1{,}600$ and $C' > 0$ when $x > 1{,}600$. With the relevant interval $0 \leq 2{,}000$, it is most economical for the cable to surface at a point 400 meters from the factory.

21. The material is $M = (2)(\pi r^2) + 2\pi r h$ and the volume $V = \pi r^2 h = 6.89\pi$.
Solving for h we get $h = \dfrac{6.89}{r^2}$

$$M = \pi\left(2r^2 + \frac{13.78}{r}\right)$$
$$M' = \pi\left(4r - \frac{13.78}{r^2}\right) = 0 \text{ at } r^3 = 3.445$$

This leads to $r \approx 1.51$ and $h \approx 3.02$ inches. $M'' > 0$ indicates that these can dimensions minimize the material needed.
The actual dimensions of a can of cola differ because of ease of handling and packaging.

23. The amount of material is

$$27\pi = \pi r^2 + 2\pi r h.$$

Dividing by π and solving for h leads to $h = \dfrac{27}{2}r^{-1} - \dfrac{1}{2}r.$

The volume is $V(r) = \pi r^2 h = \dfrac{\pi}{2}(27r - r^3)$

3.5. PRACTICAL OPTIMIZATION

$V' = \dfrac{\pi}{2}(27 - 3r^2) = 0$ when $r = 3$

$V''(3) < 0$ from which we deduce that $r = 3$ indicates a maximum value for the volume. $V(3) = 27\pi$ cubic inches.

25. The material will cost
$$C = (3)(\pi r^2) + (2)(2\pi rh).$$
The volume is $V = \pi r^2 h$ which leads to $h = \dfrac{V}{\pi r^2}$. Thus

$$C = 3\pi r^2 + \frac{4V}{r}$$
$$C' = 6\pi r - \frac{4V}{r^2} = 0 \text{ if } 3\pi r^3 = 2V \text{ or } r = \frac{2h}{3}$$

27. (a) Let x denote the number of machines used and $C(x)$ the corresponding total cost. Then

$$\begin{aligned} C(x) &= \text{set-up cost } + \\ & \quad \text{operating cost} \\ &= 20 \text{ (number of machines)} \\ & \quad + 15(\text{number of hours}). \end{aligned}$$

Since each machine produces 30 kickboards per hour, x machines produce $30x$ kickboards per hour and the number of hours required to produce 8,000 kickboards is $\dfrac{8,000}{30x}$.

Hence $C(x) = 20x + 15\left(\dfrac{8,000}{30x}\right) = 20x + \dfrac{4,000}{x}$

Since the firm owns only 10 machines, the relevant interval is $1 \le x \le 10$.

$$\begin{aligned} C'(x) &= 20 - \frac{4,000}{x^2} = \frac{20(x^2 - 200)}{x^2} \\ &= 0 \text{ or } x \approx 14 \\ C''(x) &= \frac{4,000}{x^3} > 0 \end{aligned}$$

hence the total cost C has an absolute minimum when $x = 14$ (the graph of $C(x)$ is concave upward), that is, when 14 machines are used. Unfortunately the firm owns only 10 machines, so we'll use all of them. Note that $C(1) = 4,020$ and $C(10) = 600$.

(b) When $x = 10$, the supervisor earns $\dfrac{4,000}{10} = \$400$.

(c) The cost of setting up 14 machines is $20(10) = \$200$

29. (a) Let x denote the number of bottles in each shipment. The number of shipments is $\dfrac{800}{x}$ and the ordering cost is
$$10 \times \frac{800}{x} = \frac{8,000}{x}$$

The purchase cost is $800 \times 20 = 16,000$. The storage cost is $\dfrac{x}{2} \times 0.4 = 0.2x$.

As a result the total cost is
$$\begin{aligned} C &= 0.2x + \frac{8,000}{x} + 16,000 \\ C' &= 0.2 - \frac{8,000}{x^2} = 0 \end{aligned}$$

when $x^2 = \dfrac{8,000}{0.2} = 40,000$. Thus $x = 200$ which is a minimum for the cost since $C'' > 0$.

(b) $\dfrac{800}{200} = 4$ orders per year.

31. $\eta = \dfrac{p}{q}\dfrac{dq}{dp}$.

(a) $\eta = \dfrac{p}{200 - 2p^2}(-4p) = -\dfrac{2p^2}{100 - p^2}$.

(b) $\eta|_{p=6} = -\dfrac{2 \times 36}{64} = -1.125$.

A 1 % increase in price will produce a decrease in demand of 1.125 %.

(c) $-1 = \dfrac{-2p^2}{100 - p^2}$ or $p = 5.77$

33. $q = 120 - 0.1p^2$.

(a) $\eta = \dfrac{p}{q}\dfrac{dp}{dq} = \dfrac{p}{120 - 0.1p^2}(-0.2p)$
$= -\dfrac{2p^2}{1,200 - p^2}.$

The demand is of unit elasticity when
$$|\eta| = 1$$

or $2p^2 = 1,200 - p^2$, $p = 20$.
If $0 \leq p < 20$ then

$$|\eta| = \frac{2p^2}{1,200 - p^2}$$
$$< \frac{2 \times 20^2}{1,200 - 400} = 1$$

the demand is inelastic.
If $20 < p \leq \sqrt{1,200}$ then $|\eta| > 1$
hence the demand is elastic.

(b) The total revenue increases when the demand is inelastic, that is when $0 \leq p < 20$.
The total revenue decreases when the demand is elastic, that is when $20 < p \leq \sqrt{1,200}$.
When $p = 20$, the revenue is maximized.

(c) $R(p) = pq = 120p - 0.1p^3$.

$$R'(p) = 120 - 0.3p^2 = 0$$

when $p^2 = 400$, that is when $p = 20$, for which the revenue is maximized, since $R''(p) < 0$.

(d)

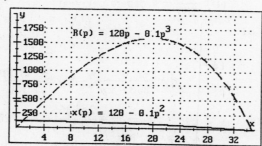

35. Let x be the dimension of one side of the square base and y the length.

The volume is $V = x^2 y$.

The cross sectional perimeter plus the length is

$$4x + y = 108 \text{ (maximum)}.$$

Thus $y = 4(27 - x)$ and

$$V = 4x^2(27 - x) = 4(27x^2 - x^3)$$
$$V' = 4(54x - 3x^2) = 12x(18 - x) = 0$$

if $x = 18$ and $y = 4(27 - 18) = 36$.

$$V''(x) = 4(54 - 6x)$$

$V''(18) = -216$, thus $V(18)$ is a maximum.
The largest volume is $V = 4(18)^2(9) = 11,664$ cubic inches.

37. Let x be the number of units and $C(x)$ the cost. Total cost

$$= \text{wages} + \text{production cost}$$
$$+ \text{maintenance cost}$$
$$= 1,200 + 1.2x + \frac{100}{x^2}$$
$$C'(x) = 1.2 - \frac{200}{x^3} = 0$$

if $x = \sqrt[3]{\frac{200}{1.2}} = 5.503 \approx 6$, so $C(6) = 1209.98$,
$C(0)$ is not defined, and $C(100) = 1,320.01$.
Thus, the minimum cost occurs when 6 units are produced.

39. The rate of change of value is $V' = 53 - 10x$.
The profit P after x years is the present value V less storage less initial outlay I.

$$P = V - 3x - I$$
$$P' = V' - 3 - I' = 53 - 10x - 3 = 0$$

when $x = 5$ years.
(The initial cost is a constant.) $P'' < 0$ indicates that the profit is maximized.

41. $$R(p) = p\, x(p) = 72\sqrt{p}$$

If the price per unit p increases, so will the revenue $R(p)$. although the rate of increases $R'(p) = \dfrac{36}{\sqrt{p}}$ will slow down.

43. Let x be the number of additional days beyond 80 before the club takes all its glass to the recycling center. Let's assume that the same quantity of glass is collected daily, namely $\dfrac{24,000}{80} = 300$ lbs. The daily revenue for the first 80 days would be 300 cents. The reduction in daily revenue for x days beyond 80 is $3x$ or 1 cent per 100 lbs per day.

3.5. PRACTICAL OPTIMIZATION

The club's revenue on day $80 + x$ would be

$$R(x) = (300 - 3x)(80 + x) = 3(80 + x)(100 - x)$$

The key to this problem is understanding that all the glass is taken to the recycling center on day $80 + x$.

$$\begin{aligned} R'(x) &= 3[(80 + x)(-1) + (100 - x)] \\ &= 3[-80 - x + 100 - x] = 0 \end{aligned}$$

when $x = 10$. The number of additional days is 10.

45.
$$\begin{aligned} S(r) &= ar^2(r_0 - r) \\ F(r) &= \pi r^2 S(r) = a\pi(r_0 r^4 - r^5) \quad 0 \le r \le r_0 \\ F'(r) &= a\pi r^3(4r_0 - 5r) = 0 \text{ when } r = \frac{4r_0}{5} \end{aligned}$$

$F(0) = F(r_0) = 0$, $F\left(\dfrac{4r_0}{5}\right) > 0$ and

$F(r)$ is maximized for $r = \dfrac{4r_0}{5}$.

47. Let v be the speed at which the truck is driven and k_1 as well as k_2 positive constants of proportionality.
The driver's wages added to the cost of fuel lead to a cost function

$$\begin{aligned} C &= \frac{k_1}{v} + k_2 v = k_1 v^{-1} + k_2 v. \\ C' &= -k_1 v^{-2} + k_2 = 0 \end{aligned}$$

when $\dfrac{k_1}{v^2} = k_2$ or $\dfrac{k_1}{v} = k_2 v$ (driver's wages equal cost of fuel).
With $C' = -k_1 v^{-2} + k_2$ falling to the left of the critical point and rising to the right, the critical point is a minimum for the cost function.

49. (a) Let x be the number of machines and t the number of hours required to produce q units.
The setup cost is $C_s = xs$ and the operating cost (for all x machines) is $C_o = pt$.

Since n units can be produced per machine per hour, $q = nxt$ or $t = \dfrac{q}{nx}$.
The total cost is

$$\begin{aligned} C &= xs + \frac{pq}{nx} \\ C' &= s - \frac{pq}{nx^2} = 0 \end{aligned}$$

when $x = \sqrt{\dfrac{pq}{ns}}$. $C'' > 0$ which flags a minimum.

(b)
$$\begin{aligned} C_s &= xs = \sqrt{\frac{pqs}{n}} \\ C_o &= \frac{pq}{nx} = \sqrt{\frac{pqs}{n}} \end{aligned}$$

51. We have $w^2 + h^2 = 15^2$ and $S(w) = kwh^3$ or

$$S(w) = kw(225 - w^2)^{3/2}$$

since $h = \sqrt{225 - w^2}$. The stiffness is

$$S'(w) = k(225 - w^2)^{1/2}(225 - 4w^2) = 0$$

at $w = \dfrac{15}{2}$ inches, $h \approx 13$ inches.

53. (a) Let x be the number of units produced, $p(x)$ the price per unit, t the tax per unit, and $C(x)$ the total cost. With

$$\begin{aligned} C(x) &= \frac{7x^2}{8} + 5x + 100 \\ p(x) &= 15 - \frac{3x}{8}, \text{ the revenue is} \\ R(x) &= xp(x) = 15x - \frac{3x^2}{8} \end{aligned}$$

Now profit

$$\begin{aligned} P(x) &= \text{revenue} - \text{taxation} - \text{cost} \\ P(x) &= 15x - \frac{3x^2}{8} - tx \\ &\quad - \frac{7x^2}{8} - 5x - 100 \\ P'(x) &= 15 - \frac{3x}{4} - t - \frac{7x}{4} - 5 = 0 \end{aligned}$$

when $\dfrac{5x}{2} = 10 - t$ or $x = \dfrac{2(10 - t)}{5}$.

(b) The government share is

$$G(x) = tx = (\frac{2}{5})(10t - t^2)$$

$$G'(t) = (\frac{2}{5})(10 - 2t) = 0 \text{ if } t = 5$$

This represents an absolute maximum since $G''(t) < 0$ and there is only one critical point.

(c) From part (a), with $t = 0$,
$x = \dfrac{2(10-0)}{5} = 4$, and with $t = 5$,
$x = \dfrac{2(10-5)}{5} = 2$.
The price per unit for the two quantities produced is, respectively,

$$p(4) = 15 - \frac{3(4)}{8} = \$13.50 \text{ and}$$

$$p(2) = 15 - \frac{3(2)}{8} = \$14.25$$

The difference between the two unit prices is $14.25 - 13.50$ or 75 cents.

(d) Writing Exercise —
Answers will vary.

Review Problems

1.
$$f(x) = -2x^3 + 3x^2 + 12x - 5$$
$$f'(x) = -6x^2 + 6x + 12$$
$$= -6(x+1)(x-2) = 0$$

when $x = -1$ or $x = 2$.
The function rises on $-1 < x < 2$
it falls on $x < -1$ and $2 < x$.
$f(-1) = -12$ and $f(2) = 15$.
$(2, 15)$ is a relative maximum
and $(-1, -12)$ is a relative minimum.

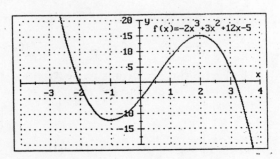

2.
$$f(x) = 3x^3 - 4x^2 - 12x + 17$$
$$f'(x) = 9x^2 - 8x - 12 = 0$$

when $x = 1.6817$ or $x = -0.7928$.
The function rises on $x < -0.7928$ and $1.6817 < x$
it falls on $-0.7928 < x < 1.6817$.
A relative minimum is $(1.6817, -0.2247)$
and a relative maximum $(-0.7928, 22.5046)$.

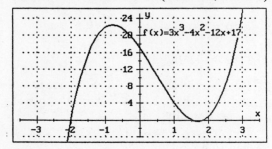

3.
$$f(t) = 3t^5 - 20t^3$$
$$f'(t) = 15t^4 - 60t^2$$
$$= 15t^2(t-2)(t+2) = 0$$

when $t = -2$, $t = 0$, and $t = 2$.
The function rises when $t < -2$ and $2 < t$,
it falls when $-2 < t < 2$.
$f(-2) = 64$, $f(0) = 0$, and $f(2) = -64$.
$(-2, 64)$ is a relative maximum and
$(2, -64)$ is a relative minimum.
$(0,0)$ is a point of inflection.

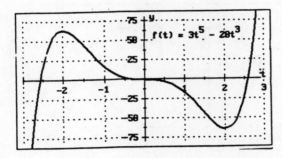

4.
$$g(t) = \frac{t^2}{t+1}$$
$$g'(t) = \frac{(t+1)(2t) - (t^2)(1)}{(t+1)^2}$$
$$= \frac{t(t+2)}{(t+1)^2}$$
$$= 0 \text{ when } t = -2 \text{ or } t = 0$$

The function rises when $x < -2$
or $0 < x$,
it falls when $-2 < t < -1$
or $-1 < t < 0$.
$g(-2) = -4$ and $g(0) = 0$.
$(-2, -4)$ is a relative maximum
and $(0, 0)$ is a relative minimum.

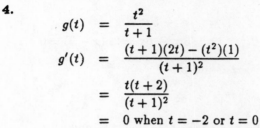

5.
$$F(x) = 2x + \frac{8}{x} + 2$$
$$F'(x) = 2 - \frac{8}{x^2}$$
$$= \frac{2(x-2)(x+2)}{x^2} = 0$$

when $= -2$ and $x = -2$
The function rises when $x < -2$ and $2 < x$,
it falls when $-2 < x < 0$ and $0 < x < 2$.
$F(-2) = -6$ and $F(2) = 10$.
$(-2, -6)$ is a relative maximum

and $(2, 10)$ is a relative minimum.

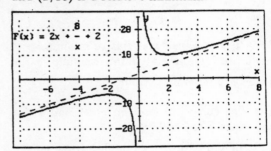

6.
$$G(x) = (2x-1)^2(x-3)^3$$
$$G'(x) = 3(x-3)^2(2x-1)^2$$
$$\quad + 4(2x-1)(x-3)^3$$
$$= 5(2x-1)(x-3)^2(2x-3) = 0$$

at $x = \frac{1}{2}, \frac{3}{2}$, and $x = 3$.
The function increases on $x < \frac{1}{2}$, $\frac{3}{2} < x$
it decreases on $\frac{1}{2} < x < \frac{3}{2}$. $G\left(\frac{1}{2}\right) = 0$,
$G\left(\frac{3}{2}\right) = -13.5$, and $G(3) = 0$. $\left(\frac{1}{2}, 0\right)$ is a
relative maximum,
$(1.5, -13.5)$ is a relative minimum,
$(3, 0)$ is not an extremum.

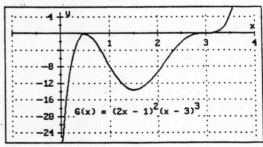

7.
$$f'(x) = x^3(2x-3)^2(x+1)^5(x-7) = 0$$

at $x = -1$, $x = 0$, $x = \frac{3}{2}$, and $x = 7$.
The sign of $f'(x)$ changes at all these critical
numbers excepts $x = \frac{3}{2}$ because of the even
exponent.
If $7 < x$ all the factors are positive.
The pattern of increase/decrease follows:

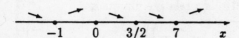

$x = -1$, $x = 7$ produce relative minima,
$x = 0$ produces a relative maximum,
$x = \dfrac{3}{2}$ produces no extremum.

8. (a) The curve increases on $x < 0$ and $5 < x$ since $f'(x) > 0$,

   (b) the curve decreases on $0 < x < 5$ since $f'(x) < 0$,

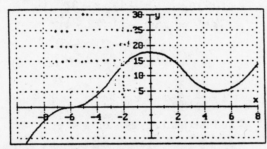

   (c) the curve is concave upward on $-6 < x < -3$ and $2 < x$ since $f''(x) > 0$,

   (d) the curve is concave downward on $x < -6$ and $-3 < x < 2$ since $f''(x) < 0$.

9. $f(x) = x^2 - 6x + 1$
   $f'(x) = 2(x-3) = 0$ when $x = 3$

   $f''(x) = 2 > 0$, $f(3) = -8$.
   Thus $(3, -8)$ is a relative minimum (actually the absolute minimum).

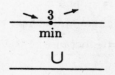

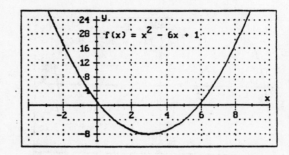

10. $f(x) = x^3 - 3x^2 + 2$
    $f'(x) = 3x^2 - 6x = 3x(x-2) = 0$

    when $x = 0$ and $x = 2$.

    $f''(x) = 6x - 6 = 6(x-1) = 0$

    when $x = 1$. $f(0) = 2$, $f(1) = 0$, and $f(2) = -2$.
    Thus $(0, 2)$ is a relative maximum,
    $(2, -2)$ is a relative minimum,
    and $(1, 0)$ is a point of inflection.

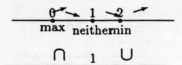

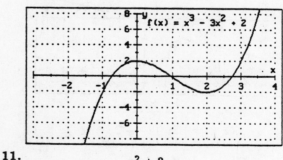

11. $f(x) = \dfrac{x^2 + 3}{x - 1}$

    $f'(x) = \dfrac{(x-1)(2x) - (x^2+3)(1)}{(x-1)^2}$

    $= \dfrac{(x-3)(x+1)}{(x-1)^2}$

    $= 0$ when $x = -1$, $x = 3$

REVIEW PROBLEMS

$$f''(x) = \frac{1}{(x-1)^4}[(x-1)^2(2x-2)$$
$$-(x^2 - 2x - 3)(2)(x-1)]$$
$$= \frac{8}{(x-1)^3} \neq 0.$$

$f(-1) = -2$ and $f(3) = 6$.
Thus $(-1, -2)$ is a relative maximum,
$(3, 6)$ is a relative minimum,
and the graph changes concavity at $x = 1$ even though this is not in the domain of the function.

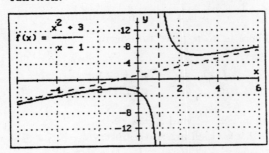

12.
$$f(x) = \frac{1}{x^3} + \frac{2}{x^2} + \frac{1}{x}$$
$$= x^{-3} + 2x^{-2} + x^{-1}.$$
$$f'(x) = -3x^{-4} - 4x^{-3} - x^{-2}$$
$$= -\frac{3 + 4x + x^2}{x^4} = 0$$

when $x = -3$ or $x = -1$.

$$f''(x) \frac{2}{x^5}(6 + 6x + x^2) = 0$$

when $x = -4.7321$ or $x = -1.2679$
$f(-3) = -0.148$ and $f(-1) = 0$.
$f(x)$ increases on $-3 < x < -1$
decreases on $x < -3$ and $-1 < x$.
It is concave up on $-4.7321 < x < -1.268$ and $0 < x$,

concave down on $x < -4.7321$ and
$-1.268 < x < 0$.
Thus $(-3, -0.148)$ is a relative minimum,
and $(-1, 0)$ is a relative maximum.
$(-4.7321, -0.1436)$ and $(-1.268, -0.0353)$ are points of inflection.

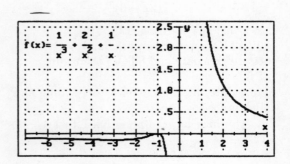

13.
$$f(x) = -2x^3 + 3x^2 + 12x - 5$$
$$f'(x) = -6(x+1)(x-2) = 0, \ x = -1, \ x = 2$$
$$f''(x) = -12x + 6 = -6(2x - 1)$$

$f(-1) = -12$, $f(2) = 15$. $f''(-1) = 18 > 0$,
so f has a relative minimum at $(-1, -12)$.
Similarly $f''(2) = -18 < 0$,
so f has a relative maximum at $(2, 15)$.

14.
$$f(x) = \frac{x^2}{x+1}$$
$$f'(x) = \frac{(x+1)(2x) - (x^2)(1)}{(x+1)^2}$$
$$= \frac{x(x+2)}{(x+1)^2} = 0$$

when $x = -2$ and $x = 0$.
$$f''(x) = \frac{1}{(x+1)^4}[(x+1)^2(2x+2) - (x^2+2x)(2)(x+1)]$$
$$= \frac{2}{(x+1)^3}$$

$f(0) = 0$, $f(-2) = -4$.
$f''(0) = 2 > 0$, so f has a relative minimum at $(0,0)$.
Similarly $f''(-2) = -2 < 0$, so f has a relative maximum at $(-2, -4)$.

15.
$$f(x) = x(2x-3)^2$$
$$f'(x) = 2x(2x-3)(2) + (2x-3)^2 = 12x^2 - 24x + 9$$

Get this result from your graphing utility by STOring
$$x(2x-3)^2$$
into a memory location 'EQ', (the EQUATION writer is quite handy) then differentiate and STOre the result in 'FPR'
Manipulate the expression by expanding and collecting a few times
OK (you may have to press NXT first)
STOre
$$12x^2 - 24x + 9$$
in 'FPR'
differentiate again and store in 'FPP'
Manipulate by collecting to get the form
$$-24 + 24x$$
STOre it in 'FPP'
Plot $f'(x)$ and find $x = 0.5$ and $x = 1.5$ as roots.
$f''(0.5) = -12$ (you may have to edit 'FPP' to store 'FPP(X)=-24+24X')
Similarly $f''(1.5) = 12$
Thus $(0.5, 2)$ is a relative maximum
$(1.5, 0)$ is a relative minimum.

16.
$$f(x) = \frac{1}{x} - \frac{1}{x+3}$$
$$f'(x) = -\frac{1}{x^2} + \frac{1}{(x+3)^2}$$
$$= -\frac{3(2x+3)}{x^2(x+3)^2}$$
$$f''(x) = \frac{2}{x^3} - \frac{2}{(x+3)^3}$$

$f'(x) = 0$ at $x = -\frac{3}{2}$.
$f''(-1.5) = -1.1852 < 0$ so $\left(-\frac{3}{2}, -\frac{4}{3}\right)$ is a relative maximum.
There is no relative minimum.

17.
$$f(x) = -2x^3 + 3x^2 + 12x - 5$$
$$f'(x) = -6(x+1)(x-2) = 0$$

when $x = -1$ and $x = 2$,
both of which are in the interval $-3 \le x \le 3$.
$f(-1) = -12$, $f(2) = 15$, $f(-3) = 40$, $f(3) = 4$.
Thus $f(-3) = 40$ is the absolute maximum and $f(-1) = -12$ the absolute minimum.

18.
$$f(t) = -3t^4 + 8t^3 - 10$$
$$f'(t) = -12t^2(t-2) = 0$$

when $t = 0$ and $t = 2$,
both of which are in the interval $0 \le t \le 3$.
$f(0) = -10$, $f(2) = 6$, and $f(3) = -37$.
Thus $f(2) = 6$ is the absolute maximum and $f(3) = -37$ the absolute minimum.

19.
$$g(s) = \frac{s^2}{s+1}$$
$$g'(s) = \frac{s(s+2)}{(s+1)^2} = 0$$

when $s = 0$ and $s = -2$, of which only $s = 0$ is in the interval $-\frac{1}{2} \le s \le 1$.
Note that the derivative is undefined at $s = -1$, which is not in the domain of g.
$g\left(-\frac{1}{2}\right) = \frac{1}{2}$, $g(0) = 0$, and $g(1) = \frac{1}{2}$.

REVIEW PROBLEMS   105

Thus $g\left(-\frac{1}{2}\right) = g(1) = \frac{1}{2}$ is the absolute maximum and
$g(0) = 0$ the absolute minimum.

20.
$$f(x) = 2x + \frac{8}{x} + 2$$
$$f'(x) = 2 - \frac{8}{x^2}$$
$$= \frac{2(x+2)(x-2)}{x^2} = 0$$

when $x = -2$ and $x = 2$, of which only $x = 2$ is in the interval $0 < x$. There are no endpoints, so the only possible absolute extremum is $f(2) = 10$.
$f'(x) < 0$ (f is decreasing) when $0 < x < 2$ and $f'(x) > 0$ (f is increasing) when $2 < x$.
Thus $f(2) = 10$ is the absolute minimum and there is no absolute maximum.

21. $\lim\limits_{x \to +\infty} \left(2 + \frac{1}{x^2}\right) = 2 + 0 = 2$

22. $\lim\limits_{x \to 0} \left(2 - \frac{1}{x^3}\right) = \pm\infty$

does not exist because of 0 in the denominator.

23. $\lim\limits_{x \to 0^+} \left(x^3 - \frac{1}{x^2}\right) = 0 - \infty = -\infty$

24. $\lim\limits_{x \to +\infty} \frac{x}{x^2 + 5} = \lim\limits_{x \to +\infty} \frac{1}{x + \frac{5}{x}} = 0$

25. $\lim\limits_{x \to -\infty} \frac{x^3 - 3x + 5}{2x + 3}$
$$= \lim\limits_{x \to -\infty} \frac{1 - \frac{3}{x^2} + \frac{5}{x^3}}{\frac{2}{x^2} + \frac{3}{x^3}} = +\infty$$

26. $\lim\limits_{x \to -\infty} \frac{x^4 + 3x^2 - 2x + 7}{x^3 + x + 1}$
$$= \lim\limits_{x \to -\infty} \frac{x - \frac{3}{x} - \frac{2}{x^2} + \frac{7}{x^3}}{1 + \frac{1}{x^2} + \frac{1}{x^3}} = -\infty$$

27. $\lim\limits_{x \to +\infty} \frac{x(x-3)}{7 - x^2} = \lim\limits_{x \to +\infty} \frac{1 - \frac{3}{x}}{\frac{7}{x^2} - 1} = -1$

28. $\lim\limits_{x \to +\infty} \frac{1 + \frac{1}{x} + \frac{1}{x^2}}{x^2 + 3x - 1} = 0$

29. $\lim\limits_{x \to 0} \sqrt{x\left(1 + \frac{1}{x^2}\right)} = \lim\limits_{x \to 0} \sqrt{x + \frac{1}{x}} = +\infty$

30. $\lim\limits_{x \to 0^-} x\sqrt{1 - \frac{1}{x}}$
$$= \lim\limits_{x \to 0^-} \sqrt{x^2\left(1 - \frac{1}{x}\right)} = \lim\limits_{x \to 0^-} \sqrt{x^2 - x} = 0$$

31. $S(t) = t^3 - 9t^2 + 15t + 45$

on the interval $0 \le t \le 7$.
$$S'(t) = 3t^2 - 18t + 15$$
$$= 3(t-1)(t-5) = 0$$

when $t = 1$ or $t = 5$.
$S(0) = 45$, $S(1) = 52$, $S(5) = 20$, and $S(7) = 52$.
Traffic is moving fastest at 1:00 p.m. and 7:00 p.m. when its speed is 52 mph and slowest at 5:00 p.m. when its speed is 20 mph.

32. Let R denote the rate at which the rumor is spreading,
Q the number of people who have heard the rumor,
and P the total population of the community.
Then $R(Q) = kQ(P - Q)$ where k is a positive constant of proportionality.
$$R'(Q) = kQ(-1) + (P - Q)(k)$$
$$= Pk - 2kQ = 0$$

when $2kQ = Pk$ or $Q = \frac{P}{2}$,
which is in the interval $0 \le Q \le P$.
$$R''(Q) = -2k < 0$$

which means that the absolute maximum is reached when half the people have heard the rumor.

33. Let x be the number of $1.00 increases.
The sales price per lamp is $6 + x$
and the corresponding number of lamps is

$$3,000 - 1,000x = 1,000(3 - x).$$

The total cost

$$C = (3,000 - 1,000x) \times 4 = 4,000(3 - x)$$

and the total revenue

$$R = 1,000(3 - x)(6 + x).$$

The profit is

$$P = R - C = 1,000(6 + x - x^2)$$
$$P' = 1,000(1 - 2x) = 0 \text{ when } x = \frac{1}{2}$$
$$P''(0.50) < 0$$

so $x = 0.50$ maximizes the profit.
The corresponding sales price is $6.50.
(Note that the graph of p is a parabola pointing downward.)

34. Let x be the number of 25 cent reductions in price.
The profit per card will be

$$10 - 0.25x - 5 = 5(1 - 0.05x)$$

while the number of cards sold will be

$$25 + 5x = 5(5 + x).$$

The total profit will be

$$P(x) = 25(5 + x)(1 - 0.05x)$$
$$= 25(5 + 0.75x - 0.05x^2)$$
$$P'(x) = 18.75 - 2.5x = 0 \text{ when } x = 7.5$$

Since $P(7) = 195 = P(8)$, the best price is $10 - 0.25(8) = \$8$ per card.

35. Label the sides of the rectangular plot as indicated in the figure and let A denote the area.
Then $A = xy + xy = 2xy$,

a function of two variables. The fact that 300 meters of fencing are to be used implies

$$4x + 3y = 300 \text{ or } y = \frac{300 - 4x}{3} = 100 - \frac{4x}{3}.$$

Thus $A(x) = 2x\left(100 - \frac{4x}{3}\right)$
$$= 200x - \frac{8x^2}{3}.$$
$$A'(x) = 200 - \frac{16x}{3} = 0 \text{ or } x = 37.5$$

$$A''(x) = -\frac{16}{3} < 0,$$

$x = 37.5$ indicates the absolute maximum.
The corresponding value of y is
$$y = 100 - \frac{4(37.5)}{3} = 50.$$
Each plot of land should be 37.5 by 50 meters.

36. (a) Let x be the length and y the width of rectangular pasture, and let A denote the area. Then

$$A = xy,$$

a function of two variables.
The fact that 320 ft of fencing are to be used implies

$$2x + 2y = 320 \text{ or } y = 160 - x$$

Thus $A(x) = x(160 - x)$
$$= 160x - x^2.$$
$$A'(x) = 160 - 2x = 0$$

or $x = 80$.

$$A''(x) = -2 < 0,$$

$x = 80$ indicates the absolute maximum.
The corresponding value of y is
$y = 160 - 80 = 80$.
The pasture should be a square with 80 ft per side.

(b) The fact that 320 ft of fencing are to be used implies

$$x + 2y = 320 \text{ or } x = 2(160 - y)$$

REVIEW PROBLEMS

Thus
$$A(y) = 2y(160 - y) = 2(160y - y^2).$$
$$A'(y) = 2(160 - 2y) = 0$$
or $y = 80$.
$$A''(y) = -4 < 0,$$
$y = 80$ indicates the absolute maximum. The corresponding value of x is
$x = 2(160 - 80) = 160$.
The pasture measures 160 ft (on the unfenced side) by 80 ft.

37. Let r denote the radius,
h the height,
C the (fixed) cost (in cents),
and V the volume.
$$V = \pi r^2 h,$$
a function of two variables.
To write h in terms of r, use the fact that the cost is to be C cents. That is

$$\begin{aligned} C &= \text{cost of bottom } + \text{ cost of side} \\ &= 3(\text{area of bottom}) \\ &\quad + 2(\text{area of side}) \end{aligned}$$

or $C = 3\pi r^2 + 4\pi rh$.

To solve for h,
$$h = \frac{C - 3\pi r^2}{4\pi r}$$
where C is a constant.

Thus
$$\begin{aligned} V(r) &= \pi r^2 \left(\frac{C - 3\pi r^2}{4\pi r} \right) \\ &= \frac{rC}{4} - \frac{3\pi r^3}{4}. \\ V'(r) &= \frac{C}{4} - \frac{9\pi r^2}{4} = 0 \end{aligned}$$

when $\frac{C}{4} = \frac{9\pi r^2}{4}$ or $C = 9\pi r^2$.
Substituting in the formula for cost leads to
$$3\pi r^2 + 4\pi rh = 9\pi r^2,$$

$$h = \frac{3r}{2}.$$
$$V''(r) = -\frac{9\pi r}{2} < 0,$$
so the volume is maximized when the height is 1.5 times the radius of the cylindrical container.

38.
$$f(t) = -t^3 + 7t^2 + 200t$$
is the number of letters the clerk can sort in t hours.
The clerk's rate of output is
$$R(t) = f'(t) = -3t^2 + 14t + 200$$
letters per hour.
The relevant interval is $0 \leq t \leq 4$.
$$R'(t) = f''(t) = -6t + 14 = 0$$
when $t = \frac{7}{3}$. $R\left(\frac{7}{3}\right) = 216.33$,
$R(0) = 200$, and $R(4) = 208$.
Thus the rate of output is greatest when $t = \frac{7}{3}$ hours, that is, after 2 hours and 20 minutes.

39. Let x be the number of houses built beyond the planned 60. The profit for per house will be $47,500 - 500x$. The total profit will be
$$\begin{aligned} P(x) &= (47,500 - 500x)(60 + x) \\ &= 2,850,000 + 17,500x - 500x^2 \\ P'(x) &= 17,500 - 1000x = 0 \end{aligned}$$
at $x = 17.5$, that is when 77 or 78 houses are built. The developer will probably choose 77, since it costs more to build one additional house and the profit is greater then.

40. Let x denote the number of machines used and $C(x)$ the corresponding cost of producing the 400,000 medals. Then
$$\begin{aligned} C(x) &= \text{set-up cost } + \text{ operating cost} \\ &= 80 \text{ (number of machines)} \\ &\quad + 5.76 \text{ (number of hours)}. \end{aligned}$$

Each machine can produce 200 medals per hour, so x machines can produce $200x$ medals

107

per hour, and it will take $\dfrac{400,000}{200x}$ hours to produce the 400,000 medals. Hence

$$\begin{aligned} C(x) &= 80x + 5.76\,\dfrac{400,000}{200x} \\ &= 80x + \dfrac{11,520}{x}. \\ C'(x) &= 80 - \dfrac{11,520}{x^2} \\ &= \dfrac{80(x-12)(x+12)}{x^2} = 0 \end{aligned}$$

when x=12
For $0 < x < 12$, $C'(x) < 0$ and $C(x)$ decreases.
For $12 < x$, $C'(x) > 0$ and $C(x)$ increases.
Thus $C(x)$ has its absolute minimum when $x = 12$ machines are used.

41. Let x denote the number of hours after 8:00 a.m. at which the 15 minute coffee break begins.
Then $f(x)$ is the number of radios assembled before the break.
After the break, $4 - x$ hours remain until lunch time at 12:15 (4 hours 15 minutes minus x hours before the break minus 15 minutes for the break).
Thus the number of radios assembled during the $4 - x$ hours after the break is $g(4 - x)$, and the total number of radios assembled between 8:00 a.m. and 12:15 p.m. is

$$\begin{aligned} N(x) &= f(x) + g(x - 4) \\ &= -x^3 + 6x^2 + 15x - \dfrac{1}{3}(4-x)^3 \\ &\quad + (4-x)^2 + 23(4-x). \end{aligned}$$

The relevant interval is $0 \le x \le 4$.

$$\begin{aligned} N'(x) &= -3x^2 + 12x + 15 - (4-x)^2(-1) \\ &\quad + 2(4-x)(-1) - 23 \\ &= -2x(x-3) = 0 \text{ at } x = 0, \ x = 3 \end{aligned}$$

$N(0) = 86.67$, $N(3) = 95.67$, and $N(4) = 92$.
Thus the worker will assemble the maximum number of radios by lunch time if the coffee break begins when $x = 3$, that is, at 11:00 a.m.

42. Let n denote the number of floors and $A(n)$ the corresponding average cost.
Since the total cost is

$$C(n) = 2n^2 + 500n + 600 \text{ (thousand dollars)}$$

$$A(n) = \dfrac{C(n)}{n} = 2n + 500 + \dfrac{600}{n}.$$

The relevant interval is $n > 0$.

$$A'(n) = 2 - \dfrac{600}{n^2} = \dfrac{2(n^2 - 300)}{n^2} = 0$$

when $n = \sqrt{300} \approx 17.32$.

$$A''(n) = \dfrac{1,200}{n^3} > 0 \text{ when } n > 0$$

$A(n)$ has its absolute minimum when $n = \sqrt{300}$.
In the context of this practical problem, the optimal value of n (which denotes the number of floors) must be an integer.
$A(17) = 569,294$ and $A(18) = 569,333$.
Thus 17 floors should be built to minimize the average cost per floor.

43. Let Q be the point on the opposite bank straight across from the starting point.
With $PQ = x$, the distance along the bank is $1 - x$.
The distance across the water is given by the pythagorean theorem to be $\sqrt{1 + x^2}$.
The time t is

$$\begin{aligned} t &= \text{time in the water} \\ &\quad + \text{time on the land} \\ &= \dfrac{\text{distance in the water}}{\text{speed in the water}} \\ &\quad + \dfrac{\text{distance on the land}}{\text{speed on the land}} \\ &= \dfrac{1}{4}(1 + x^2)^{1/2} + \dfrac{1}{5}(1 - x). \end{aligned}$$

The relevant interval is $0 \le x \le 1$ and

$$t'(x) = \dfrac{x}{4\sqrt{1+x^2}} - \dfrac{1}{5} = 0$$

if $\dfrac{x}{4\sqrt{1+x^2}} = \dfrac{1}{5}.$

REVIEW PROBLEMS

$$25x^2 = 16 + 16x^2 \text{ or } x = \pm\frac{4}{3}$$

Neither of these critical points lies in the interval $0 \leq x \leq 1$.

Hence, the absolute minimum must occur at an endpoint. $t(0) = \frac{1}{4} + \frac{1}{5} = 0.45$ hour

and $t(1) = \frac{\sqrt{2}}{4} = 0.354$ hour.

The minimum time thus occurs if $x = 1$, that is, if you row all the way to the town.

44.
$$f(x) = Ax^3 + Bx^2 + C$$
$$f'(x) = 3Ax^2 + 2Bx$$
$$\text{and } f''(x) = 6Ax + 2B$$

$E(2, 11)$ is an extremum, so

$$f'(2) = 12A + 4B = 0$$

or $B = -3A$.

$I(1, 5)$ is a point of inflection, so
$f''(1) = 6A + 2B = 0$ or $B = -3A$.
Since the points E and I are on the curve, their coordinates must satisfy its equation. Thus

$$f(2) = 8A + 4B + C = 8A - 12A + C = 11$$

and $f(1) = A + B + C = A - 3A + C = 5$.
Solving simultaneously leads to $A = -3$, $B = 9$, and $C = -1$ and

$$f(x) = -3x^3 + 9x^2 - 1.$$

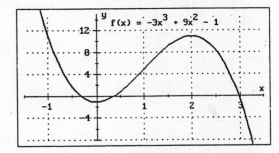

45. Let x be the dimension of one side of the equilateral triangle, which is also the dimension of the base of the rectangle.

y is the height of the rectangle.
The perimeter of the window is $3x + 2y = 20$ giving $y = \frac{20 - 3x}{2}$.

We need to find the height h of the equilateral triangle.
With x as the hypotenuse of one-half of the equilateral triangle,

$$h = \frac{\sqrt{3}x}{2}$$

The base of the right triangle is $\frac{x}{2}$.
Thus the area of the (whole) equilateral triangle is

$$A_t = \frac{\sqrt{3}x^2}{4}.$$

The area of the rectangle is

$$A_r = xy = \frac{x(20 - 3x)}{2}.$$

Since twice as much light passes through the rectangle as through the stained glass isosceles triangle, the light is

$$L = k\left[\frac{\sqrt{3}x^2}{4} + (2)\frac{x(20 - 3x)}{2}\right].$$

k is a proportionality constant which does not affect the outcome of the problem. With

$$L = k\left[\left(\frac{\sqrt{3}}{4} - 3\right)x^2 + 20x\right],$$

$$L' = k\left[\left(\frac{\sqrt{3}}{2} - 6\right)x + 20\right] = 0$$

when $x = \frac{20}{6 - \sqrt{3}/2} = 3.8956$ feet.

Then $y = \left(\frac{1}{2}\right)(20 - 11.6869) = 4.1566$ feet.

$L'' < 0$ which signifies that $x = 3.8956$ is the absolute maximum of the curve, whose graph is a parabola pointing downward.

46. (a)
$$C(x) = x^3 - 24x^2 + 350x + 400.$$
$$C'(x) = 3x^2 - 48x + 350 = 0$$

and $[C'(x)]' = 6x - 48 = 0$ if $x = 8$

This leads to a relative minimum as $[C'(x)]'' = 6 > 0$

(b)
$$A(x) = x^2 - 24x + 350 + \frac{400}{x}$$
$$A'(x) = \frac{1}{x^2}(2x^3 - 24x^2 - 400) = 0$$

when $f(x) = 2x^3 - 24x^2 - 400 = 0$.
$x = 13.156$ is the only solution (calculator solution).

47. Let $A(x, y)$ and $B(-x, y)$, $(x > 0)$, be points on the parabola

$$y = \frac{1}{14}(48 - x^2)$$

to form the isosceles triangle OAB. Since AB is horizontal, the area is

$$A = \frac{1}{2}(2xy)$$
$$A'(x) = \frac{48 - 3x^2}{14} = 0 \text{ when } x = 4 > 0$$

$A''(4) < 0$ reveals a relative maximum for $A(x)$.

Thus $y = \frac{16}{7}$ and the largest area is $A = \frac{64}{7}$ square units.

48. The relationship between the number of Moppsy dolls and Floppsy dolls is given by

$$y = \frac{82 - 10x}{10 - x}$$

with the relevant interval $0 \le x \le 8$.
Suppose Moppsy sells for $2 and Floppsy sells for $1 each. The revenue is then

$$R(x) = x + \frac{2(82 - 10x)}{10 - x}$$
$$= \frac{x^2 + 10x - 164}{x - 10}.$$
$$R'(x) = \frac{1}{(x-10)^2}[(x-10)(2x+10)$$
$$- (x^2 + 10x - 164)(1)]$$
$$= \frac{x^2 - 20x + 64}{(x-10)^2} = 0 \text{ when } x = 4$$

($x = 16$ is outside the interval $0 \le x \le 8$.)
Since $R(0) = 16.4$, $R(4) = 18$, and $R(8) = 10$, the maximum occurs at $x = 4$ and we find $y(4) = 7$. That is, revenue is maximized when 4 hundred Floppsies and 7 hundred Moppsies are sold.

49. Let Q be the point on the shore straight across from the oil rig and P the point on the shore where the pipe starts on land.
With $PQ = x$, the distance along the bank is $8 - x$.
The distance across the water is given by the pythagorean theorem to be

$$\sqrt{9 + x^2}.$$

The cost C is

$$C = \text{cost in the water} + \text{cost on the shore}$$
$$= (1.5)\sqrt{9 + x^2} + (1)(8 - x),$$

using 1 unit of price.

$$C'(x) = \frac{(1.5)(2x)}{2\sqrt{9 + x^2}} - 1 = 0$$

when $2.25x^2 = 9 + x^2$, $1.25x^2 = 9$, or $x = 2.6833 > 0$, which is in the relevant interval $0 \le x \le 8$. $C(0) = 12.5$, $C(2.6833) = 11.35$, and $C(8) = 12.81$.
The minimum thus occurs if $x = 2.6833$, that is, if $8 - x = 5.3167$ miles or 28,072 ft of pipe is laid on the shore.

50. Let x be the width and y the depth. Then the strength is

$$s = kxy^2$$

where k is the proportionality constant.
Since the beam is 15 in. in diameter $x^2 + y^2 = 225$

$$s = k(225x - x^3)$$
$$s' = k(225 - 3x^2) = 0 \text{ at } x = 5\sqrt{3}$$

and $y = \sqrt{225 - 3(25)} = 5\sqrt{6}$.
Thus the stiffest beam has dimensions $5\sqrt{3}$ by $5\sqrt{6}$.

REVIEW PROBLEMS

51. $f'(x) = x(x-1)^2 = 0$ when $x = 0$ and $x = 1$

$f''(x) = (x-1)(3x-1)$ $f''\left(\frac{1}{3}\right) = f''(1) = 0.$

```
 ↘  0  ↗  1/3  ↗  1  ↗
   min  neither neither
    ∪   1/3  ∩  1   ∪
```

(a) $f(x)$ is increasing for $0 < x$. It is decreasing for $x < 0$.

(b) $f(x)$ is concave down on $\frac{1}{3} < x < 1$ and concave up elsewhere.

(c) A minimum occurs when $x = 0$. Inflection points correspond to $x = \frac{1}{3}$ and $x = 1$.

(d)
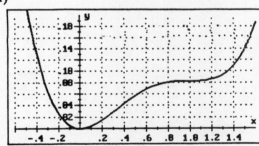

52. Let x denote the number of maps per batch and $C(x)$ the corresponding cost. Then,

$$C(x) = \text{(storage cost)} + \text{(production cost)} + \text{(set-up cost)}.$$

The storage cost is $\left(\frac{1}{2}x\right)(0.20) = 0.1x.$
(See example 5.1 for an explanation of the storage cost.)
The production cost is $0.06(16,000) = 960.$
Since x maps are produced per batch and 16,000 maps are needed, the number of batches is $\frac{16,000}{x}$.

The set-up cost is
$100\left(\frac{16,000}{x}\right) = \frac{1,600,000}{x}$. Thus

$$C(x) = 0.1x + 960 + \frac{1,600,000}{x}.$$

The relevant interval is $0 < x \leq 16,000$.

$$\begin{aligned} C'(x) &= 0.1 - \frac{1,600,000}{x^2} \\ &= 0.1\left(\frac{(x-4,000)(x+4,000)}{x^2}\right) \\ &= 0 \text{ when } x = 4,000. \end{aligned}$$

Since this is the only critical point in the relevant interval, and

$$C''(x) = \frac{3,200,000}{x^3},$$

which indicates that C has its absolute minimum when $x = 4,000$ maps per batch.

53. Let x be the number of units ordered and k_1, k_2 constants of proportionality.
Since the storage cost is $C_s = k_1 x$ and the ordering cost $C_o = \frac{k_2}{x}$, the total cost is

$$\begin{aligned} C &= k_1 x + \frac{k_2}{x} \\ C' &= k_1 - \frac{k_2}{x^2} = 0 \end{aligned}$$

when the critical point is $x_c = \sqrt{\frac{k_2}{k_1}}$.
$C'' > 0$ signifies a minimum ordering cost at the critical point, so

$$\begin{aligned} C_o &= k_2 \frac{\sqrt{k_1}}{\sqrt{k_2}} = \sqrt{k_1 k_2} \\ C_s &= k_1 \frac{\sqrt{k_2}}{\sqrt{k_1}} = \sqrt{k_1 k_2} \end{aligned}$$

This shows that $C_s = C_o$.

54. Let v be the speed at which the truck is driven and k_1 as well as k_2 positive constants of proportionality.

The driver's wages added to the cost of fuel lead to a cost function

$$C = \frac{k_1}{v} + k_2 v = k_1 v^{-1} + k_2 v$$
$$C' = -k_1 v^{-2} + k_2 = 0$$

when $\frac{k_1}{v^2} = k_2$ or $\frac{k_1}{v} = k_2 v$ (driver's wages equal cost of fuel).

With $C' = -k_1 v^{-2} + k_2$ falling to the left of the critical point and rising to the right, the critical point is a minimum for the cost function.

55. $$f(x) = \frac{x^2 + 2x - 5}{x^2 - 1}$$

According to the graphing utility,

$f'(x)$ is $\frac{(\text{num})'}{\text{den}} - \frac{(\text{num})(\text{den})'}{(\text{den})^2}$

If your viewing rectangle is long enough watch the graph's linearity when x gets large.
Note the vertical asymptotes at $x = -1$ and $x = 1$, the horizontal asymptote at $y = 1$.

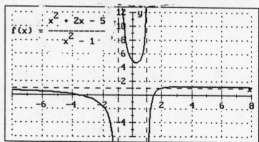

56. $$f(x) = \frac{x^2 + 3x - 5}{x^2 - 5x + 3}$$
$$f'(x) = \frac{1}{(x^2 - 5x + 3)^2}[(x^2 - 5x + 3)(2x + 3)$$
$$-(x^2 + 3x - 5)(2x - 5)]$$
$$f'(x) \neq 0$$

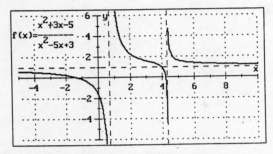

57. Note: The domain is $-1 < x < 1$.

$$f(x) = (1 - x^{2/3})^{3/2}$$
$$f'(x) = \frac{3}{2}(1 - x^{2/3})^{1/2}\left(-\frac{2}{3}x^{-1/3}\right)$$
$$= -x^{-1/3}(1 - x^{2/3})^{1/2}$$

$f'(x) = 0$ at $x = 1$.

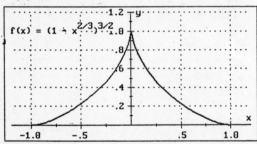

58.
$$\lim_{x \to +\infty} (\sqrt{x^2 - x} - x)$$
$$= \lim_{x \to +\infty} \frac{(\sqrt{x^2 - x} - x)(\sqrt{x^2 - x} + x)}{(\sqrt{x^2 - x} + x)}$$
$$= \lim_{x \to +\infty} \frac{x^2 - x - x^2}{\sqrt{x^2 - x} + x}$$
$$= \lim_{x \to +\infty} \frac{-x}{(\sqrt{x^2 - x} + x)}$$
$$= \lim_{x \to +\infty} \frac{-1}{(\sqrt{1 - \frac{1}{x}} + 1)}$$
$$= -\frac{1}{2}$$

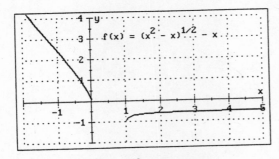

59. $R(x) = -2x^2 + 68x - 10$

(a) $\dfrac{R(x)}{x} = -2x + 68 - \dfrac{10}{x}$

(b) $R'(x) = -4x + 68$

(c)

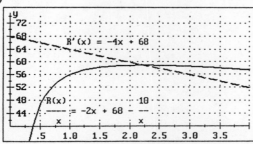

60. (a) $R(x) = -2x^2 + 68x - 10$
$C(x) = 110 + 47x - 9x^2 + 0.9x^3$
$A(x) = \dfrac{C(x)}{x} = \dfrac{110}{x} + 47 - 9x + 0.9x^2$

A graphing utility places the minimum at $(6.4630, 43.4463)$.

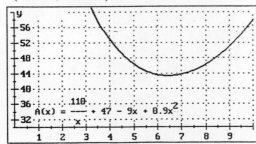

(b) $P(x) = R(x) - C(x)$
$= -2x^2 + 68x - 10 - 110$
$ -47x + 9x^2 - 0.9x^3$
$= -0.9x^3 + 7x^2 + 21x - 120$

A graphing utility places the maximum at $(6.400, 65.1904)$.

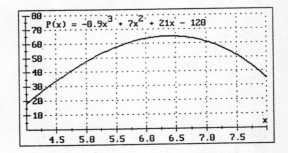

61. $f'(x) > 0$ when $x < 2$ and when $2 < x < 5$, $f'(x) < 0$ when $x > 5$, $f'(2) = 0$,
$f''(x) < 0$ when $x < 2$ and when $4 < x < 7$,
$f''(x) > 0$ when $2 < x < 4$ and when $x > 7$. A possible graph is shown below

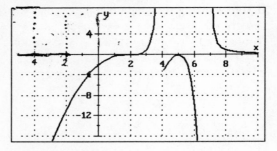

62. $f'(x) > 0$ when $x < 1$,
$f'(x) < 0$ when $x > 1$,
$f''(x) > 0$ when $x < 1$,
$f''(x) > 0$ when $1 < x$.
$f'(1)$ must be undefined. A possible graph is shown below

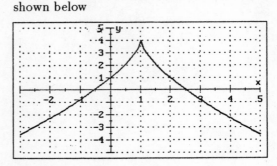

63. (a) $R(t) = \dfrac{dP}{dt} = A\left(1 - \dfrac{P}{B}\right)P - H$

$$R'(t) = \frac{d^2P}{dt^2} = A\left(1 - \frac{P}{B}\right) - \frac{AP}{B}$$

$$= A - \frac{2AP}{B} = 0 \text{ when } P = 0.5B$$

(b) If $H > \dfrac{AB}{4}$,

$$\frac{dP}{dt} < A\left(1 - \frac{P}{B}\right)P - \frac{AB}{4}$$

$$= -\frac{A}{4B}(2P - B)^2 < 0$$

Since the change in population is negative, the population will decrease continuously, until it dies out.

(c) Writing Exercise — Answers will vary.

64. (a)
$$E = \frac{\dfrac{dQ}{Q}100}{\dfrac{dI}{I}100} = \frac{I}{Q}\frac{dQ}{dI}$$

(b) Writing Exercise— Answers will vary.

Chapter 4

Exponential and Logarithmic Functions

4.1 Exponential Functions

1. Use your HP48G.
   For e^2, punch 2, then e^x, to get
   $$e^2 = 7.389.$$
   For the TI-85, press 2nd e^x then 2 ENTER. Similarly $e^{-2} = 0.135$, $e^{0.05} = 1.051$, $e^{-0.05} = 0.951$, $e^0 = 1$, $e = 2.718$, $\sqrt{e} = 1.649$, $\dfrac{1}{\sqrt{e}} = 0.607$.

3.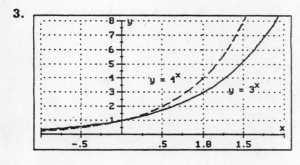

   $y = 3^x$ and $y = 4^x$.
   The x-axis is a horizontal asymptote.

5. (a) $\quad 27^{2/3} = (27^{1/3})^2 = 3^2 = 9$

   (b) $\quad (-128)^{3/5} = -18.379$

   (c) $8^{2/3} + 16^{3/4} = (8^{1/3})^2 + (16^{1/4})^3 = 4 + 8 = 12$

   (d) $\quad (2^3 - 3^2)^{11/7} = (8-9)^{11/7} = -1$

7. $\quad (3^3)(3^{-2}) = 3^{3-2} = 3$

9. $\quad \dfrac{5^2}{5^3} = 5^{2-3} = \dfrac{1}{5}$

11. $\quad (3^2)^{5/2} = [(3^2)^{1/2}]^5 = 3^5 = 243$

13. If P dollars is invested at an annual interest rate r and interest is compounded k times per year, the balance after t years will be
    $$B(t) = P\left(1 + \frac{r}{k}\right)^{kt}$$
    dollars and if interest is compounded continuously, the balance will be
    $$B(t) = Pe^{rt} \text{ dollars.}$$

    (a) If $P = 1,000$, $r = 0.07$, $t = 10$, and $k = 1$, then
    $$\begin{aligned} B(10) &= 1,000\left(1 + \frac{0.07}{1}\right)^{10} \\ &= 1,000(1.07)^{10} = \$1,967.15. \end{aligned}$$

    (b) If $P = 1,000$, $r = 0.07$, $t = 10$, and $k = 4$, then
    $$\begin{aligned} B(10) &= 1,000\left(1 + \frac{0.07}{4}\right)^{40} \\ &= \$2,001.60. \end{aligned}$$

115

(c) If $P = 1,000$, $r = 0.07$, $t = 10$, and $k = 12$, then

$$B(10) = 1,000\left(1 + \frac{0.07}{12}\right)^{120}$$
$$= \$2,009.66.$$

(d) If $P = 1,000$, $r = 0.07$, $t = 10$, and interest is compounded continuously, then

$$B(10) = 1,000e^{0.7} = \$2,013.75.$$

15. $$B(5) = P\left(1 + \frac{0.07}{4}\right)^{20} = 5,000.$$

Thus

$$P = \frac{5,000}{(1.0175)^{20}} = \frac{5,000}{1.4178} = \$3,534.12.$$

17. (a) $$B(5) = P\left(1 + \frac{0.07}{4}\right)^{20} = 5,000.$$

Thus

$$P = \frac{9,000}{(1.0175)^{20}} = \frac{9,000}{1.4148} = \$6,361.42$$

(b) $$B(5) = 9,000\, e^{-(0.07)(5)} = \$6,342.19$$

19. (a) The population in t years will be

$$P(t) = 50e^{0.02t}$$

million. Thus $P(0) = 50$ million.

(b) The population 30 years from now will be

$$P(30) = 50e^{0.02(30)} = 50e^{0.6} = 91.11$$

million.

21. $$\begin{aligned} f(x) &= e^{kx} \\ f(1) &= e^k = 20 \\ f(2) &= e^{2k} = (e^k)^2 = (20)^2 = 400 \end{aligned}$$

23. $$f(x) = A2^{kx},$$
$f(0) = A = 20$, $f(2) = 20\,(2^{2k}) = 40$, so $2^{2k} = 2$.
$$\begin{aligned} f(8) &= 20\,(2^{8k}) = 20(2^{2k})^4 \\ &= 20(2^4) = 320. \end{aligned}$$

25. $$B(15) = 2P = (P)\left(1 + \frac{r}{4}\right)^{60}$$

$$\begin{aligned} B(30) &= (P)\left(1 + \frac{r}{4}\right)^{120} \\ &= (P)\left[\left(1 + \frac{r}{4}\right)^{60}\right]^2 = (P)(2)^2 = 4P \end{aligned}$$

The money has quadrupled, which is not surprising since $30 = (2)(15)$.

27. $$P(t) = P_0 e^{kt}$$

When $t = 0$ $P_0 = 5,000$, and when $t = 10$ $P = 8,000$. $8,000 = 5,000e^{10k}$, so $e^{10k} = \frac{8}{5}$ and

$$\begin{aligned} P(30) &= 5,000(e^{10k})^3 \\ &= 5,000 \times \left(\frac{8}{5}\right)^3 = 20,480 \text{ bacteria} \end{aligned}$$

29. $$P(t) = P_0 e^{-kt}$$

When $t = 0$, $P_0 = 25,000$ copies and when $t = 1$ $P = 10,000$ copies.
$25,000e^{-k} = 10,000$, so $e^{-k} = \frac{2}{5}$ and

$$P(2) = 25,000e^{-2k} = 25,000\left(\frac{2}{5}\right)^2 = 4,000$$

copies.

31. The population density x miles from the center of the city is $D(x) = 12e^{-0.07x}$ thousand people per square mile.

(a) At the center of the city, the density is $D(0) = 12$ thousand people per square mile.

(b) Ten miles from the center, the density is

$$D(10) = 12e^{-0.07(10)} = 12e^{-0.7} = 5.959$$

that is 5,959 people per square mile.

33. Let $Q(t)$ denote the amount of the radioactive substance present after t years.
Since the decay is exponential and 500 grams were present initially,

$$Q(t) = 500e^{-kt}$$

4.1. EXPONENTIAL FUNCTIONS

Moreover, since 400 grams are present 50 years later, $400 = Q(50) = 500e^{-50k}$ or $e^{-50k} = \dfrac{4}{5}$. The amount present after 200 years will be

$$Q(200) = 500e^{-200k} = 500(e^{-50k})^4$$
$$= 500\left(\dfrac{4}{5}\right)^4 = 204.8 \text{ grams}$$

35. $f(t) = e^{-0.2t}$ toasters will still be in use after t years.

(a) After $t = 3$ years, $f(3) = e^{-0.6} = 0.5488$.

(b) The fraction of toasters still working after 2 years is
$f(2) = e^{-0.4} = 0.6703$, so the fraction of toasters failing is $1.0 - 0.6703$.
Similarly the fraction failing after 3 years will be $1 - f(3) = 1.0 - 0.5488$.
The fraction of toasters failing during the third year is
$1.0 - 0.5488 - (1.0 - 0.6703) = 0.1215$.

(c) The fraction of toasters in use before the first year is $e^{-0.1}$, so the fraction of toasters failing during the first year is $1.0 - e^{-0.1} = 0.0952$.

37. Let k be the number of compounding periods per year.

(a) $(P)(1+i)^k = (P)(1+r_e)$ from which
$$r_e = (1+i)^k - 1$$

(b) $(P)e^r = (P)(1+r_e)$ from which
$$r_e = e^r - 1$$

39. Let the period of investment be 1 year and the principal $100.
$$B(t) = 100\left(1 + \dfrac{0.082}{4}\right)^4 = 108.46$$
$$B(t) = 100e^{0.081} = 108.44$$
Thus 8.2% is the winner.

41. (a) $$N(t) = N_0 e^{-0.217t}$$
Let $t = 0$ be 200 B.C. In 2000 A.D. $t = 2.2$.
$$N(2200) = 500e^{-0.217(2.2)} = 310.2$$

(b) In 950 $t = 0$, so $t = 1$ in 1950.
$$N(1) = 210e^{-0.217(1)} = 169$$
This prediction is very close to the actual count of 167.

(c) Writing Exercise — Answers will vary.

43. $$M = \dfrac{Pi}{1 - (1+i)^{-n}}$$
$P = 150,000$, $i = \dfrac{0.09}{12} = 0.0075$, $n = 360$.
$$M = \dfrac{150,000(0.0075)}{1 - (1.0075)^{-360}} = \$1,206.93$$

45. (a) If the loan of $5,000 is amortized over 3 years, the monthly payment would be
$$5000(1 + 0.12)^{-36} = 161.75$$
In this case you would receive
$$\$1,000 + (161.75)(36) = \$6,823$$
which is
$$6,823 - 1,000 - (36)(160) = 6,823 - 6,760 = \$63$$
less than the buyer's way.

(b) Writing Exercise — Answers will vary

47. $$A(n) = \left(1 + \dfrac{1}{n}\right)^n.$$
$A(-1,000) = (1 - 0.001)^{-1,000} = 2.71964,$
$A(-2,000) = (1 - 0.0005)^{-2,000} = 2.71896,$
$A(-50,000) = \left(1 - \dfrac{1}{50,000}\right)^{-50,000} = 2.71831.$
The values approach $e \approx 2.71828$.

49. $$A(n) = \left(2 - \dfrac{5}{2n}\right)^{n/3}$$
$$A(10) = \left(2 - \dfrac{5}{20}\right)^{10/3} = 6.4584$$
$$A(100) = \left(2 - \dfrac{5}{200}\right)^{100/3} = 7.12 \times 10^9$$
which suggests $\lim\limits_{n \to +\infty}\left(2 - \dfrac{5}{2n}\right) = +\infty$

4.2 Logarithmic Functions

1. Use your calculator. For the HP48G, first enter the number a, then press the ln key. For the TI-85, press LN and the number. This gives

$$\ln 1 = 0$$
$$\ln 2 = 0.6931472$$
$$\ln 5 = 1.6094379$$
$$\ln \frac{1}{5} = \ln 0.2 = -1.6094379$$

Now, for $\ln e^n$, enter n, then press the e^x key, followed by the ln key (HP48G) or enter LN followed by e^n (TI-85). Thus

$$\ln e^1 = 1 \text{ and } \ln e^2 = 2.$$

These results should not be surprising since the natural logarithm and the exponential functions are inverses. $\ln e^n = n$.

For $\ln 0$ and $\ln(-2)$

the calculator displays an error signal (such as the letter E or a flashing light on the HP48G or a complex number – a pair of numbers in parentheses – on the TI-85.) This is due to the domain of the logarithmic function which is $x > 0$.
(Ask yourself "what exponent of e will give -2 ?")
Answer: $e^x = -2$ is impossible since $e^x > 0$ for all real x.
Remember that "log" means "exponent".)

3. $$\ln e^3 = 3\ln e = 3 \times 1 = 3$$
since $\ln u^v = v \ln u$ and $\ln e = 1$.

5. Let's give the expression a name, say A, so that we can handle it.
$$A = e^{\ln 5}$$
which can be written as
$$\ln A = \ln 5.$$
A solution (the only solution) is A=5. Thus
$$e^{\ln 5} = 5.$$
Actually this result is immediate from the inverse relationship between e^x and $\ln x$.

7. Let's call the given expression A.
$$\begin{aligned} A &= e^{3\ln 2 - 2\ln 5} \\ &= e^{\ln 2^3 - \ln 5^2} \end{aligned}$$
(since $v \ln u = \ln u^v$).
$$\begin{aligned} A &= e^{\ln 8 - \ln 25} \\ &= e^{\ln(8/25)} \end{aligned}$$
(since $\ln \frac{u}{v} = \ln u - \ln v$).
$A = \frac{8}{25}$ because of the inverse relationship between e^x and $\ln x$.

9. $$2 = e^{0.06x}$$
is equivalent to
$$\ln 2 = 0.06x,$$
from which $x = \frac{\ln 2}{0.06} = 11.55$.

11. $$3 = 2 + 5e^{-4x}$$
$$\frac{1}{5} = e^{-4x}$$
(using arithmetic),
$$-4x = \ln \frac{1}{5} = -\ln 5$$
(since $\ln \frac{u}{v} = \ln u - \ln v$ and $\ln 1 = 0$), from which $x = \frac{\ln 5}{4} = 0.402$.

13. $$-\ln x = \frac{t}{50} + C$$
or
$$\ln x = -\frac{t}{50} - C.$$
Thus
$$x = e^{-C - t/50} = (e^{-C})(e^{-t/50})$$
because $a^{r+s} = a^r a^s$ (which) certainly applies when $a = e$.

4.2. LOGARITHMIC FUNCTIONS

15.
$$\begin{aligned}\ln x &= \frac{1}{3}(\ln 16 + 2\ln 2)\\ &= \frac{1}{3}(\ln 16 + \ln 2^2)\\ &= \frac{1}{3}\ln[(16)(2^2)]\\ &= \frac{1}{3}\ln 2^6 = \ln(2^6)^{1/3}\\ &= \ln 2^2.\end{aligned}$$

Thus $\ln x = \ln 4$ which is valid when $x = 4$.

17.
$$3^x = e^2$$

is valid if the logarithms of both members are taken.

$$\begin{aligned}\ln 3^x &= \ln e^2 = 2\ln e = 2\\ x\ln 3 &= 2, \text{ or}\\ x &= \frac{2}{\ln 3} = 1.82\end{aligned}$$

19.
$$\begin{aligned}a^{x+1} &= b\\ \text{if } \ln a^{x+1} &= \ln b\\ (x+1)\ln a &= \ln b\\ x &= \frac{\ln b}{\ln a} - 1\end{aligned}$$

21.
$$\begin{aligned}\log_2 x &= 5\\ 2^5 &= x\\ \ln 2^5 &= 5\ln 2 = \ln x\\ \text{or } \ln x &= 3.4657\end{aligned}$$

23.
$$\begin{aligned}\log_5(2x) &= 7\\ 5^7 &= 2x\\ \ln 5^7 &= \ln(2x) = \ln 2 + \ln x\\ \ln x &= 7\ln 5 - \ln 2 = 10.5729\end{aligned}$$

25.
$$\begin{aligned}\ln\frac{1}{\sqrt{ab^3}} &= 0 - \ln(ab^3)^{1/2}\\ &= -\frac{1}{2}\ln(ab^3)\\ &= -\frac{1}{2}(\ln a + \ln b^3)\\ &= -\frac{1}{2}(\ln a + 3\ln b)\\ &= -\frac{1}{2}(2+9) = -5.5.\end{aligned}$$

27.
$$B(t) = Pe^{rt}.$$

After a certain time the investment will have grown to $B(t) = 2P$ at the interest rate of 0.06. Thus

$$\begin{aligned}2P &= Pe^{0.06t}\\ 2 &= e^{0.06t}\\ \ln 2 &= 0.06t\end{aligned}$$

and $t = \dfrac{\ln 2}{0.06} = 11.55$ years.

29. The balance after t years is

$$B(t) = Pe^{rt},$$

where P is the initial investment and r is the interest rate compounded continuously. Since money doubles in 13 years,

$$\begin{aligned}2P &= B(13) = Pe^{13r}\\ 2 &= e^{13r}\\ \ln 2 &= 13r\end{aligned}$$

or $r = \dfrac{\ln 2}{13} = 0.0533$. Thus the annual interest rate is 5.33%.

31.
$$\begin{aligned}Q(t) &= Q_0 e^{-kt}\\ \frac{Q_0}{2} &= Q(1{,}690) = Q_0 e^{-1{,}690k}\\ e^{-1{,}690k} &= \frac{1}{2}\\ -1{,}690k &= \ln\frac{1}{2}\\ k &= \frac{\ln 2}{1{,}690}\end{aligned}$$

Since $Q_0 = 50$, $5 = 50e^{-kt}$,
$e^{-kt} = \dfrac{1}{10}$, $-kt = \ln\dfrac{1}{10} = -\ln 10$,
$t = \dfrac{\ln 10}{k} = \dfrac{1{,}690 \ln 10}{\ln 2} \approx 5{,}614$ years

33. The number of bacteria is

$$Q(t) = Q_0 e^{kt}$$

Since 6,000 bacteria were present initially, $Q_0 = 6{,}000$ so that

$$\begin{aligned}Q(t) &= 6{,}000 e^{kt}\\ Q(20) &= 9{,}000 = 6{,}000 e^{20k}\\ 20k &= \ln\frac{3}{2}\end{aligned}$$

$k \approx 0.0203$ and $Q(t) = 6,000e^{0.0203t}$.

35.
$$Q(t) = 500 - Ae^{-kt}$$
and
$$Q(0) = 300 = 500 - A.$$
Thus $A = 200$ and
$$Q(t) = 500 - 200e^{-kt},$$
$$Q(6) = 410 = 500 - 200e^{-6k}.$$
$$e^{-6k} = \frac{9}{20} \text{ or } k = \approx 0.1331$$
It follows that
$$Q(t) = 500 - 200e^{-0.1331t}.$$

37. From the half-life of ^{14}C
$$\frac{1}{2}R_0 = R_0 e^{-5,730k}$$
$$\ln\frac{1}{2} = -5,730k \text{ or } k = \frac{\ln 2}{5,730}$$

Now $R(t) = R_0 e^{-kt} = 0.28 R_0$
$$\ln 0.28 = -\frac{\ln 2}{5,730} t$$
$$t = -\frac{5,730 \ln 0.28}{\ln 2} = 10,523 \text{ years}.$$

The artifacts at the Debert site in Nova Scotia are about 10,500 years old.

39. From the half-life of ^{14}C
$$\frac{1}{2}R_0 = R_0 e^{-5,730k}$$
$$\ln\frac{1}{2} = -5,730k \text{ or } k = \frac{\ln 2}{5,730}$$

Now $R(t) = R_0 e^{-kt} = 0.997 R_0$

$\ln 0.997 = -\dfrac{\ln 2}{5,730} t$ or $t = 24.8372$ years.

The forged Rembrandt painting is only approximately 25 years old.
In the year 2000 $t = 2,000 - 1,640 = 360$.
Let p be the percentage of ^{14}C left.
$$R(t) = pR_0$$
$$\ln p = -\frac{360 \ln 2}{5,730} \text{ or } p = 0.9574$$

The original Rembrandt painting will contain approximately 95.74 % of ^{14}C.

41. (a)
$$e^{-(\ln 2/20.9)(24)} = 0.45$$
or 45% of the original ^{133}I should be detected.

(b)
$$e^{-(\ln 2/20.9)(25)} = 0.4364$$
or 43.64% of ^{133}I remains in the thyroid. $43.64 - 41.3 = 2.34\%$ of ^{133}I still remains in the rest of the patient's body.

43.
$$T = T_a + (T_d - T_a)(0.97)^t$$
$$\text{so } 40 = 10 + (98.6 - 10)(0.97)^t,$$
$$30 = 88.6(0.97)^t,$$
$$t \ln 0.97 = \ln\left(\frac{30}{88.6}\right) \text{ or } t = 35.55$$

The murder occurred around 1:30 a.m. on Wednesday morning. Blohardt was in the slammer, so Scélerat must have done it.

45. (a)
$$R = \frac{\ln I}{\ln 10} = 8.3$$
$$\ln I = 8.3 \ln 10 = 19.1115$$
$$I_a = e^{19.1115} = 1.995 \times 10^8$$

(b)
$$\ln I_b = 7.1 \ln 10 = 16.3484$$
$$I_b = e^{16.3484} = 1.2589 \times 10^7$$
$$\frac{I_a}{I_b} = 15.8472$$

times more intense.

47. Let t denote the number of years after 1960. If the population P (measured in billions) is growing exponentially and was 3 billion in 1960 (when $t = 0$), then
$$P(t) = 3e^{kt}.$$

Since the population in 1975 (when $t = 15$) was 4 billion,
$$4 = P(15) = 3e^{15k} \text{ or } k = \frac{1}{15} \ln \frac{4}{3}$$
$$\text{For } P(t) = 40 \text{ so } 3e^{kt} = 40$$
$$e^{kt} = \frac{40}{3} \text{ or } t \approx 135$$

The population will reach 40 billion in the year $1960 + 135 = 2095$.

49. (a) $0.05 = e^{-3k}$, $\ln(0.05) = -3k$ or $k = 0.999$

$$0.01 = e^{-0.999x}$$

or $x = -\dfrac{\ln(0.01)}{0.999} = 4.61$ m.

(b) Writing exercise — Answers will vary.

51. (a) $\lambda = \dfrac{\ln 2}{k}$. Therefore

$$Q(t) = Q_0 e^{-(\ln 2/\lambda)t}$$

(b) $Q_0 e^{-(\ln 2/\lambda)t} = Q_0 (0.5)^{kt}$

$$k = \dfrac{-\ln 2}{(\ln 0.5)\lambda} = \dfrac{-\ln 2}{=\ln 2\lambda} = \dfrac{1}{\lambda}$$

53. $\log_a b = \dfrac{\log_b b}{\log_b a} = \dfrac{1}{\log_b a}$

Therefore $\log_a b \log_b a = 1$.

55. 10^x and $\log_{10} x$ are reflections about $y = x$.

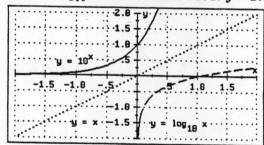

57. $3{,}500 e^{0.31x} = \dfrac{e^{-3.5x}}{1 + 257 e^{-1.1x}}$

Using a graphing utility and tracing, we find $x = -5.06$.

59. $\ln(x+3) - \ln x = 5\ln(x^2 - 4)$

$$\ln \dfrac{x+3}{x(x^2-4)^5} = 0$$

or $\dfrac{x+3}{x(x^2-4)^5} = 1$. Using a graphing utility we find $x = 2.28$.

61. $k = \dfrac{\ln 2}{46.5} = 0.015$. After 24 hours

$$100 e^{-0.015(24)} = 69.77 \text{ mg}$$

will be left. The time required for the isotope to decline to 25 mg is
$t = -\dfrac{\ln 0.25}{0.015} = 92.4$ hours.

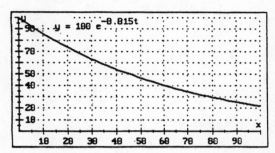

4.3 Differentiation of Logarithmic and Exponential Functions

1.
$$f(x) = e^{5x}$$
$$f'(x) = e^{5x} \dfrac{d}{dx}(5x) = 5e^{5x}$$

3.
$$f(x) = e^{x^2 + 2x - 1}$$
$$f'(x) = e^{x^2 + 2x - 1} \dfrac{d}{dx}(x^2 + 2x - 1)$$
$$= (2x + 2) e^{x^2 + 2x - 1}.$$

5.
$$f(x) = 30 + 10 e^{-0.05x}$$
$$f'(x) = 0 + 10 e^{-0.05x} \dfrac{d}{dx}(-0.05x)$$
$$= -0.5 e^{-0.05x}.$$

7.
$$f(x) = (x^2 + 3x + 5) e^{6x}$$
$$f'(x) = (x^2 + 3x + 5) \dfrac{d}{dx} e^{6x}$$
$$+ e^{6x} \dfrac{d}{dx}(x^2 + 3x + 5)$$
$$= (6x^2 + 20x + 33) e^{6x}.$$

9.
$$f(x) = (1 - 3e^x)^2$$
$$f'(x) = 2(1 - 3e^x) \dfrac{d}{dx}(1 - 3e^x)$$
$$= -6 e^x (1 - 3e^x).$$

11. $f(x) = e^{\sqrt{3x}} = e^{(3x)^{1/2}}$

$f'(x) = e^{\sqrt{3x}} \dfrac{d}{dx}(3^{1/2}x^{1/2})$

$= e^{\sqrt{3x}}\left[3^{1/2}\left(\dfrac{1}{2}\right)x^{-1/2}\right] = \dfrac{3}{2\sqrt{3x}}e^{\sqrt{3x}}.$

13. $f(x) = \ln x^3 = 3\ln x$

$f'(x) = 3\left(\dfrac{1}{x}\right)\dfrac{d}{dx}x = \dfrac{3}{x}.$

15. $f(x) = x^2 \ln x$

$f'(x) = x^2 \dfrac{d}{dx}(\ln x) + \ln x \dfrac{d}{dx}(x^2)$

$= x(1 + 2\ln x).$

17. $f(x) = \sqrt[3]{e^{2x}} = e^{2x/3}$

$f'(x) = e^{2x/3}\left(\dfrac{2}{3}\right) = \dfrac{2}{3}\sqrt[3]{e^{2x}}$

19. $f(x) = \ln\left(\dfrac{x+1}{x-1}\right)$

$f'(x) = \dfrac{x-1}{x+1}\dfrac{d}{dx}\left(\dfrac{x+1}{x-1}\right)$

$= \dfrac{x-1}{x+1}\dfrac{-2}{(x-1)^2}$

$= \dfrac{-2}{x^2-1}.$

21. $f(x) = xe^{-x}$

$f'(x) = -xe^{-x} + e^{-x}$

$= e^{-x}(1-x)$

$f'(0) = 1$ and $f(0) = 0$. An equation of the tangent line is

$y - 0 = (1)(x - 0)$ or $y = x$

23. $f(x) = \dfrac{e^{2x}}{x^2}$

$f'(x) = \dfrac{x^2(2e^{2x}) - e^{2x}(2x)}{x^4} = \dfrac{2(x-1)e^{2x}}{x^3}$

$f'(1) = 0$ and $f(1) = e^2$. An equation of the tangent line is

$y = e^2$

25. $f(x) = x^2 \ln\sqrt{x} = \dfrac{x^2 \ln x}{2}$

$f'(x) = \dfrac{x + 2x\ln x}{2}$

$f'(1) = \dfrac{1}{2}$ and $f(1) = 0$. An equation of the tangent line is

$y - 0 = \left(\left(\dfrac{1}{2}\right)(x-1)\right.$ or $x - 2y - 1 = 0$

27. $f(x) = \dfrac{(x+2)^5}{\sqrt[6]{3x-5}}.$

$\ln f(x) = 5\ln(x+2) - \dfrac{1}{6}\ln(3x-5).$

$\dfrac{f'(x)}{f(x)} = \dfrac{5}{x+2} - \dfrac{3}{6(3x-5)},$

$f'(x) = \dfrac{(x+2)^5}{\sqrt[6]{3x-5}}\left[\dfrac{5}{x+2} - \dfrac{1}{2(3x-5)}\right]$

29. $f(x) = (x+1)^3(6-x)^2(2x+1)^{1/3}.$

$\ln f(x) = \ln[(x+1)^3(6-x)^2(2x+1)^{1/3}]$

$= \ln(x+1)^3 + \ln(6-x)^2$

$\quad + \ln(2x+1)^{1/3}$

$= 3\ln(x+1) + 2\ln(6-x)$

$\quad + \dfrac{1}{3}\ln(2x+1).$

Differentiating leads to

$\dfrac{f'(x)}{f(x)} = \dfrac{3}{x+1}$

$\quad + \dfrac{2(-1)}{6-x} + \dfrac{2}{3(2x+1)},$

$f'(x) = (x+1)^3(6-x)^2\sqrt[3]{2x+1}$

$\left[\dfrac{3}{x+1} - \dfrac{2}{6-x} + \dfrac{2}{3(2x+1)}\right].$

31. $f(x) = 2^{x^2}$

$\ln f(x) = x^2 \ln 2$

$\dfrac{f'(x)}{f(x)} = 2x\ln 2$

$f'(x) = 2^{x^2+1}x\ln 2$

33. $C(x) = e^{0.2x}$

(a) $C'(x) = 0.2e^{0.2x}$

4.3. DIFFERENTIATION OF LOGARITHMIC AND EXPONENTIAL FUNCTIONS

(b)
$$A(x) = \frac{e^{0.2x}}{x}$$
$$A'(x) = \frac{e^{0.2x}(0.2x-1)}{x^2}$$

(c)
$$R(x) = xe^{-3x}$$
$$R'(x) = e^{-3x}(1-3x)$$

(d) $R'(x) = C'(x)$ when
$$e^{-3x}(1-3x) = 0.2e^{0.2x}$$
or $x = 0.2049$

(e)
$$0.2e^{0.2x} = \frac{e^{0.2x}}{x}$$
or $x = 5$.

35.
$$C(x) = x^2 + 2$$

(a) $\qquad C'(x) = 2x$

(b)
$$A(x) = x + 2x^{-1}$$
$$A'(x) = 1 - 2x^{-2}$$

(c)
$$R(x) = \frac{x\ln(x+3)}{x+3}$$
$$R'(x) = \frac{1}{(x+3)^2}\left[(x+3)\left(\frac{x}{x+3} + \ln(x+3)\right) - x\ln(x+3)\right]$$
$$= \frac{x + 3\ln(x+3)}{(x+3)^2}$$

(d) $R'(x) = C'(x)$ when
$$\frac{x + 3\ln(x+3)}{(x+3)^2} = 2x$$
or $x = 0.1805$ (obtained with a graphing utility.)

(e)
$$2x = x + \frac{2}{x}$$
or $x = \sqrt{2}$.

37. (a) The population t years from now will be
$$P(t) = 50e^{0.02t}$$
million. Hence the rate of change of the population t years from now will be
$$P'(t) = 50e^{0.02t}(0.02) = e^{0.02t}$$
and the rate of change 10 years from now will be
$$P'(t) = e^{0.2} = 1.22$$
million per year.

(b) The percentage rate of change t years from now will be
$$100\left[\frac{P'(t)}{P(t)}\right] = 100\left(\frac{e^{0.02t}}{50e^{0.02t}}\right)$$
$$= \frac{100}{50} = 2$$
% per year, which is a constant, independent of time.

39. (a) The value of the machine after t years is
$$Q(t) = 20,000e^{-0.4t}$$
dollars. Hence the rate of depreciation after t years is
$$Q'(t) = 20,000e^{-0.4t}(-0.4) = -8,000e^{-0.4t}$$
and the rate after 5 years is
$$Q'(5) = -8,000e^{-2} = -\$1,082.68 \text{ per year.}$$

(b) The percentage rate of change t years from now will be
$$100\left[\frac{Q'(t)}{Q(t)}\right] = 100\left(\frac{-8,000e^{-0.4t}}{20,000e^{-0.4t}}\right) = -40$$
% per year, which is a constant, independent of time.

41. (a) The first year sales of the text will be
$$f(x) = 20 - 15e^{-0.2x}$$
thousand copies when x thousand complementary copies are distributed. If the number of complimentary copies distributed is increased from 10,000, that is when $x = 10$, by 1,000, that is, $\Delta x = 1$, the approximate change in sales is
$$\Delta f = f'(10)\Delta x.$$
Since $f'(x) = 3e^{-0.2x}$ and $\Delta x = 1$,
$\Delta f = f'(10) = 3e^{-2} = 0.406$ thousand or 406 copies.

(b) The actual change in sales is
$$\Delta f = f(11) - f(10)$$
$$= (20 - 15e^{-2.2}) - (20 - 15e^{-2})$$
$$= 0.368 \text{ or } 368 \text{ copies.}$$

43.
$$Q(t) = Q_0 e^{-0.0015t}$$

(a) $$\frac{1}{2} = e^{-0.0015t} \text{ or } t = 462$$

Half of the ozone will be depleted in 462 years.

(b) $$0.2 = e^{-0.0015t}, \; t = -\frac{\ln 0.22}{0.0015} = 1,073$$

80% of the ozone will be depleted in 1,073 years.

45.
$$C(t) = 0.2te^{-t/2}$$

(a)
$$\begin{aligned} C'(t) &= 0.2e^{-t/2} + 0.2t\left(-\frac{1}{2}e^{-t/2}\right) \\ &= 0.2e^{-t/2}\left(1 - \frac{t}{2}\right) = 0 \end{aligned}$$

at $t = 2$. $C(2) = 0.147$ or 14.7%.

(b) We want t such that
$$C(t) = 0.3(0.147) = 0.0441$$

Using a graphing utility, we find $t = 6.9$ hours.

47.
$$f(x) = \frac{2^x}{x}$$
$$f'(x) = \frac{x(2^x)\ln 2 - 2^x}{x^2} = \frac{2^x(\ln 2^x - 1)}{x^2}$$

49.
$$f(x) = x\log_{10} x = \frac{x \ln x}{\ln 10}$$
$$f'(x) = \frac{1}{\ln 10}(1 + \ln x)$$

51. By definition, the percentage rate of change of f with respect to x is
$$100\frac{f'(x)}{f(x)}$$

Since $\frac{d}{dx}\ln f(x) = \frac{f'(x)}{f(x)}$, it follows that the percentage rate of change can be written as $100\frac{d}{dx}\ln f(x)$.

53. The population of the town x years from now will be
$$P(x) = 5,000\sqrt{x^2 + 4x + 19}$$

From problem 51, the percentage rate of change x years from now will be

$$100\frac{d}{dx}\ln 5,000\sqrt{x^2 + 4x + 19}$$
$$= 100\frac{d}{dx}\left[\ln 5,000 + \frac{1}{2}\ln(x^2 + 4x + 19)\right]$$
$$= \frac{100(x + 2)}{x^2 + 4x + 19}$$

Hence the percentage rate of change 3 years from now will be
$$\frac{100(3 + 2)}{3^2 + 12 + 19} = 12.5 \text{ \% per year}$$

55.
$$f(x) = (3.7x^2 - 2x + 1)e^{-3x+2}$$
$$f'(-2.17) = -428,640$$

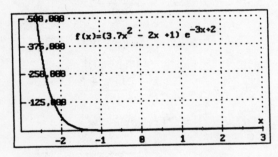

4.4 Additional Exponential Models

1.
$$y = f(t) = 2 + e^t$$

The line $y = 2$ is a horizontal asymptote.

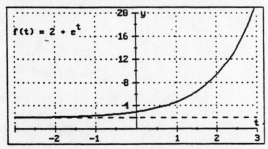

4.4. ADDITIONAL EXPONENTIAL MODELS

3. $\quad y = g(x) = 2 - 3e^x.$

The line $y = 2$ is a horizontal asymptote.

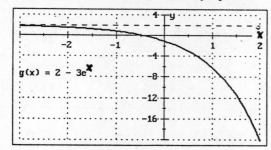

5. $\quad y = f(x) = 3 - 2(2^{-x}).$

The line $y = 3$ is a horizontal asymptote.

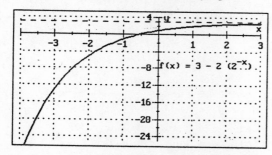

7. $\quad y = g(t) = 5 - 3e^{-t}.$

The line $y = 5$ is a horizontal asymptote.

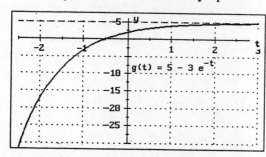

9. $\quad y = f(x) = \dfrac{2}{1 + 3e^{-2t}}.$

The lines $y = 0$ and $y = 2$ are horizontal asymptotes.

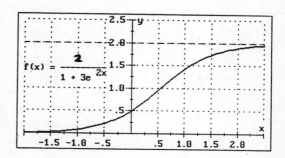

11.
$$\begin{aligned} f(x) &= xe^x \\ f'(x) &= e^x(x+1) = 0 \end{aligned}$$

when $x = -1$.

$$\begin{aligned} f''(x) &= e^x(1 + x + 1) \\ &= e^x(x+2) = 0 \end{aligned}$$

when $x = -2$.
$(-1, -0.37)$ is a minimum,
$(-2, -0.27)$ is a point of inflection.
Note that $f(x) = xe^x < 0$ if $x < 0$ and decreasing (if $x < -1$).
The line $y = 0$ is an upper bound (the curve is never above this line for $x < 0$.)
This strongly suggests that the x axis is a horizontal asymptote.

x	$-\infty$	-2	-1	∞
$f(x)$	0	-0.27	-0.37	∞
$f'(x)$	$-$	$-$	0	$+$
$f''(x)$	$-$	0	$+$	$+$

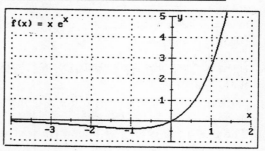

13.
$$\begin{aligned} f(x) &= xe^{2-x} \\ f'(x) &= e^{2-x}(1-x) = 0 \end{aligned}$$

when $x = 1$.

$$\begin{aligned} f''(x) &= e^{2-x}[-1 + (1-x)(-1)] \\ &= e^{2-x}(x-2) = 0 \end{aligned}$$

when $x=2$.
$(1, 2.7)$ is a maximum,
$(2, 2)$ is a point of inflection.
Note that $f(x) = xe^{2-x} > 0$ if $x > 0$ and decreasing for $x > 1$.
The line $y = 0$ is a lower bound (the curve is never below this line if $x > 0$.)
This strongly suggests that the x axis is a horizontal asymptote.

x	$-\infty$	1	2	∞
$f(x)$	$-\infty$	2.7	2	0
$f'(x)$	+	0	−	−
$f''(x)$	−	−	0	+

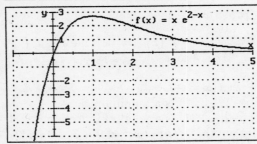

15. $f(x) = x^2 e^{-x}$.

$f'(x) = e^{-x}[-x^2 + 2x] = xe^{-x}(2-x) = 0$

when $x = 0, 2$.

$$f''(x) = e^{-x}[-2x + 2 - (-x^2 + 2x)]$$
$$= e^{-x}(x^2 - 4x + 2) = 0$$

when $x = \dfrac{4 \pm \sqrt{16-8}}{2}$,

so $x = 0.59$ or $x = 3.41$.
$(0, 0)$ is a minimum,
$(2, 0.54)$ is a maximum,
$(0.59, 0.19)$ and $(3.41, 0.38)$ are points of inflection.
Note that $f(x) = x^2 e^{-x} > 0$ if $x > 0$ and decreasing.
The line $y = 0$ is a lower bound (the curve is never below this line.)
This strongly suggests that the x axis is a horizontal asymptote.

x	$-\infty$	$\cdots$	0	$\cdots$	0.59	2	3.41	∞
$f(x)$	∞	$\cdots$	0	$\cdots$	0.19	0.5	0.38	0
$f'(x)$	−	−	0	+	+	0	−	−
$f''(x)$	+	+	+	+	0	−	0	+

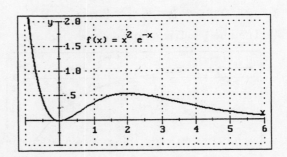

17. $f(x) = \dfrac{6}{1+e^{-x}} = 6(1+e^{-x})^{-1}.$

$f'(x) = -6(1+e^{-x})^{-2}(-e^{-x})$

$ = \dfrac{6}{e^x(1+e^{-x})^2} > 0.$

$f''(x) = \dfrac{6}{e^{2x}(1+e^{-x})^4}$
$[0 - (e^x)(2)(1+e^{-x})(-e^{-x})$
$- (1+e^{-x})^2 e^x]$

$ = -\dfrac{6(1-e^{-x})}{e^x(1+e^{-x})^3} = 0$

when $e^{-x} = 1$ or $x = 0$.
$(0, 3)$ is a point of inflection.
Note that if $x \to -\infty$ $1 + e^{-x} \to \infty$ and $y \to 0$, so the x axis is a horizontal asymptote.
Similarly if $x \to +\infty$ $1 + e^{-x} \to 1$ and $y \to 6$, so the line $y = 6$ is a horizontal asymptote also.
When $x < 0$, $f(x)$ is concave up, and when $x > 0$, $f(x)$ is concave down.

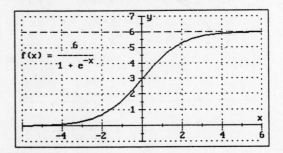

19. $f(x) = (\ln x)^2$

$f'(x) = 2\left(\dfrac{\ln x}{x}\right) = 0$

when $x = 1$.

$f''(x) = \dfrac{2}{x^2}(1 - \ln x) = 0$

4.4. ADDITIONAL EXPONENTIAL MODELS

when $x = e$.
$(1, 0)$ is a minimum,
$(e, 1)$ is a point of inflection.
Note that the curve seems to level off to the right, but this is an optical illusion.
$f(x)$ will keep increasing beyond all bounds.
For example $f(1,000) = 47.7$ and
$f(10^{98}) = 51,964$.
The y-axis is a vertical asymptote.
When $x < 1$, $f(x)$ is decreasing.
When $x > 1$, $f(x)$ is increasing.
When $x < e$, $f(x)$ is concave up.
When $x > e$, $f(x)$ is concave down.

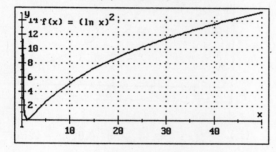

21. **(a)** The reliability function is
$$f(t) = 1 - e^{-0.03t}.$$

As t increases without bound, $e^{-0.03t}$ approaches 0 and so $f(t)$ approaches 1. Furthermore, $f(0) = 0$.
The graph is like that of a learning curve.

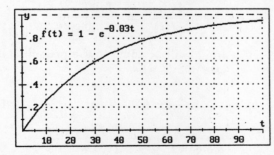

(b) The fraction of tankers that sink in fewer than 10 days is $f(10) = 1 - e^{-0.3}$.
The fraction of tankers that remain afloat for at least 10 days is therefore $1 - f(10) = e^{-0.3} = 0.7408$.

(c) The fraction of tankers that can be expected to sink between the 15th and 20th days is $f(20) - f(15)$
$$\begin{aligned} &= (1 - e^{-0.6}) - (1 - e^{-0.45}) \\ &= -e^{-0.6} + e^{-0.45} \\ &= -0.5488 + 0.6373 = 0.0888. \end{aligned}$$

23. The temperature of the drink t minutes after leaving the refrigerator is
$$f(t) = 30 - Ae^{-kt}.$$

Since the temperature of the drink when it left the refrigerator was 10 degrees Celsius,
$$10 = f(0) = 30 - A \text{ or } A = 20.$$

Thus
$$f(t) = 30 - 20e^{-kt}.$$

Since the temperature of the drink was 15 degrees Celsius 20 minutes later,
$$15 = f(20) = 30 - 20e^{-20k} \text{ or } e^{-20k} = \frac{3}{4}.$$

The temperature of the drink after 40 minutes is therefore
$$\begin{aligned} f(40) &= 30 - 20e^{-40k} = 30 - 20(e^{-20k})^2 \\ &= 30 - 20\left(\frac{3}{4}\right)^2 = 18.75 \text{ degrees.} \end{aligned}$$

25. **(a)**
$$f(t) = \frac{2}{1 + 3e^{-0.8t}}.$$

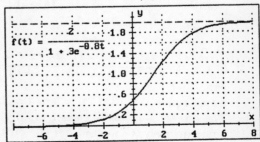

(b) $f(0) = 0.5$ thousand people (500 people).

(c) $f(3) = \dfrac{2}{1 + 3 \times 0.0907} = 1.572$, so 1,572 people have caught the disease.

(d) The highest number of people who can contract the disease is $\frac{2}{1+0} = 2$ or 2,000 people.

Note that the graph is shown only for $t \geq 0$.

27. $Q(t) = 40 - Ae^{-kt}$
$Q(0) = 20$, $20 = 40 - A$ so that $A = 20$.
$Q(1) = 30$, $30 = 40 - 20e^{-k}$ and $e^{-k} = \frac{1}{2}$.
$Q(3) = 40 - 20e^{-3k} = 37.5$ units per day.

29. $$f(x) = 15 - 20e^{-0.3x}$$
$x = 9$ thousand.
$$f(9) = 15 - 20e^{-2.7} = 13.656$$
(a) $f'(9) = 6e^{-2.7} = 0.403$ thousand
(b) $f(10) = 15 - 20e^{-3} = 14.004$
The actual increase was $14.004 - 13.656 = 0.348$ or 348 books. The estimate of 403 books was not too bad an approximation.

31. $$Q(t) = 80(4 + 76e^{-1.2t})^{-1}$$
$$Q'(t) = 80(-1)(4 + 76e^{-1.2t})^{-2}(76)e^{-1.2t}(-1.2)$$
$$= 80 \times 76 \times 1.2 \frac{e^{-1.2t}}{(4 + 76e^{-1.2t})^2}$$

After 2 weeks (at the end of the second week)
$$Q(2) = 80 \times 76 \times 1.2 \frac{e^{-2.4}}{(4 + 76e^{-2.4})^2}$$
$$= 5.576 \text{ or } 5,576 \text{ people}$$

$$Q''(t) = \frac{7296(-1.2)(4 + 76e^{-1.2t} - 152e^{-1.2t})}{e^{1.2t}(4 + 76e^{-1.2t})^3}$$

which equals 0 if $76e^{-1.2t} = 4$ or $t = 2.45$. The disease is spreading most rapidly approximately $2\frac{1}{2}$ weeks after the outbreak.

33. $$P(t) = \frac{Ce^{kt}}{1 + Ce^{kt}}$$
(a) $$P(0) = P_0 = \frac{C}{1 + C}$$
$$\text{or } C = \frac{P_0}{1 - P_0}$$

(b) $$\lim_{t \to \infty} \frac{Ce^{kt}}{1 + Ce^{kt}}$$
$$= \lim_{t \to \infty} \frac{C}{e^{-kt} + C} = 1 = 100\%$$

35. (a) The profit per VCR is $x - 125$. The number of units sold is $1,000e^{-0.02x}$. The weekly profit is
$$P(x) = 1,000(x - 125)e^{-0.02x}$$

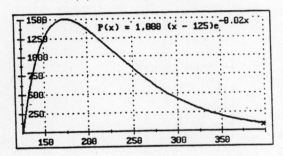

(b) $P'(x) = 1,000[-0.02(x - 125)e^{-0.02x} + e^{-0.02x}] = 0$

when $0.02(x - 125) = 1$ or $x = 175$.

37. The percentage rate of change of the market price
$$V(t) = 8,000e^{\sqrt{t}}$$
of the land (expressed in decimal form) is
$$\frac{V'(t)}{V(t)} = \frac{8,000e^{\sqrt{t}}}{8,000e^{\sqrt{t}}} \frac{1}{2\sqrt{t}} = \frac{1}{2\sqrt{t}}$$

which will be equal to the prevailing interest rate of 6 % when
$$\frac{1}{2\sqrt{t}} = 0.06 \text{ or } t = \left(\frac{1}{0.12}\right)^2 = 69.44.$$

Moreover, $\frac{1}{2\sqrt{t}} > 0.06$ when $0 < t < 69.44$ and $\frac{1}{2\sqrt{t}} < 0.06$ when $69.44 < t$.

Hence the percentage rate of growth of the value of the land is greater than the prevailing interest rate when $0 < t < 69.44$ and less than the prevailing interest rate when $69.44 < t$. Thus the land should be sold in 69.44 years.

4.4. ADDITIONAL EXPONENTIAL MODELS

39. Since the stamp collection is currently worth $1,200 and its value is increases linearly at the rate of $200 per year, its value t years from now is
$$V(t) = 1,200 + 200t.$$
The percentage rate of change of the value (expressed in decimal form) is
$$\frac{V'(t)}{V(t)} = \frac{200}{1,200 + 200t}$$
$$= \frac{200}{200(6+t)}$$
$$= \frac{1}{6+t}$$
which will be equal to the prevailing interest rate of 8% when $\frac{1}{6+t} = 0.08$ or
$$t = \frac{1}{0.08} - 6 = 6.5.$$
Moreover, $\frac{1}{6+t} > 0.08$ when $0 < t < 6.5$
and $\frac{1}{6+t} < 0.08$ when $6.5 < t$.
Hence the percentage rate of growth of the value of the collection is greater than the prevailing interest rate when $0 < t < 6.5$ and less than the prevailing interest rate when $t > 6.5$. Thus the collection should be sold in 6.5 years.

41. $y = \frac{c}{b-a}(e^{-at} - e^{-bt})$

(a) $y' = \frac{c}{b-a}(-ae^{-at} + be^{-bt}) = 0$
when $ae^{-at} = be^{-bt}$ or $\frac{a}{b} = e^{(a-b)t}$

Thus $t = \frac{1}{a-b}\ln\left(\frac{a}{b}\right)$.

In the long run both exponential terms approach zero, so $y \to 0$.

(b)
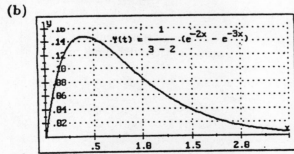

43. (a)
$$N(t) = 500(0.03)^{(0.4)^t}$$
$$N(0) = 500(0.03)^{(0.4)^0} = 15$$
$$N(5) = 500(0.03)^{(0.4)^5} \approx 482 \text{ employees.}$$
$$300 = 500(0.03)^{(0.4)^t}$$
if $(0.4)^t = \frac{\ln 0.6}{\ln 0.03} = 0.145677$
$$t = \frac{\ln 0.145677}{\ln 0.4} = 2.10 \text{ years.}$$
$$\lim_{t \to \infty} N(t) = 500(0.03)^0 = 500 \text{ employees.}$$

(b) $F(t) = 500(0.03)^{-(0.4)^{-t}}$

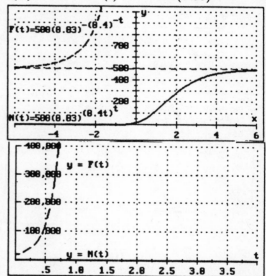

45. Let's assume continuous growth, so
$$Q(t) = Q_0 e^{0.06t}$$
Let $t = 0$ be 1947. $Q_0 = 1,139$.

(a) $Q(7) = 1,139e^{0.06(7)} = 1,733.5$
$Q(53) = 27,389.3$

(b) We want $2,000 = 1,139e^{0.06t}$ or $1.756 = e^{0.06t}$.
This leads to $t = 9.38$.
$2,278 = 1,139e^{0.06t}$ or $2 = e^{0.06t}$.
This leads to $t = 11.55$.

(c) Writing Exercise — Answers will vary.

47. $P(x) = \lambda^2 x e^{-\lambda x}, 0 < \lambda < e$

(a) $$P'(x) = \lambda^2 e^{-\lambda x} + \lambda^2 x(-\lambda e^{-\lambda x})$$
$$= \lambda^2 e^{-\lambda x}(1 - x\lambda) = 0$$

at $x = \frac{1}{\lambda}$. $P\left(\frac{1}{\lambda}\right) = \lambda e^{-1} = \frac{\lambda}{e}$

(b)

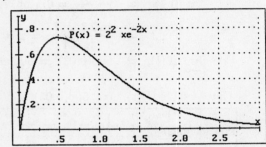

49. $f(x) = x(e^{-x} + e^{-2x})$

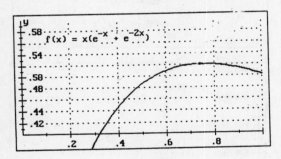

$$\lim_{x \to \infty} f(x) = 0$$

Note: The portion of the graph in the third quadrant is likely to be hidden from view on your graphing utility unless you specificaly request a negative domain.
The high point occurs at $(0.76, 0.52)$.

51. $P(t) = 20,000 t e^{\sqrt{0.4t} - 0.07t}$

A graphing utility indicates a maximum present value of \$150,543 at $t = 44.38$ years.

then $f(x)$ approaches 0 as x increases without bound, and $f(x)$ increases without bound as x decreases without bound. The y-intercept is $f(0) = 5$.

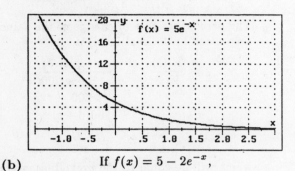

(b) If $f(x) = 5 - 2e^{-x}$,

then $f(x)$ approaches 5 as x increases without bound, and $f(x)$ decreases without bound as x decreases without bound. The y-intercept is $f(0) = 3$.

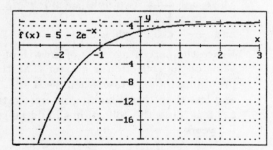

(c) If $f(x) = 1 - \dfrac{6}{2 + e^{-3x}}$,

the y-intercept is
$$f(0) = 1 - \frac{6}{2 + e^0} = 1 - \frac{6}{2 + 1} = -1.$$
As x increases without bound, e^{-3x} approaches 0, and so
$$\lim_{x \to \infty} f(x) = 1 - \frac{6}{2 + 0} = -2.$$
As x decreases without bound, e^{-3x} increases without bound, and so
$$\lim_{x \to -\infty} f(x) = 1 - 0 = 1.$$

Review Problems

1. (a) If $f(x) = 5e^{-x}$,

REVIEW PROBLEMS

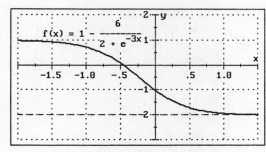

(d) If $f(x) = \dfrac{3 + 2e^{-2x}}{1 + e^{-2x}}$,

then $f(x)$ approaches 3 as x increases without bound, since e^{-2x} approaches 0. Rewriting the function, by multiplying numerator and denominator by e^{2x} yields

$$f(x) = \frac{3e^{2x} + 2}{e^{2x} + 1}.$$

As x decreases without bound, e^{2x} approaches 0 and thus $f(x)$ approaches 2. The y-intercept is $f(0) = \dfrac{3+2}{1+1} = \dfrac{5}{2}$. Applying the quotient rule to determine $f'(x)$ yields

$$f'(x) = \frac{2e^{2x}}{(e^{2x} + 1)^2}.$$

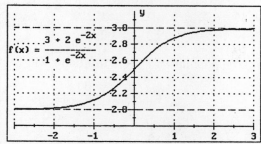

Since $2e^{2x} > 0$, there are no critical points and $f(x)$ increases for all x.
To find $f''(x)$, differentiate $f'(x)$ using the quotient rule obtaining

$$f''(x) = \frac{4e^{2x}(1 - e^{2x})}{(e^{2x} + 1)^3}$$

There is one inflection point $(0, f(0)) = (0, \dfrac{5}{2})$ since $f''(0) = 0$. The function $f(x)$ is concave down for $x > 0$ and concave up for $x < 0$.

2. (a) If $f(x) = Ae^{-kx}$

and $f(0) = 10$, then $10 = Ae^0$.
Hence $f(x) = 10e^{-kx}$.
Since $f(1) = 25$, $25 = 10e^{-k}$ or $e^{-k} = \dfrac{5}{2}$.

Then $f(4) = 10e^{-4k} = 10(e^{-k})^4$
$= 10\left(\dfrac{5}{2}\right)^4 = 390.625.$

(b) If $f(x) = Ae^{kx}$ and $f(1) = 3$ as well as $f(2) = 10$,
then $3 = Ae^k$ and $10 = Ae^{2k}$, two equations in two unknowns.
Division eliminates A, so $\dfrac{3}{10} = \dfrac{Ae^k}{Ae^{2k}}$,
$\dfrac{3}{10} = \dfrac{e^k}{e^{2k}}$, or $e^k = \dfrac{10}{3}$.
Since $e^k = \dfrac{10}{3}$, $3 = A\left(\dfrac{10}{3}\right)$
and so $A = \dfrac{9}{10}$. Thus

$$f(3) = \frac{9}{10}e^{3k} = \frac{9}{10}(e^k)^3 = \frac{100}{3}$$

(c) If
$$f(x) = 30 + Ae^{-kx}$$

and $f(0) = 50$, then $50 = 30 + Ae^0$ or $A = 20$.
Hence, $f(x) = 30 + 20e^{-kx}$.
Since $f(3) = 40$, $40 = 30 + 20e^{-3k}$,
$10 = 20e^{-3k}$, or $e^{-3k} = \dfrac{1}{2}$.

Thus $f(9) = 30 + 20e^{-9k}$
$= 30 + 20(e^{-3k})^3$
$= 30 + 20\left(\dfrac{1}{2}\right)^3 = 32.5.$

(d) If $f(x) = \dfrac{6}{1 + Ae^{-kx}}$

and $f(0) = 3$ then $3 = \dfrac{6}{1 + Ae^0}$, or $A = 1$.
Hence, $f(x) = \dfrac{6}{1 + e^{-kx}}$.

Since $f(5) = 2$, $2 = \dfrac{6}{1 + e^{-5k}}$, $2 + 2e^{-5k} = 6$, or $e^{-5k} = 2$. Then,

$$f(10) = \dfrac{6}{1 + e^{-10k}} = \dfrac{6}{1 + (e^{-5k})^2}$$
$$= \dfrac{6}{1 + (2)^2} = \dfrac{6}{5}.$$

3. Let $V(t)$ denote the value of the machine after t years.
   Since the value decreases exponentially and was originally $50,000, it follows that
   $$V(t) = 50,000e^{-kt}.$$

   Since the value after 5 years is $20,000,
   $$20,000 = V(5) = 50,000e^{-5k},$$
   $$e^{-5k} = \dfrac{2}{5},$$
   $$\text{or } k = -\dfrac{1}{5}\ln\dfrac{2}{5}.$$

   Hence $V(t) = 50,000e^{[1/5\ln(2/5)]t}$ and so
   $$V(10) = 50,000e^{2\ln(2/5)}$$
   $$= 50,000e^{\ln(4/25)}$$
   $$= 50,000\left(\dfrac{4}{25}\right) = \$8,000.$$

4. The sales function is
   $$Q(x) = 50 - 40e^{-0.1x}$$
   units, where x is the amount (in thousands) spent on advertising.

   (a) As x increases without bound, $Q(x)$ approaches 50. The vertical axis intercept is $Q(0) = 10$. The graph is like that of a learning curve.

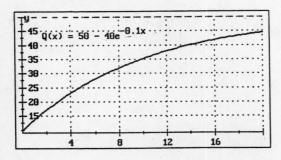

   (b) If no money is spent on advertising, sales will be $Q(0) = 10$ thousand units.

   (c) If $8,000 is spent on advertising, sales will be $Q(8) = 50 - 40e^{-0.8} = 32.027$ thousand or 32,027 units.

   (d) Sales will be 35 thousand if $Q(x) = 35$, that is, if
   $$50 - 40e^{-0.1x} = 35,$$
   $$e^{-0.1x} = \dfrac{3}{8},$$
   $$x = -\dfrac{\ln(3/8)}{0.1} = 9.808$$
   thousand or $9,808.

   (e) Since $Q(x)$ approaches 50 as x increases without bound, the most optimistic sales projection is 50,000 units.

5. The output function is
   $$Q(t) = 120 - Ae^{-kt}.$$

   Since $Q(0) = 30$, $30 = 120 - A$ or $A = 90$.
   Since $Q(8) = 80$, $80 = 120 - 90e^{-8k}$,
   $-40 = -90e^{-8k}$ or $e^{-8k} = \dfrac{4}{9}$. Hence,
   $$Q(4) = 120 - 90e^{-4k} = 120 - 90(e^{-8k})^{1/2}$$
   $$= 120 - 90\left(\dfrac{4}{9}\right)^{1/2} = 60 \text{ units.}$$

6. The population t years from now will be
   $$P(t) = \dfrac{30}{1 + 2e^{-0.05t}}.$$

   (a) The vertical axis intercept is
   $$P(0) = \dfrac{30}{1 + 2} = 10 \text{ million.}$$
   As t increases without bound, $e^{-0.05t}$ approaches 0. Hence,
   $$\lim_{t \to \infty} P(t) = \lim_{t \to \infty} \dfrac{30}{1 + 2e^{-0.05t}} = 30$$

   As t decreases without bound, $e^{-0.05t}$ increases without bound.
   Hence, the denominator $1 + 2e^{-0.05t}$ increases without bound and
   $$\lim_{t \to \infty} P(t) = \lim_{t \to \infty} \dfrac{30}{1 + 2e^{-0.05t}} = 30.$$

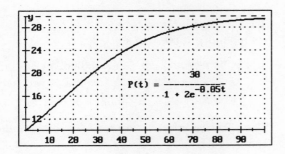

(b) The current population is $P(0) = 10$ million.

(c) The population in 20 years will be

$$P(20) = \frac{30}{1 + 2e^{-0.05(20)}}$$
$$= \frac{30}{1 + 2e^{-1}} = 17.2835$$

million or 17,283,500 people.

(d) In the long run (as t increases without bound), $e^{-0.05t}$ approaches 0 and so the population $P(t)$ approaches 30 million.

7. (a) $\ln e^5 = 5$ since $n = \ln e^n$.

(b) $e^{\ln 2} = 2$ since $n = e^{\ln n}$.

(c) $e^{3\ln 4 - \ln 2} = e^{\ln 4^3 - \ln 2}$
$$= e^{\ln \frac{64}{2}} = e^{\ln 32} = 32.$$

(d) $\ln(9e^2) + \ln(3e^{-2}) = \ln[(9e^2)(3e^{-2})]$
$$= \ln 27 = 3\ln 3.$$

8. (a) $8 = 2e^{0.04x},$
$e^{0.04x} = 4,$
$0.04x = \ln 4$
$x = 34.657.$

(b) $5 = 1 + 4e^{-6x},$
$4e^{-6x} = 4$
$-6x = \ln 1 = 0,$ or $x = 0.$

(c) $4\ln x = 8,$ $\ln x = 2,$
or $x = e^2 = 7.389.$

(d) $5^x = e^3,$
$\ln 5^x = \ln e^3,$
$x \ln 5 = 3,$ or
$x = \frac{3}{\ln 5} = 1.864.$

9. Let $Q(t)$ denote the number of bacteria after t minutes. Since $Q(t)$ grows exponentially and 5,000 bacteria were present initially,

$$Q(t) = 5,000e^{kt}.$$

Since 8,000 bacteria were present after 10 minutes,

$8,000 = Q(10) = 5,000e^{10k}$, $e^{10k} = \frac{8}{5}$, or

$k = \frac{1}{10} \ln \frac{8}{5}.$

The bacteria will double when

$$Q(t) = 10,000,$$

that is, when

$5,000e^{kt} = 10,000$
$kt = \ln 2$
or $t = \frac{\ln 2}{k} = \frac{10 \ln 2}{\ln(8/5)}$
$= 14.75$ or 14. min. and 45 seconds.

10. (a) $f(x) = 2e^{3x+5},$
$f'(x) = 2e^{3x+5} \frac{d}{dx}(3x+5)$
$= 6e^{3x+5}.$

(b) $f(x) = x^2 e^{-x},$
$f'(x) = e^{-x}\frac{d}{dx}(x^2) + x^2\frac{d}{dx}e^{-x}$
$= x(2-x)e^{-x}.$

(c) $g(x) = \ln \sqrt{x^2 + 4x + 1}$
$= \frac{1}{2}\ln(x^2 + 4x + 1),$
$g'(x) = \frac{1}{2}\frac{1}{x^2 + 4x + 1}$
$\frac{d}{dx}(x^2 + 4x + 1)$
$= \frac{x+2}{x^2 + 4x + 1}.$

(d) $\quad h(x) = x\ln x^2 = 2x\ln x,$
$\quad h'(x) = 2\left(x\dfrac{d}{dx}\ln x + \ln x\dfrac{d}{dx}x\right)$
$\quad\quad\quad = 2(1+\ln x).$

(e) $\quad f(t) = \dfrac{t}{\ln 2t},$
$\quad f'(t) = \dfrac{(\ln 2t)(1) - t\left(\dfrac{1}{2t}\right)(2)}{(\ln 2t)^2}$
$\quad\quad\quad = \dfrac{\ln 2t - 1}{(\ln 2t)^2}.$

(f) $\quad g(t) = t\log_3 t^2,$
$\quad\quad\quad = \dfrac{2}{\ln 3}\, t\ln t$
$\quad g'(t) = \dfrac{2}{\ln 3}(1+\ln t).$

11. (a) Using the formula
$$B(t) = P\left(1+\dfrac{r}{k}\right)^{kt}$$
with $P = 2,000$, $B = 5,000$, $r = 0.08$, and $k = 4$,
$$5,000 = 2,000\left(1+\dfrac{0.08}{4}\right)^{4t}$$
$$(1.02)^{4t} = \dfrac{5}{2}$$
or $t = \dfrac{1}{4}\dfrac{\ln(5/2)}{\ln 1.02} = 11.57$ years.

(b) Using the formula
$$B(t) = Pe^{rt}$$
with $P = 2,000$, $B = 5,000$, and $r = 0.08$,
$5,000 = 2,000e^{0.08t}$,
$$e^{0.08t} = \dfrac{5}{2},$$
or $t = \dfrac{\ln(5/2)}{0.08} = 11.45$.

12. Compare the effective interest rates.

The effective interest rate for 8.25 % compounded quarterly is
$$\left(1+\dfrac{r}{k}\right)^k - 1 = \left(1+\dfrac{0.0825}{4}\right)^4 - 1$$
$$= 0.0851 \text{ or } 8.51\ \%.$$

The effective interest rate for 8.20 compounded continuously is
$$e^r - 1 = e^{0.082} - 1 = 0.0855 \text{ or } 8.55\ \%.$$

13. (a) Using the present value formula
$$P = B\left(1+\dfrac{r}{k}\right)^{-kt}$$
with $B = 2,000$, $t = 10$, $r = 0.0625$, and $k = 12$,
$$P = 2,000\left(1+\dfrac{0.0625}{12}\right)^{-120}$$
$$= \dfrac{2,000}{1.8652182} = \$1,072.26.$$

(b) Using the present value formula
$$P = Be^{-rt}$$
with $B = 2,000$, $t = 10$, and $r = 0.0625$,
$$P = 2,000e^{-0.0625(10)} = \$1,070.52.$$

14. (a) $\quad 8,000 = (P)\left(1+\dfrac{0.0625}{2}\right)^{2\times 10}$
$\quad\quad P = \dfrac{8,000}{(1.03125)^{20}} = 4,323.25$

(b) $\quad 8,000 = Pe^{0.0625\times 10}$
$\quad\quad P = 8,000e^{-0.625} = 4,282.09$

15. $\quad B(t) = 2,054.44 = 1,000e^{12r}$
$\quad r = \dfrac{\ln 2.05444}{12} = 0.06$, or $r = 6\%$

16. At 6% compounded annually, the effective interest rate is
$$\left(1+\dfrac{r}{k}\right)^k - 1 = \left(1+\dfrac{0.06}{1}\right)^1 - 1$$
$$= 0.06.$$

At r % compounded continuously, the effective interest rate is $e^r - 1$. Setting the two effective rates equal to each other yields

$$e^r - 1 = 0.06,$$

$e^r = 1.06$, $r = \ln 1.06 = 0.0583$ or 5.83 %.

17. The average level of carbon monoxide in the air t years from now is $Q(t) = 4e^{0.03t}$ parts per million.

(a) The rate of change of the carbon monoxide level t years from now is $Q'(t) = 0.12e^{0.03t}$, and the rate two years from now is $Q'(2) = 0.12e^{0.06} = 0.13$ parts per million per year.

(b) The percentage rate of change of the carbon monoxide level t years from now is

$$100\left[\frac{Q'(t)}{Q(t)}\right] = 100\left(\frac{.12e^{.03t}}{4e^{.03t}}\right) = 3\% \text{ per year}$$

which is a constant, independent of time.

18. Let $F(p)$ denote the profit, where p is the price per camera. Then

$$\begin{aligned} F(p) &= \text{(number of cameras sold)} \\ &\quad \text{(profit per camera)} \\ &= 800(p-40)e^{-0.01p}. \\ F'(p) &= 800[e^{-0.01p}(1) \\ &\quad + (p-40)e^{-0.01p}(-0.01)] \\ &= 8e^{-0.01p}(140-p) = 0 \end{aligned}$$

when $p = 140$.
Since $F'(p) > 0$ (and F is increasing) for $0 < p < 140$,
and $F'(p) < 0$ (and F is decreasing) for $p > 140$, it follows that $F(p)$ has its absolute maximum at $p = 140$.
Thus the cameras should be sold for $140 apiece to maximize the profit.

19. (a) $f(x) = xe^{-2x}$,
$f'(x) = xe^{-2x}(-2) + e^{-2x}(1)$
$\quad = e^{-2x}(1 - 2x) = 0$ when $x = \frac{1}{2}$.
$f\left(\frac{1}{2}\right) = \frac{1}{2e}.$
$f''(x) = 4e^{-2x}(x - 1) = 0$

when $x = 1$. $f(1) = \dfrac{1}{e^2}$.

x	$-\infty$		$\frac{1}{2}$		1		∞
$f(x)$	$-\infty$		$\frac{1}{2e}$		$\frac{1}{e^2}$		0
$f'(x)$	∞	+	0	−		−	0
		↑		↓		↓	
$f''(x)$		−		−		0	+
		↓		↓			↑

$\left(\dfrac{1}{2}, \dfrac{1}{2e}\right)$ is the absolute maximum while $\left(1, \dfrac{1}{e^2}\right)$ is a point of inflection.

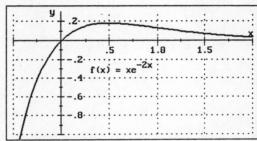

(b) $f(x) = e^x - e^{-x}$
$f'(x) = e^x + e^{-x} > 0$
$f''(x) = e^x - e^{-x} = 0$ when $x = 0$

There are no extrema.

x	$-\infty$		0		∞
$f(x)$	$-\infty$		0		∞
$f'(x)$	∞	+	+	+	
		↑	↑	↑	
$f''(x)$		−	0	+	
		↓		↑	

$(0,0)$ is a point of inflection.

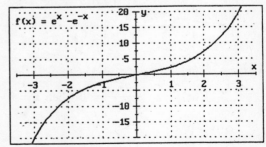

(c) $f(x) = \dfrac{4}{1+e^{-x}} = 4(1+e^{-x})^{-1}$,
$f'(x) = 4(-1)(1+e^{-x})^{-2}\dfrac{d}{dx}(1+e^{-x})$

$$= \frac{4e^{-x}}{(1+e^{-x})^2} > 0.$$

$$f''(x) = \frac{1}{(1+e^{-x})^4}[(1+e^{-x})^2$$
$$(4e^{-x})(-1) - 4e^{-x}(2)$$
$$(1+e^{-x})(e^{-x})(-1)]$$
$$= \frac{4e^{-x}(e^{-x} - 1)}{(1+e^{-x})^3} = 0 \text{ when } x = 0$$

x	$-\infty$		0		∞
$f(x)$	0		2		4
$f'(x)$	1	+	+	+	.0707
		↑	↑	↑	
$f''(x)$		+	0	−	
		up		↓	

$(0,2)$ is a point of inflection.

$$\lim_{x \to -\infty} \frac{4}{1+e^{-x}} = 0$$

because the denominator increases beyond all bounds.

$$\lim_{x \to \infty} \frac{4}{1+e^{-x}} = 4$$

because $e^{-x} \to 0$.
So $y = 0$ and $y = 4$ are horizontal asymptotes.

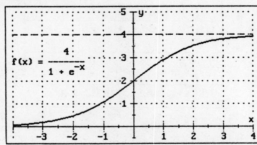

(d) $\quad f(x) = \ln(x^2 + 1)$
$\quad\quad f'(x) = \dfrac{2x}{x^2+1} = 0$ when $x = 0$

$f(0) = 0$.

$$f''(x) = \frac{2(1-x)(1+x)}{(x^2+1)^2} = 0$$

when $x = \pm 1$. $f(\pm 1) = \ln 2$.

x	$-\infty$		-1		0		1		∞
$f(x)$	∞		$\ln 2$		0		$\ln 2$		∞
$f'(x)$	0	−	−	−	0	+	+	+	0
		↓	↓	↓		↑	↑	↑	
$f''(x)$	0	−	0	+	2	+	0	−	0
		↓		↑	↑			↓	

$(0,0)$ is the absolute minimum
while $(\pm 1, \ln 2)$ are points of inflection.

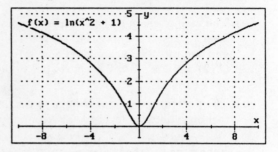

20. The value of the coin collection in t years is

$$V(t) = 2{,}000 e^{\sqrt{t}}.$$

Hence, the percentage rate of change of the value of the collection (expressed in decimal form) is

$$\frac{V'(t)}{V(t)} = \frac{2{,}000 e^{\sqrt{t}}}{2{,}000 e^{\sqrt{t}}} \left(\frac{1}{2\sqrt{t}}\right) = \frac{1}{2\sqrt{t}}$$

which will be equal to the prevailing interest rate of 7 % when $\dfrac{1}{2\sqrt{t}} = 0.07$ or

$$t = \left(\frac{1}{0.14}\right)^2 = 51.02.$$

Moreover, $\dfrac{1}{2\sqrt{t}} > 0.07$ when $0 < t < 51.02$
and $\dfrac{1}{2\sqrt{t}} < 0.07$ when $51.02 < t$.
Hence the percentage rate of growth of the collection is greater than the prevailing interest rate when $0 < t < 51.02$ and less than the prevailing interest rate when $51.02 < t$.
Thus the coin collection should be sold in 51.02 years.

21. Let a be the desired ratio. Then

$$R(t) = R_0 e^{-kt},$$

$$aR_0 = R_0 e^{-kt}, a = e^{-15,000k}.$$

Since the half-life is 5,730, $\frac{1}{2} = e^{-5,730k}$ or

$k = \dfrac{\ln 2}{5,730}$ and

$a = e^{-15,000 \ln 2/5,730} = e^{-1.8145} = 0.1629.$

22. (a)
$$\begin{aligned} R(t) &= R_0 e^{-(\ln 2)t/5,730} \\ R(1,960) &= R_0 e^{-(\ln 2)(1,960)/5,730} \\ &= R_0 e^{-0.2371} = 0.7889 R_0 \end{aligned}$$

Thus about 78.89% of $_{14}C$ should be left in the shroud.

(b) Since 92.3% of $_{14}C$ is left in the forgery,

$$\begin{aligned} 0.923 &= \exp\left[-\frac{(\ln 2)t}{5,730}\right] \\ t &= -\frac{(\ln 0.923)(5,730)}{\ln 2} = 662 \end{aligned}$$

1988 − 662 = 1326 so Pierre d'Arcis's suspicions were well founded.

23. $\quad f(t) = 70 - Ae^{-kt}$

$f(0) = 70 - A = 212$ or $A = -142$. Thus $f(t) = 70 + 142 e^{-kt}$. Let T_0 be the ideal temperature. Then

$$T_0 + 15 = 70 + 142 e^{-k(5)} \text{ and}$$
$$T_0 = 70 + 142 e^{-k(7)}$$

Solving (with the SOLVE utility of our calculator) we get $k = 0.091$ and $T_0 = 145^0$.

24. (a) $D(t) = (D_0 - 0.00046)e^{-0.162t} + 0.00046$

With $D_0 = 0.008$,

$$\begin{aligned} D(10) &= (0.008 - 0.00046)e^{-1.62} + 0.00046 \\ &= 0.00195 \end{aligned}$$

and $D(25) = 0.00590$.

(b) $\quad \lim_{t \to \infty} D(t) = 0$

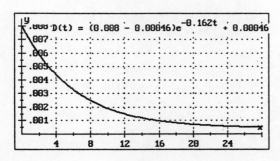

25. (a) To compute doubling time in terms of the rate set $2P = Pe^{rt}$ and solve for t.

$$2 = e^{rt} \text{ or } t = \frac{\ln 2}{r}$$

In the chart below find a comparison of the rules of 69, 70, 72 with the true doubling time.

r	0.04	0.06	0.09	0.10	0.12
69	17.25	11.50	7.67	6.9	5.75
70	17.5	11.67	7.78	7.0	5.83
72	18	12	8	7.2	6
True	17.33	11.55	7.7	6.93	5.78

The rule of 69 is closest because $\dfrac{69}{100}$ is closest to $\ln 2$.

(b) Writing Exercise — Answers will vary.

26. $\quad A(x) = 110 \dfrac{\ln x - 2}{x}$, for $10 \leq x$.

$\quad A'(x) = 110 \dfrac{3 - \ln x}{x^2} = 0$

when $x = e^3 = 20.0855$ years. $A(10) = 3.3284$ so a person's aerobic capacity is maximized at about age 20.

27.
$$f(x) = \frac{4e^{-(\ln x)^2}}{\sqrt{\pi} x} \text{ for } x > 0$$

$$f'(x) = \frac{4}{\sqrt{\pi}} \frac{e^{-(\ln x)^2}}{x^2}[-2\ln x - 1] = 0$$

when $x = \dfrac{1}{\sqrt{e}} = 0.6065.$

According to the graphing utility, the most common age is at $(0.6065, 2.8977)$.

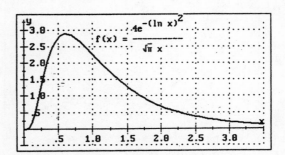

28. The Bronze age began about 5,000 years ago (around 3,000 B.C.).
The maximum percentage is

$$P(5,000) = e^{-(\ln 2)5,000/5,730} = 54.62\% \approx 55\%$$

29. (a)
$$f(t) = C(1 - e^{-kt}) = 0.008C$$
$$0.992 = e^{-2k} \text{ or } k = .00402$$

(b)
$$df = Cke^{-kt}$$
$$100\frac{df}{f} = \frac{100k}{e^{kt} - 1}$$

(c) Writing Exercise —
Answers will vary.

30.
$$T(t) = 35e^{-0.32t}$$
$$27 = 35e^{-0.32t} \text{ or } t = 0.811 \text{ min.}$$

Rescuers have about 49 seconds before the girl looses consciousness.

$$\frac{dT}{dt} = -35(0.32)e^{-0.32t}$$

At $T = 27$ or $e^{-0.32t} = \frac{27}{35}$

$$\frac{dT}{dt} = (-35)(0.32)\left(\frac{27}{35}\right) = -8.64$$

Thus the girl's temperature is dropping at 8.64 degrees per minute.

31. (a)
$$y = 0.125(e^{4x} + e^{-4x})$$
$$y' = 0.5(e^{4x} - e^{-4x}) = 0$$

when $e^{8x} = 1$ at $x = 0$.

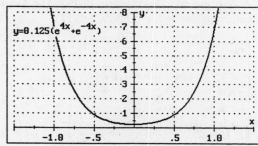

(b) Writing exercise —
Answers will vary.

32.
$$k_1 = A\exp\left[-\frac{E_0}{RT_1}\right]$$
$$k_2 = A\exp\left[-\frac{E_0}{RT_2}\right]$$
$$\frac{k_1}{k_2} = \exp\left[\frac{E_0}{RT_2} - \frac{E_0}{RT_1}\right]$$
$$\ln\frac{k_1}{k_2} = \frac{E_0}{R}\left(\frac{1}{T_2} - \frac{1}{T_1}\right)$$

33. (a)
$$V(t) = V_0\left(1 - \frac{2}{L}\right)^t$$

$V_0 = 875$, $L = 8$, $t = 5$,

$$V(5) = 875\left(1 - \frac{2}{8}\right)^5 = 207.64.$$

The refrigerator will be worth $207.64 after 5 years.

$$\frac{875 - 207.64}{5} = \$133.47 \text{ and}$$
$$\frac{133.47}{875} = 0.15 \text{ or } 15\%$$

(b)
$$V'(t) = V_0\left(1 - \frac{2}{L}\right)^t \ln\left(1 - \frac{2}{L}\right)$$

The percentage rate of change is

$$100\frac{V'(t)}{V(t)} = 100\frac{V_0\left(1 - \frac{2}{L}\right)^t \ln\left(1 - \frac{2}{L}\right)}{V_0\left(1 - \frac{2}{L}\right)^t}$$
$$= 100\ln\left(1 - \frac{2}{L}\right)$$

34.
$$f(t) = 30 - Ae^{-kt}$$

Solve for $-Ae^{-kt} = f(t) - 30$.

$$f'(t) = -Ae^{-kt}(-k)$$
$$= kAe^{-kt} = k[30 - f(t)]$$

where k is a constant of proportionality and $f(t)$ is the temperature of the drink.

35.
$$PH = -\log_{10}[H_3O^+]$$

For milk and lime, $PH_m = 3PH_l$.
For lime and orange, $PH_l = 0.5PH_o$.

$$PH_l = \frac{3.2}{2} = 1.6$$
$$[H_3O^+]_l = 10^{-1.6} = 0.0251$$

REVIEW PROBLEMS

36. $$P(t) = \frac{202.31}{1 + e^{3.938 - 0.314t}}$$

Hint: Use the equation writer on the HP48G to enter the function.
For the TI-85, press 2nd CALC
EVALF($202.31/(1 + e^{(3.938-0.314x)}), x, 0$), etc.

(a)

year	t	$P(t)$
1790	0	3,867,087
1800	1	5,256,550
1830	4	12,956,719
1860	7	30,207,500
1880	9	50,071,364
1900	11	77,142,427
1920	13	108,425,601
1940	15	138,370,607
1960	17	162,289,822
1980	19	178,782,499
1990	20	184,566,652
2000	21	189,034,385

(b) This model predicts that the population will be increasing most rapidly when $t = 12.5$ or in 1915.

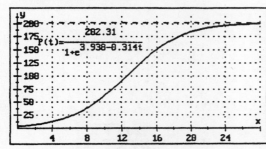

(c) Writing exercise —
Answers will vary.

37.
$$y = 2^{-x} = \left(\frac{1}{2}\right)^x$$
$$y = 3^{-x} = \left(\frac{1}{3}\right)^x$$
$$y = 5^{-x} = \left(\frac{1}{5}\right)^x$$
$$y = (0.5)^{-x} = 2^x$$

The graphs of $y = b^x$ and $y = \left(\frac{1}{b}\right)^x$ are reflections of each other in the y-axis $(0 < b < 1)$.
The larger b the steeper the curve.

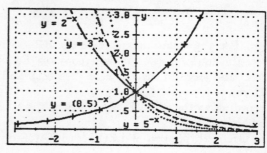

38.
$$y = \sqrt{3^x} = 3^{x/2}$$
$$y = \sqrt{3^{-x}} = 3^{-x/2}$$
$$y = 3^{-x} = 3^{-x}$$

The graphs of $y = 3^{bx}$ and $y = 3^{-bx}$ are reflections of each other in the y-axis $(0 < b)$.
The larger b the steeper the curve.

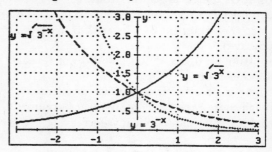

39. Hint: Use the equation writer on the HP48G to enter the function or $y(x) =$ on the TI-85.

$$y = 3^x$$

intersects

$$y = 4 - \ln\sqrt{x}$$

at $(1.2373, 3.8935)$ according to the graphing utility.

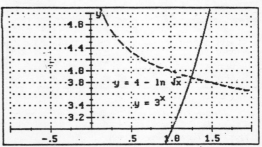

40. $$\log_5(x+5) - \log_2 x = 2\log_{10}(x^2+2x)$$

can be rewritten as

$$\frac{1}{\ln 5}\ln(x+5) - \frac{1}{\ln 2}\ln x = \frac{1}{\ln 10}\ln(x^2+2x)^2$$

Hint: Use the equation writer on the HP48G to enter the equation (and make a copy on the stack) and store in the EQ field (use 'EQ' and STOre).
According to the SOLVE capability of the graphing utility.
Use GRAPH, $y(x) =$, and ZOOM/TRACE on the TI-85.

$$x = 1.06566543483$$

is a solution of this equation. Plot

$$f(x) = \log_5(x+5) - \log_2 x - 2\log_{10}(x^2+2x)$$

because the message "sign reversal" (on the HP48G) indicates another root.
$x < 0$ leads to a complex solution (as it should since the argument of a logarithm must be positive).
Plotting on $0 < x < 500{,}000$ did not reveal another root.
There is no other root because x^2 increases much more rapidly than any other argument, making $f(x)$ monotonically decreasing.

41. $$y = f(x) = \ln(1+x^2) \text{ and}$$
$$y = g(x) = \frac{1}{x}$$

intersect at $(1.166, 0.858)$ according to the graphing utility.

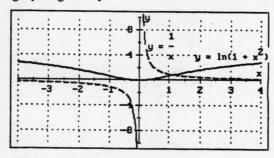

42.

n	$(\sqrt{n})^{\sqrt{n+1}}$	$(\sqrt{n+1})^{\sqrt{n}}$
8	22.63	22.36
9	32.27	31.62
12	88.21	85.00
20	957.27	904.84
25	3,665	3,447
31	16,528	15,494
37	68,159	63,786
38	85,679	80,166
43	261,578	244,579
50	1,165,565	1,089,362
100	1.12×10^{10}	1.05×10^{10}
1,000	2.87×10^{47}	2.76×10^{47}

Thus $(n+1)^{\sqrt{n}} \leq n^{\sqrt{n+1}}$

Mathematically one could reason as follows:

$$\lim_{x\to\infty} \frac{(x+1)^{\sqrt{x}}}{x^{\sqrt{x+1}}}$$

$$\leq \lim_{x\to\infty} \left(\frac{x+1}{x}\right)^{\sqrt{x+1}}$$

$$= \lim_{x\to\infty} \exp\left[\ln\left(\frac{x+1}{x}\right)\right]^{\sqrt{x+1}}$$

$$= \lim_{x\to\infty} \exp\left[\frac{\ln\left(\frac{x+1}{x}\right)}{(x+1)^{-1/2}}\right]$$

$$= \exp\left[\lim_{x\to\infty} \frac{x(-1/x^2)}{(x+1)(-1/2)(x+1)^{-3/2}}\right]$$

$$= \exp\left[\lim_{x\to\infty} \frac{2\sqrt{x+1}}{x}\right]$$

$$= \exp\left[\lim_{x\to\infty} \frac{2}{2\sqrt{x+1}}\right] = e^0 = 1$$

Thus $(n+1)^{\sqrt{n}} \leq n^{\sqrt{n+1}}$

even when n is large.

Chapter 5

Integration

5.1 Antidifferentiation; the Indefinite Integral

1. $I = \int x^5 dx = \dfrac{x^6}{6} + C.$

3.
$$\begin{aligned} I &= \int \dfrac{1}{x^2} dx = \int x^{-2} dx \\ &= -x^{-1} + C = -\dfrac{1}{x} + C. \end{aligned}$$

5. $I = \int 5 dx = 5x + C.$

7.
$$\begin{aligned} I &= \int (3t^2 - \sqrt{5}t + 2) dt \\ &= 3\int t^2 dt - \sqrt{5} \int t^{1/2} dt + 2\int dt \\ &= 3\left(\dfrac{t^3}{3}\right) - \sqrt{5}\left(\dfrac{t^{3/2}}{3/2}\right) + 2t + C \\ &= t^3 - \dfrac{2\sqrt{5}t^{3/2}}{3} + 2t + C. \end{aligned}$$

9.
$$\begin{aligned} I &= \int (3\sqrt{y} - \dfrac{2}{y^3} + \dfrac{1}{y}) dy \\ &= 3\int y^{1/2} dy - 2\int y^{-3} dy + \int y^{-1} dy \\ &= 3\left(\dfrac{y^{3/2}}{3/2}\right) - 2\left(\dfrac{y^{-2}}{-2}\right) + \ln|y| + C \\ &= 2y^{3/2} + y^{-2} + \ln|y| + C \\ &= 2y^{3/2} + \dfrac{1}{y^2} + \ln|y| + C. \end{aligned}$$

11.
$$\begin{aligned} I &= \int (\dfrac{e^x}{2} + x\sqrt{x}) dx \\ &= \dfrac{1}{2} \int e^x dx + \int x^{3/2} dx \\ &= \dfrac{1}{2} e^x + \dfrac{x^{5/2}}{5/2} + C \\ &= \dfrac{e^x}{2} + \dfrac{2x^{5/2}}{5} + C. \end{aligned}$$

13.
$$\begin{aligned} I &= \int \left(\dfrac{1}{3u} - \dfrac{3}{2u^2} + e^2 + \dfrac{\sqrt{u}}{2}\right) du \\ &= \dfrac{1}{3} \int u^{-1} du - \dfrac{3}{2} \int u^{-2} du \\ &\quad + e^2 \int du + \dfrac{1}{2} \int u^{1/2} du \\ &= \dfrac{1}{3} \ln|u| - \dfrac{3}{2}\left(\dfrac{u^{-1}}{-1}\right) + e^2 u + \dfrac{1}{2}\left(\dfrac{u^{3/2}}{3/2}\right) + C \\ &= \dfrac{1}{3} \ln|u| + \dfrac{3}{2u} + e^2 u + \dfrac{u^{3/2}}{3} + C. \end{aligned}$$

15.
$$\begin{aligned} I &= \int \dfrac{x^2 + 2x + 1}{x^2} dx \\ &= \int \left(1 + \dfrac{2}{x} + \dfrac{1}{x^2}\right) dx \\ &= x + 2\ln|x| - \dfrac{1}{x} + C. \end{aligned}$$

17.
$$\begin{aligned} I &= \int (x^3 - 2x^2)\left(\dfrac{1}{x} - 5\right) dx \\ &= \int (x^2 - 2x - 5x^3 + 10x^2) dx \\ &= \dfrac{11x^3}{3} - x^2 - \dfrac{5x^4}{4} + C. \end{aligned}$$

19.
$$I = \int \sqrt{t}(t^2 - 1)dt$$
$$= \int t^{5/2}dt - \int t^{1/2}dt$$
$$= \frac{2t^{7/2}}{7} - \frac{2t^{3/2}}{3} + C.$$

21. The slope of the tangent is the derivative. Thus
$$f'(x) = 4x + 1$$
and so the function $f(x)$ is
$$f(x) = \int (4x + 1)dx = 2x^2 + x + C.$$
Since the graph contains $(1, 2)$, $2 = f(1) = 2 \times 1^2 + 1 + C$ or $C = -1$. Thus
$$f(x) = 2x^2 + x - 1.$$

23. The slope of the tangent is the derivative. Thus $f'(x) = x^3 - \frac{2}{x^2} + 2$
and
$$f(x) = \int (x^3 - \frac{2}{x^2} + 2)dx$$
$$= \int (x^3 - 2x^{-2} + 2)dx$$
$$= \frac{x^4}{4} + \frac{2}{x} + 2x + C.$$
Since the graph contains $(1, 3)$,
$$3 = f(1) = \frac{1}{4} + 2 + 2 + C \text{ or } C = -\frac{5}{4}.$$
Thus $f(x) = \frac{x^4}{4} + \frac{2}{x} + 2x - \frac{5}{4}$.

25. Let $P(t)$ denote the population of the town t months from now. Since
$$\frac{dP}{dt} = 4 + 5t^{2/3},$$
$P(t)$ is an antiderivative of $4 + 5t^{2/3}$. Thus
$$P(t) = \int (4 + 5t^{2/3})dt$$
$$= 4t + 5\left(\frac{3t^{5/3}}{5}\right) + C$$
$$= 4t + 3t^{5/3} + C.$$

Since the current population is 10,000, it follows that
$$10,000 = P(0) = 4(0) + 3(0)^{5/3} + C$$
or $C = 10,000$.
$$P(t) = 4t + 3t^{5/3} + 10,000$$
and
$$P(8) = 4(8) + 3(8^{5/3}) + 10,000$$
$$= 32 + 3(2^5) + 10,000$$
$$= 32 + 96 + 10,000$$
$$= 10,128 \text{ people.}$$

27. $$\frac{dy}{dt} = 3 + 2t + 6t^2.$$
The distance is
$$y(t) = 3t + t^2 + 2t^3 + S_0.$$
$$y(2) = 6 + 4 + 16 + S_0,$$
$$y(1) = 3 + 1 + 2 + S_0,$$
so the distance traveled during the second minute is
$$|y(2) - y(1)| = 26 - 6 = 20 \text{ meters.}$$

29. The rate is $N'(t)$ people per hour. In a time period dt the number of people is
$$\int N'(t)dt = N(t) + C$$
The number of people entering the fair from 11:00 a.m. to 1:00 p.m. is
$$N(4) + C - N(2) - C = N(4) - N(2).$$

31. The rate at which population changes is
$$r(t) = 0.6t^2 + 0.2t + 0.5$$
thousand people per year and the rate of pollution 5 units per thousand people.
In a time period dt the pollution increases by
$$5(0.6t^2 + 0.2t + 0.5)dt$$

5.1. ANTIDIFFERENTIATION; THE INDEFINITE INTEGRAL

and in t years it will be
$$t^3 + \frac{t^2}{2} + 2.5t + C \text{ units}$$
The change in two years will be
$$2^3 + \frac{2^2}{2} + 2.5(2) + C - C = 15 \text{ units}.$$

33. Let $C(q)$ denote the total cost of producing q units.
Since marginal cost is
$$\frac{dC}{dq} = 6q + 1$$
dollars per unit, $C(q)$ is an antiderivative of $6q + 1$.
$$C(q) = \int (6q+1)dq = 3q^2 + q + C_1.$$
Since the cost of producing the first unit is $130,
$$130 = C(1) = 3 + 1 + C_1 \text{ or } C_1 = 126$$
$$C(q) = 3q^2 + q + 126 \text{ and } C(10) = \$436$$

35. Let $P(q)$ denote the profit from the production and sale of q units. Since
$$\frac{dP}{dq} = 100 - 2q$$
dollars per unit, $P(q)$ is an antiderivative of $100 - 2q$.
$$P(q) = \int (100 - 2q)dq = 100q - q^2 + C.$$
Since the profit is $700 when 10 units are produced,
$$700 = P(10) = 100(10) - (10)^2 + C$$
or $C = -200$.
$$P(q) = 100q - q^2 - 200$$
which attains its maximum value when its derivative,
$$\frac{dP}{dq} = 100 - 2q = 0$$
or $q = 50$. Hence the maximum profit is
$$P(50) = 100(50) - (50)^2 - 200 = \$2,300$$

37. Let $v(r)$ denote the velocity of blood through the artery. Since
$$v'(r) = -ar$$
cm per sec, $v(r)$ is an antiderivative of $-ar$.
$$v(r) = \int (-ar)dr = -\frac{ar^2}{2} + C.$$
Since $v(R) = 0$,
$$0 = v(R) = -\frac{aR^2}{2} + C \text{ or } C = \frac{aR^2}{2}$$
$$v(R) = \frac{a}{2}(R^2 - r^2).$$

39. (a) Let $P(t)$ denote the population of the endangered species. Since the species is growing at
$$P'(t) = 0.51e^{-0.03t}$$
per year, $P(t)$ is an antiderivative of $0.51e^{-0.03t}$.
$$P(t) = \int (0.51e^{-0.03t}dt = -17e^{-0.03t} + C.$$
Since $P(0) = 500$, $500 = -17 + C$ or $C = 517$.
$P(10) = 517 - 17e^{-0.3} = 504.41$ so the species will be 504 strong 10 years from now.

(b) Writing Exercise — Answers will vary.

41.
$$f''(x) = -0.12(x - 10)$$
which vanishes when $x = 10$.

(a) $f'(10) = 7$ items per minute.

(b)
$$\begin{aligned} f(x) &= \int f'(x)dx \\ &= \int (-0.12)(x-10)dx \\ &= x + 0.6x^2 - 0.02x^3 + C \end{aligned}$$

$f(0) = 0 + C = 0$ so $C = 0$.
$f(x) = x + 0.6x^2 - 0.02x^3$.

(c) $f'(x) = 0$ when $x = 20.8$.
$f(20.8) \approx 100$.

43. The distance covered during the reaction time of 0.7 seconds is

$$\frac{60 \text{ miles}}{\text{hr}} \cdot \frac{5,280 \text{ ft}}{\text{mile}} \cdot \frac{\text{hr}}{3600 \text{ sec.}} \times 0.7 \text{ sec.} = 61.6$$

ft. The speed is

$$\frac{dD}{dt} = -28t + 88 = 0 \text{ ft/sec}$$

when the car comes to rest, so $t = \frac{22}{7}$ sec. The distance traveled in this time will be

$$D\left(\tfrac{22}{7}\right) = -14 \times \left(\tfrac{22}{7}\right)^2 + \frac{88 \times 22}{7} + 61.6 = 199.89$$

ft. The camel may or may not get hit.
If the camel sits on the road in the car's path, it will make contact with the car.
If the camel is standing and its legs are in the car's path, the car will hit it.
If, on the other hand, the car is positioned between the camel's front and rear legs, and if the hood of the car is more than 0.89 ft. (10.7 inches) in length, the camel will escape undamaged.

45. $H(x) = C$ because the tangent line is horizontal at every point. Two horizontal lines differ by a constant C but are parallel.
If $F'(x) = G'(x)$ the tangent lines to $F(x)$ and $G(x)$ are parallel at every point. One curve can be translated a distance C from the other.

47.
$$\begin{aligned}
\int b^x dx &= \int e^{(\ln b)x} dx \\
&= \frac{1}{\ln b} \int e^{(\ln b)x} \ln b\, dx \\
&= \frac{e^{\ln b^x}}{\ln b} + C \\
&= \frac{b^x}{\ln b} + C
\end{aligned}$$

49. (a) Let $v(t)$ be the velocity of the car. Then

$$\frac{dv}{dt} = -23$$

and

$$v = -23t + v_0.$$

Since the velocity was 67 ft/sec when the brakes were applied, $v_0 = 67$ and

$$v(t) = -23t + 67.$$

Since $v = \dfrac{dS}{dt}$ the distance traveled is

$$s(t) = \int (-23t + 67)dt = -\frac{23}{2}t^2 + 67t + s_0.$$

For convenience $s_0 = 0$. Then

$$s(t) = -\frac{23}{2}t^2 + 67t.$$

(b)

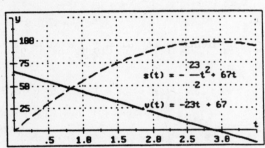

(c) The car will stop when $v(t_1) = 0$ or $t_1 = \dfrac{67}{23} = 2.91$ sec.
The distance traveled is
$s(2.91) = 97.5869$ ft.
According to the graphing utility (after zooming)
$s(0.775) = 45.0$ and $v(0.775) = 49.2$.

5.2 Integration by Substitution

1. Let $u = 2x + 6$. Then $du = 2dx$ or $dx = \dfrac{du}{2}$.

Hence $\int (2x+6)^5 dx = \dfrac{1}{2} \int u^5 du$

$= \dfrac{(2x+6)^6}{12} + C$

5.2. INTEGRATION BY SUBSTITUTION

3. Let $u = 4x - 1$. Then $du = 4dx$ or $dx = \dfrac{du}{4}$.

    Hence $\displaystyle\int \sqrt{4x-1}\,dx \;=\; \dfrac{1}{4}\int u^{1/2}du$
    $\qquad\qquad\qquad = \dfrac{1}{4}\dfrac{2u^{3/2}}{3} + C$
    $\qquad\qquad\qquad = \dfrac{(4x-1)^{3/2}}{6} + C$

5. Let $u = 1 - x$. Then $du = -dx$ or $dx = -du$.

    Hence $\displaystyle\int e^{1-x}dx = -\int e^u du = -e^{1-x} + C$

7. Let $u = x^2$. Then $du = 2x\,dx$ or $x\,dx = \dfrac{1}{2}du$.

    Hence $\displaystyle\int xe^{x^2}dx = \dfrac{1}{2}\int e^u du = \dfrac{1}{2}e^{x^2} + C$

9. Let $u = t^2 + 1$. Then $du = 2t\,dt$ or $t\,dt = \dfrac{1}{2}du$.

    Hence $\displaystyle\int t(t^2+1)^5 dt \;=\; \dfrac{1}{2}\int u^5 du$
    $\qquad\qquad\qquad = \dfrac{(t^2+1)^6}{12} + C.$

11. Let $u = x^3 + 1$. Then $du = 3x^2 dx$ or $x^2 dx = \dfrac{1}{3}du$. Hence,

    $\displaystyle\int x^2(x^3+1)^{3/4}dx \;=\; \dfrac{1}{3}\int u^{3/4} du$
    $\qquad\qquad\qquad = \dfrac{1}{3}\dfrac{4u^{7/4}}{7} + C$
    $\qquad\qquad\qquad = \dfrac{4(x^3+1)^{7/4}}{21} + C.$

13. Let $u = y^5 + 1$. Then $du = 5y^4 dy$ or $y^4 dy = \dfrac{1}{5}du$. Hence,

    $\displaystyle\int \dfrac{2y^4}{y^5+1}dy \;=\; \dfrac{2}{5}\int \dfrac{du}{u}$
    $\qquad\qquad\qquad = \dfrac{2}{5}\ln|u| + C$
    $\qquad\qquad\qquad = \dfrac{2}{5}\ln|y^5+1| + C.$

15. Let $u = x^2 + 2x + 5$. Then $du = (2x+2)dx$ or $(x+1)dx = \dfrac{1}{2}du$. Hence,

    $\displaystyle\int (x+1)(x^2+2x+5)^{12}dx$
    $\qquad = \dfrac{1}{2}\int u^{12}du = \dfrac{u^{13}}{26} + C$
    $\qquad = \dfrac{(x^2+2x+5)^{13}}{26} + C.$

17. Let $u = x^5 + 5x^4 + 10x + 12$. Then $du = (5x^4 + 20x^3 + 10)dx$ or $(x^4 + 4x^3 + 2)dx = \dfrac{1}{5}du$. Hence,

    $\displaystyle\int \dfrac{3x^4 + 12x^3 + 6}{x^5 + 5x^4 + 10x + 12}dx$
    $\qquad = 3\displaystyle\int \dfrac{x^4 + 4x^3 + 2}{x^5 + 5x^4 + 10x + 12}dx$
    $\qquad = \dfrac{3}{5}\displaystyle\int \dfrac{1}{u}du = \dfrac{3}{5}\ln|u| + C$
    $\qquad = \dfrac{3}{5}\ln|x^5 + 5x^4 + 10x + 12| + C.$

19. Let $t = u^2 - 2u + 6$. Then $dt = (2u-2)du$ or $(3u-3)du = \dfrac{3}{2}dt$. Hence,

    $\displaystyle\int \dfrac{3u-3}{(u^2-2u+6)^2}du \;=\; \dfrac{3}{2}\int \dfrac{1}{t^2}dt$
    $\qquad\qquad = \dfrac{3}{2}\int t^{-2}dt$
    $\qquad\qquad = \left(\dfrac{3}{2}\right)\left(-\dfrac{1}{t}\right) + C$
    $\qquad\qquad = -\dfrac{3}{2(u^2-2u+6)} + C.$

21. Let $u = \ln 5x$. Then $du = \dfrac{5}{5x}dx = \dfrac{1}{x}dx$.

    Hence $\displaystyle\int \dfrac{\ln 5x}{x}dx \;=\; \int u\,du$
    $\qquad\qquad\qquad = \dfrac{(\ln 5x)^2}{2} + C.$

23. Let $u = \ln x$. Then $du = \frac{1}{x}dx$. Hence,

$$\int \frac{1}{x(\ln x)^2}dx = \int \frac{1}{u^2}du$$
$$= -\frac{1}{u} + C = -\frac{1}{\ln x} + C.$$

25. Let $u = \ln(x^2 + 1)$. Then $du = \frac{2x}{x^2+1}dx$.

Hence $\int \frac{2x\ln(x^2+1)}{x^2+1}dx$
$$= \int u\, du = \frac{u^2}{2} + C$$
$$= \frac{[\ln(x^2+1)]^2}{2} + C.$$

27. $\quad \frac{dy}{dx} = x\sqrt{x^2+5}$

and $(2, 10)$ is on the curve.

$$y = \int x(x^2+5)^{1/2}dx$$
$$= \frac{1}{2}\int (x^2+5)^{1/2}d(x^2+5)$$
$$= \frac{1}{2}\frac{(x^2+5)^{3/2}}{3/2} + C = \frac{(x^2+5)^{3/2}}{3} + C.$$

From the given point we obtain
$10 = \frac{1}{3}(4+5)^{3/2} + C = 9 + C$ or $C = 1$.

$$y = \frac{1}{3}(x^2+5)^{3/2} + 1.$$

29. (a)
$$x'(t) = -2(3t+1)^{1/2}$$
$$x(t) = -\frac{2}{3}\int (3t+1)^{1/2}(3dt)$$
$$= -\frac{4}{9}\sqrt{(3t+1)^3} + C$$

Since $x(0) = 4$, $C = \frac{40}{9}$ and

$$x(t) = -\frac{4}{9}\sqrt{(3t+1)^3} + \frac{40}{9}$$

(b) $\quad x(4) = -\frac{4}{9}\sqrt{13^3} + \frac{40}{9} \approx -16.4$

(c)
$$3 = -\frac{4}{9}\sqrt{(3t+1)^3} + \frac{40}{9}$$
Thus $t \approx 0.4$

31. (a)
$$x'(t) = \frac{t}{(t+1)^2} = \frac{1}{t+1} - \frac{1}{(t+1)^2}$$
$$x(t) = \ln(t+1) + \frac{1}{t+1} + C$$

Since $x(0) = 0$, $C = -1$ and

$$x(t) = -\ln(t+1) + \frac{1}{t+1} - 1$$

(b) $\quad x(4) = -\ln 5 - \frac{5}{6} \approx -2.4$

(c) $\quad 3 = \ln(t+1) + \frac{1}{t+1} - 1$

Thus $t \approx 53$.

33. The rate of growth is

$$G' = 1 + \frac{1}{(x+1)^2}$$

and $(2, 5)$ satisfies the function.

$$G = \int [1 + (x+1)^{-2}]dx$$
$$= x + \int (x+1)^{-2}d(x+1)$$
$$= x + \frac{(x+1)^{-1}}{-1} + C$$
$$= x - \frac{1}{x+1} + C.$$

From the given point we obtain $5 = 2 - \frac{1}{3} + C$
or $C = \frac{10}{3}$.

$$y = 0 - \frac{1}{0+1} + \frac{10}{3} = \frac{7}{3} \text{ meters.}$$

35. The rate of growth is $P'(t) = e^{0.02t}$ and $(0, 50)$ satisfies the function.

$$P(t) = \int e^{0.02t}dt$$
$$= \frac{1}{0.02}\int e^{0.02t}d(0.02t)$$
$$= \frac{e^{0.02t}}{0.02} + C.$$

5.2. INTEGRATION BY SUBSTITUTION

From the given point we obtain $50 = 50 + C$ or $C = 0$.

$P(10) = 50e^{0.2} = 61.07$ million people.

37. **(a)** Let $L(t)$ denote the ozone level t hours after 7:00 a.m. Since

$$\frac{dL}{dt} = \frac{0.24 - 0.03t}{\sqrt{36 + 16t - t^2}}$$
$$= (0.24 - 0.03t)(36 + 16t - t^2)^{-1/2}$$

parts per million per hour, then by substituting $u = 36 + 16t - t^2$,

$$L(t) = \int (0.24 - 0.03t)(36 + 16t - t^2)^{-1/2} dt$$
$$= 0.03 \int (8 - t)(36 + 16t - t^2)^{-1/2} dt$$
$$= 0.03(36 + 16t - t^2)^{1/2} + C.$$

Since the ozone level was 0.25 at 7:00 a.m.,

$$0.25 = L(0) = 0.03\sqrt{36} + C = 0.18 + C$$

or $C = 0.07$. Hence,

$$L(t) = 0.03(36 + 16t - t^2)^{1/2} + 0.07.$$

The peak ozone level occurs when $\frac{dL}{dt} = 0$, that is, when $0.24 - 0.03t = 0$ or $t = 8$.
Note that $L(t)$ has its absolute maximum at this critical point since $\frac{dL}{dt} > 0$ (L is increasing) for $0 < t < 8$, and $\frac{dL}{dt} < 0$ (L is decreasing) for $8 < t$. Thus the peak ozone level is

$$L(8) = 0.03(36 + 16(8) - 8^2)^{1/2} + 0.07$$
$$= 0.03(10) + 0.07 = 0.37$$

parts per million, which occurs at 3:00 p.m. (8 hours after 7:00 a.m.).

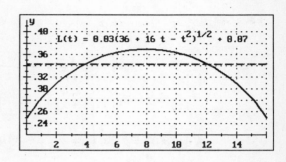

(b) According to the graphing utility $L(4) = 0.345$ and $L(12) = 0.345$, so the ozone level is the same at 11:00 a.m. as it is at 7:00 p.m.

39. **(a)** Let $V(t)$ denote the value of the machine after t years. Since

$$\frac{dV}{dt} = -960e^{-t/5}$$

dollars per year,

$$V(t) = \int (-960e^{-t/5}) dt$$
$$= 4,800e^{-t/5} + C.$$
If $V_0 = V(0) = 4,800e^0 + C,$
$C = V_0 - 4,800.$

Thus,

$$V(t) = 4,800e^{-t/5} + V_0 - 4,800.$$

(b) If $V_0 = \$5,200$, then

$$V(10) = 4,800e^{-2} + 5,200 - 4,800$$
$$\approx \$1,049.61.$$

41. $\frac{dP}{dx} = 3\sqrt{x + 1}.$

$P_0 = 300$ cents.

$$P(x) = \frac{3(x + 1)^{3/2}}{3/2} + P_0.$$

$P(8) = 2(8 + 1)^{3/2} + 300$ or $\$3.54$.

43. (a)

$$p'(x) = \frac{30x}{(3+x)^2}$$
$$= 30\left(\frac{3+x}{(3+x)^2} - \frac{3}{(3+x)^2}\right)$$
$$p(x) = 30\left(\ln|3+x| + \frac{x}{3+x}\right) + C$$

With $p(0) = 2.25$, $C = 27.75 - 30\ln 3$ and

$$p(x) = 30\left(\ln|3+x| + \frac{3}{3+x}\right) - 27.75 - 30\ln 3$$

(b)

$$p(4) = 30\left(\ln 7 + \frac{3}{7}\right) - 27.75 - 30\ln 3)$$
$$\approx \$10.50 \text{ per Weenie}$$

(c) $p(0.785) \approx 3$ Weenies will be supplied.

45. $R'(x) = \frac{11-x}{\sqrt{14-x}}$, $C'(x) = 2 + x + x^2$

Let $14 - x = u$, $du = -dx$. Then

$$R(x) = -\int \frac{u-3}{\sqrt{u}} du$$
$$= -\int (u^{1/2} - 3u^{-1/2})du$$
$$= -\left(\frac{2}{3}u^{3/2} - 6u^{1/2}\right)$$
$$= 6\sqrt{14-x} - \frac{2}{3}(14-x)^{3/2} + C_1$$

We also have

$$C(x) = \int (2 + x + x^2)dx = 2x + \frac{x^2}{2} + \frac{x^3}{3} + C_2$$

$$P(9) - P(5)$$
$$= R(9) - C(9) - [R(5) - C(5)]$$
$$= -295.54 - (-64.167) = -231.373$$

5.3 Introduction to Differential Equations

1. $\frac{dy}{dx} = 3x^2 + 5x - 6$,

then $y = \int (3x^2 + 5x - 6)dx$
$$= x^3 + \frac{5x^2}{2} - 6x + C.$$

3. Separate the variables of

$$\frac{dy}{dx} = 3y$$

and integrate to get

$$\int \frac{1}{y}dy = \int 3dx,$$
$$\ln|y| = 3x + C_1,$$
$$|y| = e^{3x+C_1} = e^{C_1}e^{3x} \text{ or } y = Ce^{3x}$$

where C is the constant $\pm e^{C_1}$.

5. Separate the variables of

$$\frac{dy}{dx} = e^y$$

and integrate to get

$$\int e^{-y}dy = \int dx,$$
$$-e^{-y} = x + C_1 \text{ or } e^{-y} = C - x$$

where C is the constant $-C_1$. Hence,

$$\ln e^{-y} = \ln(C - x),$$
$$-y = \ln(C - x), \text{ or } y = -\ln(C - x)$$

7. Separate the variables of

$$\frac{dy}{dx} = \frac{x}{y}$$

and integrate to get

$$\int y\,dy = \int x\,dx,$$
$$\frac{y^2}{2} = \frac{x^2}{2} + C_1 \text{ or } y^2 = x^2 + C$$

$y = \pm\sqrt{x^2 + C}$, where C is the constant $2C_1$.

5.3. INTRODUCTION TO DIFFERENTIAL EQUATIONS

9.
$$\frac{dy}{dx} = \sqrt{xy} \text{ or}$$
$$y^{-1/2}dy = x^{1/2}dx$$
$$2y^{1/2} = \frac{2}{3}x^{3/2} + C$$

11.
$$\frac{dy}{dx} = \frac{y}{x-1}$$
$$\frac{dy}{y} = \frac{dx}{x-1}$$
$$\ln|y| = \ln|x-1| + C_1$$
$$\ln\frac{|y|}{|x-1|} = C_1, \quad \frac{y}{x-1} = e^{C_1} = C$$

where C is any real number.

13.
$$\frac{dy}{dx} = \frac{y+3}{(2x-5)^6}$$
$$\frac{dy}{y+3} = (2x-5)^{-6}dx$$
$$\ln|y+3| = -\frac{1}{10}(2x-5)^{-5} + C_1$$
$$y = -3 + C\exp\left[-\frac{1}{10(2x-5)^5}\right]$$

15.
$$\frac{dx}{dt} = \frac{xt}{2t+1}$$
$$\frac{dx}{x} = \frac{1}{2}\left[1 - \frac{1}{2t+1}\right]dt$$
$$\ln|x| = \frac{t}{2} - \frac{1}{4}\ln|2t+1| + C_1$$
$$\ln|x| + \ln\sqrt[4]{2t+1} = \frac{t}{2} + C_1$$
$$x\sqrt[4]{2t+1} = e^{C_1}e^{t/2} = Ce^{t/2}$$
$$x = \frac{Ce^{t/2}}{\sqrt[4]{2t+1}}$$

Note: $x = \dfrac{Ce^{(2t+1)/4}}{\sqrt[4]{2t+1}}$

is correct also, since
$Ce^{(2t+1)/4} = Ce^{t/2}e^{1/4} = C_2e^{t/2}$.

17. If $\dfrac{dy}{dx} = e^{5x}$,

then $y = \displaystyle\int e^{5x}dx = \dfrac{e^{5x}}{5} + C$.

Since $y = 1$ when $x = 0$,
$$1 = \frac{e^0}{5} + C$$
or $C = \dfrac{4}{5}$. Hence,
$$y = \frac{e^{5x}}{5} + \frac{4}{5}.$$

19.
$$\frac{dy}{dx} = \frac{x}{y^2},$$
$$\int y^2 dy = \int x dx,$$
$$\frac{y^3}{3} = \frac{x^2}{2} + \frac{C}{6},$$

(remember the form of the constant of integration is arbitrary)

$$2y^3 = 3x^2 + C.$$

Since $y = 3$ when $x = 2$, $54 = 12 + C$ or $C = 42$.
Thus $2y^3 = 3x^2 + 42$ or
$$y = \left(\frac{3x^2 + 42}{2}\right)^{1/3}.$$

21.
$$\frac{dy}{dx} = y^2\sqrt{4-x}$$
$$y^{-2}dy = (4-x)^{1/2}dx$$
$$-\frac{1}{y} = -\frac{2}{3}(4-x)^{3/2} - \frac{C}{3}$$

Note: $-\dfrac{C}{3}$ is just for convenience, but make sure you introduce a constant of integration immediately after your last integration.

$$y = \frac{3}{C + 2(4-x)^{3/2}}$$

Since $y = 2$ when $x = 4$, $C = \dfrac{3}{2}$ and
$$y = \frac{6}{3 + 4(4-x)^{3/2}}$$

23.
$$\frac{dy}{dt} = \frac{y+1}{t(y-1)}$$
$$\frac{y-1}{y+1} dy = \frac{dt}{t}$$
$$\left(1 - \frac{2}{y+1}\right) dy = \frac{dt}{t}$$
$$y - 2\ln|y+1| = \ln|t| + C$$
$$2 - 2\ln 3 = C$$
$$y - 2\ln|y+1| - \ln|t| = 2 - 2\ln 3$$
$$y - 2 = \ln(y+1)^2 + \ln t - \ln 9$$
$$y = 2 + \ln\frac{t(y+1)^2}{9}$$

25. $Q = B - Ce^{-kt}$, so $Ce^{-t} = B - Q$.
$$\begin{aligned}\frac{dQ(t)}{dt} &= -Ce^{-kt}(-k) \\ &= kCe^{-kt} = k(B-Q).\end{aligned}$$

27. Let Q denote the number of bacteria. Then, $\frac{dQ}{dt}$ is the rate of change of Q, and since this rate of change is proportional to Q, it follows that
$$\frac{dQ}{dt} = kQ$$
where k is a positive constant of proportionality.

29. Let Q denote the investment. Then $\frac{dQ}{dt}$ is the rate of change of Q, and since this rate of change is equal to 7 % of the size of Q, it follows that
$$\frac{dQ}{dt} = 0.07Q$$

31. Let P denote the population. Then $\frac{dP}{dt}$ is the rate of change of P, and since this rate of change is the constant 500, it follows that
$$\frac{dP}{dt} = 500$$

33. Let N be the total population and $Q(t)$ the number of people who have caught the disease. For positive proportionality constant k,
$$\frac{dQ(t)}{dt} = kQ(t)[N - Q(t)]$$

35. (a) Let $Q(t)$ be the amount of fluoride in the reservoir at time t. Since there are 200 million gallons of solution in the reservoir, the concentration of fluoride is $\frac{Q(t)}{200}$. 4 million gallons of solution flow out of the tank per day. Thus
$$\left(\frac{Q(t) \text{ pounds}}{200 \text{ million gallons}}\right) \left(\frac{4 \text{ million gallons}}{\text{day}}\right)$$
flows out of the reservoir per day, or
$$\frac{dQ}{dt} = -\frac{Q}{50}$$

(b) Solving the differential equation leads to
$$\begin{aligned}\frac{dQ}{Q} &= -\frac{dt}{50} \\ \ln Q &= -\frac{t}{50} + C_1 \\ Q &= Ce^{-t/50}\end{aligned}$$
Since $Q(0) = 1,600$, the equation becomes
$$Q(t) = 1,600 e^{-t/50}$$

37. $D(p) = a - bp$ and $S(p) = r + sp$
$$\begin{aligned}\frac{dp}{dt} &= k(a - bp - r - sp) \\ &= k[a - r - (b+s)p]\end{aligned}$$
$$\int \frac{dp}{a - r - (b+s)p} = \int k\, dt$$
$$-\frac{1}{b+s} \ln|a - r - (b+s)p| = kt + C_1$$
$$a - r - (b+s)p = Ce^{-(b+s)kt}$$
$$p = \frac{a - r + Ce^{-(b+s)kt}}{b+s}$$
$$\lim_{t \to \infty} P(t) = \frac{a-r}{b+s}$$

5.4 Integration by Parts

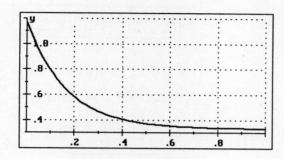

1.
$$I = \int xe^{-x}dx$$

$f(x) = x$	$g(x) = e^{-x}$
$f'(x) = 1$	$G(x) = -e^{-x}$

$$\begin{aligned}I &= -xe^{-x} + \int e^{-x}dx \\ &= -(x+1)e^{-x} + C.\end{aligned}$$

3.
$$I = \int (1-x)e^x dx.$$

$f(x) = 1-x$	$g(x) = e^x$
$f'(x) = -1$	$G(x) = e^x$

$$\begin{aligned}I &= (1-x)e^x - \int(-e^x)dx \\ &= (1-x)e^x + \int e^x dx \\ &= (2-x)e^x + C.\end{aligned}$$

5.
$$I = \int t\ln 2t\, dt.$$

$f(t) = \ln 2t$	$g(t) = t$
$f'(t) = \dfrac{1}{t}$	$G(t) = \dfrac{t^2}{2}$

$$\begin{aligned}I &= \frac{t^2}{2}\ln 2t - \int \frac{1}{t}\frac{t^2}{2}dt \\ &= \frac{t^2}{2}\ln 2t - \frac{1}{2}\int t\, dt \\ &= \frac{t^2}{2}\left(\ln 2t - \frac{1}{2}\right) + C.\end{aligned}$$

7.
$$I = \int ve^{-v/5}dv$$

$f(v) = v$	$g(v) = e^{-v/5}$
$f'(v) = 1$	$G(v) = -5e^{-v/5}$

$$\begin{aligned}I &= -5ve^{-v/5} + 5\int e^{-v/5}dv \\ &= -5(v+5)e^{-v/5} + C.\end{aligned}$$

39. Writing Exercise — Answers will vary.

41. (a)
$$\frac{dP}{dt} = P(k-mP)$$
$$\frac{dP}{P(k-mP)} = dt = \frac{A}{P} + \frac{B}{k-mP}$$

Multiplying by the Least Common Denominator leads to
$1 = A(k-mP) + BP$.
With $P = 0$, $A = \dfrac{1}{k}$, and
with $P = \dfrac{k}{m}$, $B = \dfrac{m}{k}$

(b)
$$\begin{aligned}t &= \frac{1}{k}\int \frac{dp}{P} + \frac{1}{k}\int \frac{m\, dP}{k-mP} \\ &= \frac{1}{k}(\ln|P| - \ln|k-mP|) + C_1\end{aligned}$$

(c)
$$\begin{aligned}\ln\frac{P}{k-mP} &= kt - C_2 \\ \frac{P}{k-mP} &= C_3 e^{kt} \\ (mC_3 e^{kt} + 1)P &= kC_3 e^{kt} \\ P &= \frac{\dfrac{k}{m}}{1 + \dfrac{1}{mC_3}e^{-kt}} \\ &= \frac{C}{1 + De^{-kt}}\end{aligned}$$

where $C = \dfrac{k}{m}$ and $D = \dfrac{1}{mC_3}$, with C_3 an arbitrary constant.

9. $$I = \int x\sqrt{x-6}\,dx = \int x(x-6)^{1/2}dx.$$

$f(x) = x$	$g(x) = (x-6)^{1/2}$
$f'(x) = 1$	$G(x) = \frac{2}{3}(x-6)^{3/2}$

$$\begin{aligned} I &= \frac{2}{3}x(x-6)^{3/2} - \frac{2}{3}\int (x-6)^{3/2}dx \\ &= \frac{2}{3}x(x-6)^{3/2} - \frac{4}{15}(x-6)^{5/2} + C \\ &= \frac{2}{5}(x-6)^{3/2}(x+4) + C \end{aligned}$$

11. $$I = \int x(x+1)^8 dx.$$

$f(x) = x$	$g(x) = (x+1)^8$
$f'(x) = 1$	$G(x) = \dfrac{(x+1)^9}{9}$

$$\begin{aligned} I &= \frac{1}{9}x(x+1)^9 - \int \frac{1}{9}(x+1)^9 dx \\ &= \frac{1}{9}x(x+1)^9 - \frac{(x+1)^{10}}{90} + C. \end{aligned}$$

13. $$I = \int \frac{x}{\sqrt{x+2}}dx.$$

$f(x) = x$	$g(x) = \dfrac{1}{\sqrt{x+2}} = (x+2)^{-1/2}$
$f'(x) = 1$	$G(x) = 2(x+2)^{1/2}$

$$\begin{aligned} I &= 2x(x+2)^{1/2} - \int 2(x+2)^{1/2}dx \\ &= 2x(x+2)^{1/2} - 2\left(\frac{2}{3}\right)(x+2)^{3/2} + C. \end{aligned}$$

15. $$I = \int x^2 e^{-x} dx.$$

$f_1(x) = x^2$	$g_1(x) = e^{-x}$
$f_1'(x) = 2x$	$G_1(x) = -e^{-x}$

$$\begin{aligned} I &= -x^2 e^{-x} - \int 2x(-e^{-x})dx \\ &= -x^2 e^{-x} + 2\int xe^{-x}dx. \end{aligned}$$

$f_2(x) = x$	$g_2(x) = e^{-x}$
$f_2'(x) = 1$	$G_2(x) = -e^{-x}$

$$\begin{aligned} I &= -x^2 e^{-x} + 2(-xe^{-x} + \int e^{-x}dx) \\ &= -e^{-x}(x^2 + 2x + 2) + C \end{aligned}$$

17. $$I = \int x^3 e^x dx.$$

$f(x) = x^3$	$g(x) = e^x$
$f'(x) = 3x^2$	$G(x) = e^x$

$$\begin{aligned} I &= x^3 e^x - 3\int x^2 e^x dx \\ &= x^3 e^x - 3(x^2 e^x - 2\int xe^x dx) \\ &= x^3 e^x - 3x^2 e^x + 6(xe^x - \int e^x dx) \\ &= (x^3 - 3x^2 + 6x - 6)e^x + C \end{aligned}$$

19. $$I = \int x^2 \ln x\, dx.$$

$f(x) = \ln x$	$g(x) = x^2$
$f'(x) = \dfrac{1}{x}$	$G(x) = \dfrac{x^3}{3}$

$$\begin{aligned} I &= \frac{x^3}{3}\ln x - \frac{1}{3}\int x^2 dx \\ &= \frac{x^3}{3}\ln x - \frac{x^3}{9} + C \end{aligned}$$

21. $$I = \int \frac{\ln x}{x^2}dx = \int \frac{1}{x^2}\ln x\, dx$$

$f(x) = \ln x$	$g(x) = \dfrac{1}{x^2}$
$f'(x) = \dfrac{1}{x}$	$G(x) = -\dfrac{1}{x}$

$$\begin{aligned} I &= -\frac{1}{x}\ln x + \int \frac{1}{x^2}dx \\ &= -\frac{1}{x}(\ln x + 1) + C. \end{aligned}$$

23. $$I = \int x^3 e^{x^2} dx = \int x^2 (xe^{x^2})dx.$$

$f(x) = x^2$	$g(x) = xe^{x^2}$
$f'(x) = 2x$	$G(x) = \dfrac{1}{2}e^{x^2}$

5.4. INTEGRATION BY PARTS

$$I = \frac{1}{2}x^2 e^{x^2} - \int 2x\left(\frac{1}{2}e^{x^2}\right)dx$$

$$= \frac{x^2}{2}e^{x^2} - \frac{1}{2}e^{x^2} + C.$$

25. $$I = \int x^7(x^4+5)^8 dx$$

$f(x) = x^4$	$g(x) = x^3(x^4+5)^8$
$f'(x) = 4x^3$	$G(x) = \frac{1}{36}(x^4+5)^9$

$$I = \frac{1}{36}x^4(x^4+5)^9 - \frac{1}{9}\int x^3(x^4+5)^9 dx$$

$$= \frac{x^4}{36}(x^4+9)^9 - \frac{1}{360}(x^4+5)^{10} + C$$

$$= \frac{1}{360}(x^4+5)^9(9x^4-5) + C$$

27. Let $f(x)$ be the function whose tangent has slope $x \ln \sqrt{x}$.
Then $f'(x) = x \ln \sqrt{x}$ for $x > 0$ and

$$f(x) = \int x \ln \sqrt{x}\, dx = \frac{1}{2}\int x \ln x\, dx$$

To integrate by parts,

$f(x) = \ln x$	$g(x) = x$
$f'(x) = \frac{1}{x}$	$G(x) = \frac{x^2}{2}$

$$f(x) = \frac{1}{2}\int x \ln x\, dx$$

$$= \frac{1}{2}\left[\left(\frac{x^2}{2}\right)\ln x - \int \left(\frac{x^2}{2}\right)\left(\frac{1}{x}\right)dx\right]$$

$$= \frac{1}{2}\left[\frac{x^2 \ln x}{2} - \frac{1}{2}\int x\, dx\right]$$

$$= \frac{x^2 \ln x}{4} - \frac{x^2}{8} + C$$

Since $(2, f(2)) = (2, -3)$, that is, when $x = 2$, $f = -3$,

$$-3 = \frac{2^2 \ln 2}{4} - \frac{2^2}{8} + C$$

$-3 = \ln 2 - \frac{1}{2} + C$, or $C = -\frac{5}{2} - \ln 2$. Thus

$$f(x) = \frac{x^2 \ln x}{4} - \frac{x^2}{8} - \frac{5}{2} - \ln 2$$

29. Let t denote time and $Q(t)$ the number of units produced. Then,

$$\frac{dQ}{dt} = 100t e^{-0.5t} \text{ and}$$

$$Q(t) = 100 \int t e^{-0.5t} dt$$

$f(t) = t$	$g(t) = e^{-0.5t}$
$f'(t) = 1$	$G(t) = -\frac{e^{-0.5t}}{0.5} = -2e^{-0.5t}$

Thus

$$Q(t) = 100[-2te^{-0.5t} - \int (-2)e^{-0.5t} dt]$$

$$= -200te^{-0.5t} - 400e^{-0.5t} + C$$

$$= -200(t+2)e^{-0.5t} + C$$

Since no units are produced when $t = 0$, $Q(0) = 0 = -200(2) + C$ or $C = 400$. Hence,

$$Q(t) = -200(t+2)e^{-0.5t} + 400$$

and the number of units produced during the first three hours is $Q(3) = 176.87$.

31. Let q denote the number of units produced and $C(q)$ the cost of producing the first q units. Then, $\frac{dC}{dq} = (0.1q+1)e^{0.03q}$ and

$$C(q) = \int (0.1q+1)e^{0.03q} dq.$$

$f(q) = 0.1q+1$	$g(q) = e^{0.03q}$
$f'(q) = 0.1$	$G(q) = \frac{1}{0.03}e^{0.03q}$

Thus

$$C(q) = \frac{1}{0.03}[(0.1q+1)e^{0.03q} - \frac{10}{3}\int e^{0.03q} dq]$$

$$= \frac{1}{0.03}(0.1q+1)e^{0.03q} - \frac{10}{0.09}e^{0.03q} + C_1$$

When $q = 10$, $C = 200$ and

$$200 = \frac{1}{0.03}(2)e^{.3} - \frac{10}{0.03}e^{.3} + C_1$$

or $C_1 = 177.4$. Thus

$$C(q) = \frac{1}{0.03}(0.1q+1)e^{0.03q} - \frac{10}{0.09}e^{0.03q} + 177.4$$

and

$$C(20) = \frac{1}{0.03}(3)e^{0.6} - \frac{10}{0.09}e^{0.6} + 200 - \frac{2300e^3}{9}$$
$$< 0$$

33. (a) $$I = \int x^n e^{ax} dx$$

$f(x) = x^n$	$g(x) = e^{ax}$
$f'(x) = nx^{n-1}$	$G(q) = \dfrac{1}{a}e^{ax}$

$$I = \frac{x^n}{a}e^{ax} - \frac{n}{a}\int x^{n-1}e^{ax}dx$$

(b)
$$\int x^3 e^{5x} dx$$
$$= \frac{1}{5}x^3 e^{5x} - \frac{3}{5}\int x^2 e^{5x} dx$$
$$= \frac{1}{5}x^3 e^{5x} - \frac{3}{5}\left[\frac{1}{5}x^2 e^{5x} - \frac{2}{5}\int x e^{5x} dx\right]$$
$$= \frac{1}{5}x^3 e^{5x} - \frac{3}{25}x^2 e^{5x}$$
$$\quad + \frac{6}{25}\left[\frac{1}{5}x e^{5x} - \frac{1}{5}\int e^{5x} dx\right]$$
$$= \frac{e^{5x}}{625}(125x^3 - 75x^2 + 30x - 6) + C$$

Review Problems

1. $$\int \left(x^5 - 3x^2 + \frac{1}{x^2}\right) dx$$
$$= \int (x^5 - 3x^2 + x^{-2}) dx$$
$$= \frac{x^6}{6} - x^3 - \frac{1}{x} + C.$$

2. $$\int \left(x^{2/3} - \frac{1}{x} + 5 + \sqrt{x}\right) dx$$
$$= \frac{3x^{5/3}}{5} - \ln|x| + 5x + \frac{2x^{3/2}}{3} + C.$$

3. Let $u = 3x + 1$. Then $du = 3dx$ or $dx = \frac{1}{3}du$.
$$\int \sqrt{3x+1}\, dx = \frac{1}{3}\int u^{1/2} du$$
$$= \frac{1}{3}\frac{2u^{3/2}}{3} + C$$
$$= \frac{2(3x+1)^{3/2}}{9} + C.$$

4. Let $u = 3x^2 + 2x + 5$. Then $du = (6x+2)dx$ or $(3x+1)dx = \frac{1}{2}du$. Hence
$$I = \int (3x+1)\sqrt{3x^2 + 2x + 5}\, dx$$
$$= \frac{1}{2}\int u^{1/2} du = \frac{1}{2}\frac{2u^{3/2}}{3} + C$$
$$= \frac{(3x^2 + 2x + 5)^{3/2}}{3} + C.$$

5. Let $u = x^2 + 4x + 2$. Then $du = (2x+4)dx$ or $(x+2)dx = \frac{1}{2}du$. Hence
$$I = \int (x+2)(x^2+4x+2)^5 dx$$
$$= \frac{1}{2}\int u^5 du = \frac{(x^2+4x+2)^6}{12} + C.$$

6. Let $u = x^2 + 4x + 2$. Then $du = (2x+4)dx$ or $(x+2)dx = \frac{1}{2}du$. Hence
$$I = \int \frac{x+2}{x^2+4x+2} dx$$
$$= \frac{1}{2}\int \frac{1}{u} du = \frac{1}{2}\ln|u| + C$$
$$= \frac{1}{2}\ln|x^2+4x+2| + C.$$

7. Let $u = 2x^2 + 8x + 3$. Then $du = (4x+8)dx$ or $(3x+6)dx = \frac{3}{4}du$. Hence
$$I = \int \frac{3x+6}{(2x^2+8x+3)^2} dx$$
$$= \frac{3}{4}\int \frac{1}{u^2} du = \frac{3}{4}\int u^{-2} du$$
$$= -\frac{3}{4}u^{-1} + C$$
$$= -\frac{3}{4(2x^2+8x+3)} + C.$$

REVIEW PROBLEMS

8. Let $u = x - 5$. Then $du = dx$. Hence

$$\begin{aligned} I &= \int (x-5)^{12} dx \\ &= \int u^{12} du = \frac{(x-5)^{13}}{13} + C. \end{aligned}$$

9. Method 1:
Let $u = x - 5$. Then $du = dx$ and $x = u + 5$.

$$\begin{aligned} \text{Hence } I &= \int x(x-5)^{12} dx \\ &= \int (u^{13} + 5u^{12}) du \\ &= \frac{(x-5)^{14}}{14} + \frac{5(x-5)^{13}}{13} + C. \end{aligned}$$

Method 2:

$f(x) = x$	$g(x) = (x-5)^{12}$
$f'(x) = 1$	$G(x) = \dfrac{(x-5)^{13}}{13}$

$$\begin{aligned} \text{So } I &= \int x(x-5)^{12} dx \\ &= \frac{x(x-5)^{13}}{13} - \int \frac{(x-5)^{13}}{13} dx \\ &= \frac{x(x-5)^{13}}{13} - \frac{1}{13}\left[\frac{(x-5)^{14}}{14}\right] + C \\ &= \frac{x(x-5)^{13}}{13} - \frac{(x-5)^{14}}{182} + C. \end{aligned}$$

10. Let $u = 3x$. Then $du = 3dx$ or $dx = \frac{1}{3} du$.

$$\begin{aligned} \text{Hence } I &= 5 \int e^{3x} dx \\ &= \frac{5}{3} \int e^u du \\ &= \frac{5e^{3x}}{3} + C. \end{aligned}$$

11. $I = \int 5xe^{3x} dx.$

$f(x) = x$	$g(x) = e^{3x}$
$f'(x) = 1$	$G(x) = \dfrac{e^{3x}}{3}$

$$\begin{aligned} \text{Hence } I &= 5\left(\frac{xe^{3x}}{3} - \int \frac{e^{3x}}{3} dx\right) \\ &= \left(\frac{5}{3}x - \frac{5}{9}\right) e^{3x} + C. \end{aligned}$$

12. $I = \int xe^{-x/2} dx.$

$f(x) = x$	$g(x) = e^{-x/2}$
$f'(x) = 1$	$G(x) = -2e^{-x/2}$

$$\begin{aligned} \text{Hence } I &= -2xe^{-x/2} - \int (-2e^{-x/2}) dx \\ &= -2xe^{-x/2} + 2\int e^{-x/2} dx \\ &= -2xe^{-x/2} - 4e^{-x/2} + C \\ &= -2(x+2)e^{-x/2} + C. \end{aligned}$$

13. Rewrite $I = \int x^5 e^{x^3} dx = \int x^3(x^2 e^{x^3}) dx$

$f(x) = x^3$	$g(x) = x^2 e^{x^3}$
$f'(x) = 3x^2$	$G(x) = \dfrac{e^{x^3}}{3}$

$$\begin{aligned} \text{Hence } I &= \frac{x^3 e^{x^3}}{3} - \int x^2 e^{x^3} dx \\ &= \frac{x^3 e^{x^3}}{3} - \frac{1}{3} e^{x^3} + C \\ &= \frac{1}{3}(x^3 - 1)e^{x^3} + C. \end{aligned}$$

14. $I = \int (2x+1) e^{0.1x} dx.$

$f(x) = 2x+1$	$g(x) = e^{0.1x}$
$f'(x) = 2$	$G(x) = 10e^{0.1x}$

$$\begin{aligned} \text{Hence } I &= 10e^{0.1x}(2x+1) - 20\int e^{0.1x} dx \\ &= 10(2x+1)e^{0.1x} - 200e^{0.1x} + C \\ &= 10(2x - 19)e^{0.1x} + C. \end{aligned}$$

15. $I = \int x \ln 3x \, dx.$

$f(x) = \ln 3x$	$g(x) = x$
$f'(x) = \dfrac{1}{x}$	$G(x) = \dfrac{x^2}{2}$

Hence $I = \dfrac{x^2 \ln 3x}{2} - \dfrac{1}{2}\int x\,dx$

$= \dfrac{x^2 \ln 3x}{2} - \dfrac{x^2}{4} + C.$

16. $I = \int \ln 3x\,dx.$

$f(x) = \ln 3x$	$g(x) = 1$
$f'(x) = \dfrac{1}{x}$	$G(x) = x$

Hence $I = x\ln 3x - \int dx$

$= x\ln 3x - x + C.$

17. Let $u = \ln 3x$. Then $du = \dfrac{1}{x}dx$ and so

$I = \int \dfrac{\ln 3x}{x}dx = \int u\,du$

$= \dfrac{(\ln 3x)^2}{2} + C.$

18. $I = \int \dfrac{\ln 3x}{x^2}dx.$

$f(x) = \ln 3x$	$g(x) = \dfrac{1}{x^2}$
$f'(x) = \dfrac{1}{x}$	$G(x) = -\dfrac{1}{x}$

Hence $I = -\dfrac{\ln 3x}{x} + \int \dfrac{1}{x^2}dx$

$= -\dfrac{\ln 3x}{x} - \dfrac{1}{x} + C$

$= -\dfrac{1}{x}(\ln 3x + 1) + C.$

19. Rewrite

$I = \int x^3(x^2+1)^8 dx = \int x^2[x(x^2+1)^8]dx.$

$f(x) = x^2$	$g(x) = x(x^2+1)^8$
$f'(x) = 2x$	$G(x) = \dfrac{(x^2+1)^9}{18}$

So $I = \dfrac{x^2(x^2+1)^9}{18} - \dfrac{1}{9}\int x(x^2+1)^9 dx$

$= \dfrac{x^2(x^2+1)^9}{18} - \dfrac{(x^2+1)^{10}}{180} + C.$

20. Let $u = x^2 + 1$. Then $du = 2x\,dx$ and so

$I = \int 2x\ln(x^2+1)dx = \int (1)(\ln u)du$

$f(u) = \ln u$	$g(u) = 1$
$f'(u) = \dfrac{1}{u}$	$G(u) = u$

Hence $I = u\ln u - \int u\dfrac{1}{u}du$

$= u\ln u - \int du$

$= u\ln u - u + C$

$= u(\ln u - 1) + C$

$= (x^2+1)[\ln(x^2+1) - 1] + C.$

21. The slope of the tangent is the derivative.

Hence $f'(x) = x(x^2+1)^3$

and so f is an antiderivative of $x(x^2+1)^3$.

That is $f(x) = \int x(x^2+1)^3 dx$

$= \dfrac{(x^2+1)^4}{8} + C.$

Since the graph of f passes through the point $(1,5)$, $5 = f(1) = \dfrac{2^4}{8} + C = 2 + C$ or $C = 3$.

Hence $f(x) = \dfrac{(x^2+1)^4}{8} + 3$

22. Let $Q(x)$ denote the number of commuters using the new subway line x weeks from now. It is given that

$$\dfrac{dQ}{dx} = 18x^2 + 500$$

commuters per week. Hence, $Q(x)$ is an antiderivative of $18x^2 + 500$. That is,

$Q(x) = \int (18x^2 + 500)dx = 6x^3 + 500x + C.$

Since 8,000 commuters currently use the subway, $8,000 = Q(0) = C$. Hence, $Q(x) = 6x^3 + 500x + 8,000$, and the number of commuters who will be using the subway in 5 weeks is $Q(5) = 11,250$.

REVIEW PROBLEMS

23. Let $Q(x)$ denote the number of inmates in county prisons x years from now.
It is given that

$$\frac{dQ}{dx} = 280e^{0.2x}$$

inmates per year. Hence, $Q(x)$ is an antiderivative of $280e^{0.2x}$. That is,

$$Q(x) = \int 280e^{0.2x}dx = 1{,}400e^{0.2x} + C$$

Since the prisons currently house 2,000 inmates, $2{,}000 = Q(0) = 1{,}400 + C$ or $C = 600$. Hence, $Q(x) = 1{,}400e^{0.2x} + 600$, and the number of inmates 10 years from now will be $Q(10) = 1{,}400e^2 + 600 = 10{,}945$.

24. $\frac{dy}{dx} = x^3 - 3x^2 + 5$,

$$y = \int (x^3 - 3x^2 + 5)dx$$

$$= \frac{x^4}{4} - x^3 + 5x + C.$$

25. Separate the variables of

$$\frac{dy}{dx} = 0.02xy$$

and integrate to get

$$\int \frac{1}{y}dy = 0.02xdx,$$
$$\ln|y| = 0.01x^2 + C_1,$$
$$|y| = e^{0.01x^2 + C_1} = e^{C_1}e^{0.01x^2},$$

or $y = Ce^{0.01x^2}$ where $C = \pm e^{C_1}$.

26. Separate the variables of

$$\frac{dy}{dx} = k(80 - y)$$

and integrate to get

$$\int \frac{1}{80-y}dy = \int kdx,$$
$$-\ln|80-y| = kx + C_1,$$
$$|80-y| = e^{-kx-C_1} = e^{-C_1}e^{-kx},$$
$$y - 80 = Ce^{-kx},$$

or $y = 80 + Ce^{-kx}$ where $C = \pm e^{-C_1}$

27. Separate the variables of

$$\frac{dy}{dx} = e^{2x-y} = \frac{e^x}{e^y}$$

and integrate to get

$$\int e^y dy = \int e^{2x} dx.$$

$$e^y = \frac{e^{2x}}{2} + C_1$$

or

$$y = \ln\left(\frac{e^{2x}}{2} + C\right)$$

28.

$$\frac{dy}{dx} = 5x^4 - 3x^2 - 2,$$

$$y = \int (5x^4 - 3x^2 - 2)dx = x^5 - x^3 - 2x + C.$$

Since $y = 4$ when $x = 1$, $4 = 1 - 1 - 2 + C$ or $C = 6$.
Hence

$$y = x^5 - x^3 - 2x + 6$$

29. $\frac{dy}{dx} = \frac{\ln x}{y}$

$$ydy = \ln x dx$$

$$\frac{y^2}{2} = x\ln|x| - x + C$$

Since $y = 100$ when $x = 1$, $5{,}000 = -1 + C$ or $C = 5{,}001$.

$$y = \sqrt{2(x\ln|x| - x + 5{,}001)}.$$

30. $\frac{dy}{dx} = \frac{xy}{\sqrt{1-x^2}}$

$$\frac{dy}{y} = \frac{xdx}{\sqrt{1-x^2}}$$

$$\ln|y| = -\sqrt{1-x^2} + C$$

Since $y = 2$ when $x = 0$ $C = 1 + \ln 2$ and

$$\ln|y| = -\sqrt{1-x^2} + 1 + \ln 2$$

$$\ln\left|\frac{y}{2}\right| = 1 - \sqrt{1-x^2}$$

$$y = 2e^{1-\sqrt{1-x^2}}$$

31. If
$$\frac{d^2y}{dx^2} = 2,$$
then
$$\begin{aligned}\frac{dy}{dx} &= \int \frac{d^2y}{dx^2}dx \\ &= \int 2dx \\ &= 2x + C_1.\end{aligned}$$

Since $\frac{dy}{dx} = 3$ when $x = 0$, $3 = 2(0) + C_1$ or $C_1 = 3$. Hence
$$\frac{dy}{dx} = 2x + 3$$
and $y = \int (2x + 3)dx = x^2 + 3x + C.$

Since $y = 5$ when $x = 0$, $5 = 0^2 + 3(0) + C$ or $C = 5$. Hence
$$y = x^2 + 3x + 5.$$

32. Let $V(t)$ denote the value of the machine after t years. The rate of change of V is
$$\frac{dV}{dt} = k(V - 5,000),$$
where k is a positive constant of proportionality.
Separate the variables and integrate to get
$$\int \frac{1}{V - 5,000}dV = \int kdt,$$
$$\ln(V - 5,000) = kt + C_1,$$
$$V - 5,000 = e^{kt+C_1} = e^{C_1}e^{kt},$$
or
$$V(t) = 5,000 + Ce^{kt}$$
where $C = e^{C_1}$ and the absolute values can be dropped since $V - 5,000 > 0$.
Since the machine was originally worth $40,000$,
$$40,000 = V(0) = 5,000 + C$$
or $C = 35,000$. Hence,
$$V(t) = 5,000 + 35,000e^{kt}.$$

Since the machine was worth $30,000 after 4 years,
$$30,000 = V(4) = 5,000 + 35,000e^{4k},$$
$35,000e^{4k} = 25,000$ or
$$e^{4k} = \frac{25,000}{35,000} = \frac{5}{7}.$$
The value of the machine after 8 years is
$$\begin{aligned}V(8) &= 5,000 + 35,000e^{8k} \\ &= 5,000 + 35,000(e^{4k})^2 \\ &= 5,000 + 35,000\left(\frac{5}{7}\right)^2 = \$22,857.\end{aligned}$$

33. Let $P(q)$ denote the profit, $R(q)$ the revenue, and $C(q)$ the cost when the level of production is q units. Since the marginal revenue is
$$R'(q) = 200q^{-1/2}, C'(q) = 0.4q,$$
and profit is revenue minus cost,
$$\frac{dP}{dq} = \frac{dR}{dq} - \frac{dC}{dq} = 200q^{-1/2} - 0.4q$$
dollars per unit. The profit function $P(q)$ is an antiderivative of the marginal profit. That is,
$$\begin{aligned}P(q) &= \int (200q^{-1/2} - 0.4q)dq \\ &= 400q^{1/2} - 0.2q^2 + C.\end{aligned}$$

Since profit is $2,000 when the level of production is 25 units,
$$2,000 = P(25) = 400(5) - 0.2(25)^2 + C$$
or $C = 125$. Hence,
$$P(q) = 400q^{1/2} - 0.2q^2 + 125,$$
and the profit when 36 units are produced is $P(36) = \$2,265.80$.

34. (a) The rate of change of price is
$$P'(x) = 0.2 + 0.003x^2.$$
$$\begin{aligned}P(x) &= 0.2x + 0.001x^3 + C \\ &= 0.2x + 0.001x^3 + 250\end{aligned}$$

where $C = 250$ cents ($2.50) since "now" means $x = 0$. $P(10) = 2 + 1 + 250 = 253$ or \$2.53.

(b)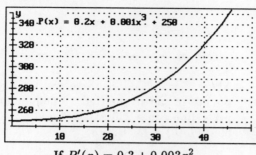

If $P'(x) = 0.3 + 0.003x^2$

then the price would be 0.1 more cents per week or $P(10) = \$2.54$.

35. Let $V(t)$ denote the value of the machine t years from now. Since

$$\frac{dV}{dt} = 220(t - 10)$$

dollars per year, the function $V(t)$ is an antiderivative of $220(t - 10)$. Thus,

$$V(t) = \int 220(t - 10)dt$$
$$= 110t^2 - 2,200t + C.$$

Since the machine was originally worth \$12,000, it follows that $V(0) = 12,000 = C$. Thus, the value of the machine after t years will be

$$V(t) = 110t^2 - 2,200t + 12,000$$

and the value after 10 years will be $V(10) = \$1,000$.

36. $$h(x) = 0.5 + \frac{1}{(x+1)^2}$$

meters per year. The growth per year is

$$h(x) = \int f(x)dx = .5x - \frac{1}{x+1} + h_0$$

During the second year the tree will grow

$$h(2) - h(1) = \frac{2}{3} \text{ meter}$$

37. Let $N(t) =$ denote the number of bushels t days from now. The number of bushels will be an antiderivative of

$$\frac{dN}{dt} = 0.3t^2 + 0.6t + 1.$$

so $N(t) = 0.1t^3 + 0.3t^2 + t + C$

and the revenue is $R(t) = 3N(t)$.
No revenue is generated initially, so $R(0) = 0 = C$.
In 5 days, the revenue will be $N(5) = \$75$.

38. Let $P(x)$ denote the population x months from now. Then

$$\frac{dP}{dx} = 10 + 2\sqrt{x},$$

and the amount by which the population will increase during a month is

$$P(x) = 10x + \frac{4x^{3/2}}{3} + C$$

The initial population is $P(0) = C$ The population at the end of 9 months is

$$P(9) = 10(9) + \frac{(4)9^{3/2}}{3} + C$$

and the population during the next 9 months is $P(9) - P(0) = 126$ people.

39. Let $N(t)$ denote the size of the crop (in bushels) t days from now. Then

$$\frac{dN}{dt} = 0.5t^2 + 4(t + 1)^{-1}$$

bushels per day. The increase in size of the crop over t days is

$$N(t) = \int [0.5t^2 + 4(t + 1)^{-1}]dt$$
$$= \frac{0.5}{3}t^3 + 4\ln|t + 1| + C$$

$N(0) = C$. The size of the crop in 6 days is $N(6)$ and the increase will be

$$N(6) - N(0) = \frac{0.5}{3}(6^3) + 4\ln 7 + C - C = 43.78$$

bushels. Hence, at \$2 per bushel, the value of the crop will increase by $2(43.78) = \$87.56$.

40. Let $Q(t)$ denote the total consumption (in billion-barrel units) of oil over the next t years. Then the demand (billion barrels per year) is the rate of change $\dfrac{dQ}{dt}$ of total consumption with respect to time.
The fact that this demand is increasing exponentially at the rate of 10 percent per year and is currently equal to 40 (billion barrels per year) implies that the demand is

$$\frac{dQ}{dt} = 40e^{0.1t}$$

billion barrels per year.
Hence, the total yearly consumption will be

$$Q(t) = \int 40e^{0.1t}dt = 400e^{0.1t} + C$$

and $Q(0) = 400 + C$. At the end of 5 years the consumption will be

$$Q(4) = 400e^{0.5} + C$$

and during the next 5 years it is
$Q(5) - Q(0) = 259.49$ billion barrels per year.

41. Let $Q(t)$ denote the number of pounds of salt in the tank after t minutes.
Then $\dfrac{dQ}{dt}$ is the rate of change of salt with respect to time (measured in pounds per minute). Thus,

$$\frac{dQ}{dt} = \text{(rate at which salt enters)}$$
$$-\text{(rate at which salt leaves)}$$
$$= \frac{\text{pounds entering}}{\text{gallon}} \cdot \frac{\text{gallons entering}}{\text{minute}}$$
$$-\frac{\text{pounds leaving}}{\text{gallon}} \cdot \frac{\text{gallons leaving}}{\text{minute}}.$$

Now

$$\frac{\text{pounds leaving}}{\text{gallon}}$$
$$= \frac{\text{pounds of salt in the tank}}{\text{gallons of brine in the tank}}$$
$$= \frac{Q}{200}.$$

Hence $\dfrac{dQ}{dt} = -\dfrac{Q}{200}(4) = -\dfrac{Q}{50}$.

Separate the variables and integrate to get

$$\int \frac{1}{Q} dQ = -\int \frac{1}{50} dt,$$
$$\ln |Q| = -\frac{t}{50} + C_1,$$
$$Q = e^{C_1} e^{-t/50} = Ce^{-t/50},$$

where $C = e^{C_1}$. Since there are initially 600 pounds of salt in the tank (3 pounds of salt per gallon times 200 gallons),
$600 = Q(0) = C$. Hence,

$$Q(t) = 600e^{-t/50}.$$

The amount of salt in the tank after 100 minutes is $Q(100) = 600e^{-2} = 81.2012$ pounds.

42. Let $Q(t)$ denote the population in millions t years after 1990. The differential equation describing the population growth is

$$\frac{dQ}{dt} = 10k(10 - Q),$$

where k is a positive constant of proportionality.

$$\frac{dQ}{10 - Q} = 10k\,dt$$

$$-\ln|10 - Q| = 10kt + C_1$$

$$Q = 10 - Ce^{-10kt}$$

Since the population was 4 million in 1990, $Q(0) = 4 = 10 - C$ and so $C = 6$,

$$Q = 10 - 6e^{-10kt}$$

Since the population was 4.74 million in 1995, $Q(5) = 4.74$ and

$$4.74 = 10 - 6e^{-50k} \text{ or } k = 0.002633$$

$$Q(t) = 10 - 6e^{-0.02633t}$$

43. Let $Q(t)$ denote the amount (in million of dollars) of new currency in circulation at time t. Then $\dfrac{dQ}{dt}$ is the rate of change of the new

REVIEW PROBLEMS

currency with respect to time (measured in million dollars per day). Thus

$$\frac{dQ}{dt} = \text{(rate at which new currency enters)} - \text{(rate at which new currency leaves)}.$$

Now, the rate at which new currency enters is 18 million per day. The rate at which new currency leaves is

$$\frac{\text{new currency at time } t}{\text{total currency}} \text{ times}$$
(rate at which new currency enters)
$$= \frac{Q(t)}{5,000}(18)$$

million per day. Putting it all together,

$$\frac{dQ}{dt} = 18 - \frac{18Q}{5,000} = 18\left(1 - \frac{Q}{5,000}\right)$$

Separate variables to obtain

$$\frac{dQ}{1 - Q/5,000} = 18\,dt \text{ and integrate}$$

$$-5,000 \ln\left|1 - \frac{Q}{5,000}\right| = 18t + C.$$

When $t = 0$, $Q(0) = 0$ which yields

$$-5,000 \ln\left|1 - \frac{0}{5,000}\right| = 18(0) + C$$

or $C = 0$. Therefore, the solution becomes

$$\ln\left|1 - \frac{Q}{5,000}\right| = -\frac{18t}{5,000}.$$

Since Q is a part of $5,000$, $1 - \frac{Q}{5,000} > 0$

and so $\ln\left(1 - \frac{Q}{5,000}\right) = -\frac{18t}{5,000}$,

$$1 - \frac{Q}{5,000} = e^{-18t/5,000}.$$

Now to find t so that $Q(t) = 0.9(5,000)$ substitute into the last solution

$$1 - \frac{4,500}{5,000} = e^{-18t/5,000},$$

$$\ln \frac{1}{10} = -\frac{18t}{5,000},$$

thus $t = \frac{5,000}{18} \ln 10 = 640$ days.

44. Let P denote the number of people, x the income. The rate of change of the number of people

$$\frac{dP}{dt} = -kP\frac{1}{x}$$

where k is a positive constant of proportionality.
Separation of variables leads to

$$\frac{dP}{P} = \frac{-k}{x}dt$$

and integrating yields

$$\ln P = \frac{-k}{x}t + C_1$$

$$P = e^{-kt/x + C_1} = e^{-kt/x}e^{C_1} = Ce^{-kt/x}$$

where $C = e^{C_1}$. Note that $P > 0$ and $x > 0$ in the context of this problem.

45.
$$\frac{dP}{dt} = P(\ln P_0)(\ln \beta)\beta^t,$$

$$\frac{dP}{P} = (\ln P_0)(\ln \beta)\beta^t dt,$$

integrating leads to

$$\ln P = (\ln P_0)\beta^t + C_1,$$

$$P = e^{(\ln P_0)\beta^t + C_1} = e^{C_1}e^{(\ln P_0)\beta^t}$$
$$= Ce^{\ln(P_0)\beta^t} = C(P_0)^{\beta^t}$$

where $C = e^{C_1}$. Absolute values were dispensed with because by context $P > 0$.

46. Let P be the number of people involved and Q the number of people implicated.

$$\frac{dQ}{dt} = kQ(P - Q)$$

$$\frac{dQ}{Q(P - Q)} = k\,dt$$

Before proceeding let's break up the fraction (by the method of partial fractions).

$$\frac{1}{Q(P-Q)} = \frac{A}{Q} + \frac{B}{P-Q}$$

Now multiply by the least common denominator. $1 = A(P-Q) + BQ$ which must be an identity (that is true **for all values**), so when

$$Q = 0, A = \frac{1}{P}$$

and when $Q = P$, $B = \frac{1}{P}$.

$$\int \frac{dQ}{Q(P-Q)} = \frac{1}{P}\int \left(\frac{dQ}{Q} + \frac{dQ}{P-Q}\right)$$
$$= k\int dt$$

$$\frac{1}{P}[\ln|Q| - \ln|P-Q|] = kt + C_1$$
$$\ln\left|\frac{Q}{P-Q}\right| = kPt + C_1$$
$$\frac{Q}{P-Q} = Ce^{kPt}$$

Since $Q = 7$ when $t = 0$, $C = \frac{7}{P-7}$.
When $t = 3$ $Q = 16$,

$$\frac{16}{P-16} = \frac{7}{P-7}e^{3Pk}$$

When $t = 6$ $Q = 28$ and

$$\frac{28}{P-28} = \frac{7}{P-7}e^{6Pk}$$
$$= \frac{7}{P-7}\left[\frac{16(P-7)}{7(P-16)}\right]^2$$

after substituting for

$$e^{3kP} = \frac{16(P-7)}{7(P-16)}$$

$$(28)(7)(P^2 - 32P + 256) = 256(P-28)(P-7)$$
$$= 256(P^2 - 35P + 196)$$

$196P^2 - 6,272P + 50,176 = 256P^2 - 8,960P + 50,176$

$(256 - 196)P^2 + (6,272 - 8,960)P = 0$, and

$P = 0$ (to be rejected in the context of this problem) or

$$P = \frac{2,688}{60} = \frac{672}{15} = 44.8 \approx 45 \text{ people}$$

47. Let S be the concentration of the solute inside the cell, S_0 that of the solute outside the cell, and A the area of the cell wall.
The rate of change of the inside solute is jointly proportional to the area of the cell surface and the difference between the solute inside and outside the wall, so

$$\frac{dS}{dt} = kA(S - S_0)$$

where k is a positive constant of proportionality, S_0 is constant, and so is A. Separation of variables and integration leads to

$$\int \frac{1}{S - S_0}dS = \int kA\,dt$$
$$\ln|S - S_0| = kAt + C_1$$
$$S - S_0 = \pm e^{C_1}e^{kAt} = Ce^{kAt}$$

The absolute sign was dropped since C can conveniently be positive or negative as the need prescribes. Thus $S = S_0 + Ce^{kAt}$.

Chapter 6

Further Topics in Integration

6.1 Definite Integration

1. $\int_0^1 (x^4+3x^3+1)dx = \left(\frac{x^5}{5} + \frac{3x^4}{4} + x\right)\Big|_0^1 = \frac{39}{20}$

3.
$$\int_2^5 (2 + 2t + 3t^2)dt$$
$$= (2t + t^2 + t^3)\Big|_2^5 = 144.$$

5.
$$\int_1^3 \left(1 + \frac{1}{x} + \frac{1}{x^2}\right) dx$$
$$= \left(x + \ln|x| - \frac{1}{x}\right)\Big|_1^3 = \frac{8}{3} + \ln 3.$$

7.
$$\int_{-3}^{-1} \frac{t+1}{t^3} dt$$
$$= \int_{-3}^{-1} (t^{-2} + t^{-3})dt$$
$$= \left(-\frac{1}{t} - \frac{1}{2t^2}\right)\Big|_{-3}^{-1} = \frac{2}{9}.$$

9. Let $u = 2x - 4$. Then $du = 2dx$ or $dx = \frac{1}{2}du$.
When $x = 1$, $u = -2$, and when $x = 2$, $u = 0$.
Hence,
$$\int_1^2 (2x-4)^4 dx = \frac{1}{2}\int_{-2}^0 u^4 du$$
$$= \frac{u^5}{10}\Big|_{-2}^0 = \frac{16}{5}.$$

11. Let $u = 6t + 1$. Then $du = 6dt$ or $dt = \frac{1}{6}du$.
When $t = 0$, $u = 1$, and when $t = 4$, $u = 25$.
Hence,
$$\int_0^4 \frac{1}{\sqrt{6t+1}}dt = \frac{1}{6}\int_1^{25} u^{-1/2}du$$
$$= \frac{u^{1/2}}{3}\Big|_1^{25} = \frac{4}{3}.$$

13. Let $u = t^4 + 2t^2 + 1$. Then $du = (4t^3 + 4t)dt$
or $(t^3 + t)dt = \frac{1}{4}du$.
When $t = 0$, $u = 1$, and when $t = 1$, $u = 4$.

Hence $\int_0^1 (t^3 + t)\sqrt{t^4 + 2t^2 + 1}\,dt$
$$= \frac{1}{4}\int_1^4 u^{1/2}du = \frac{u^{3/2}}{6}\Big|_1^4$$
$$= \frac{1}{6}(8 - 1) = \frac{7}{6}.$$

15. Let $u = x - 1$. Then $du = dx$ and $x = u + 1$.
When $x = 2$, $u = 1$, and when $x = e + 1$, $u = e$.

Hence $\int_2^{e+1} \frac{x}{x-1}dx$
$$= \int_1^e \frac{u+1}{u}du = \int_1^e \left(1 + \frac{1}{u}\right) du$$
$$= (u + \ln|u|)\Big|_1^e = e.$$

163

17. Let $u = \ln x$. Then $du = \dfrac{1}{x}dx$. When $x = 1$, $u = 0$, and when $x = e^2$, $u = 2$. Hence

$$\int_1^{e^2} \dfrac{(\ln x)^2}{x}dx = \int_0^2 u^2 du = \dfrac{u^3}{3}\Big|_0^2 = \dfrac{8}{3}$$

19.
$$I = \int_0^1 xe^{-x}dx$$

$f(x) = x$	$g(x) = e^{-x}$
$f'(x) = 1$	$G(x) = -e^{-x}$

$$\begin{aligned} I &= -xe^{-x}\Big|_0^1 + \int_0^1 1e^{-x}dx \\ &= -(x+1)e^{-x}\Big|_0^1 = 1 - 2e^{-1} \end{aligned}$$

21. The element of area has a height y and a base dx. Thus

$$dA = ydx = (4-3x)dx.$$

Summing up these infinitesimal elements leads to the integral

$$\begin{aligned} A &= \int_0^{4/3}(4-3x)dx \\ &= \left(4x - \dfrac{3x^2}{2}\right)\Big|_0^{4/3} \\ &= \dfrac{4(4)}{3} - \dfrac{3(4^2)}{2(3^2)} = \dfrac{8}{3} \text{ square units} \end{aligned}$$

23. The element of area has a height y and a base dx. Thus

$$dA = ydx = 5dx.$$

Summing up these infinitesimal elements in $-2 < x < 1$ leads to the integral

$$\begin{aligned} A &= \int_{-2}^1 (5)dx \\ &= 5x\Big|_{-2}^1 = 15 \text{ square units} \end{aligned}$$

25. The element of area has a height y and a base dx. Thus

$$dA = ydx = \sqrt{x}dx.$$

Summing up these infinitesimal elements in $4 < x < 9$ leads to the integral

$$\begin{aligned} A &= \int_4^9 \sqrt{x}dx = \int_4^9 x^{1/2}dx \\ &= \dfrac{2x^{3/2}}{3}\Big|_4^9 = \dfrac{38}{3} \text{ square units} \end{aligned}$$

27. The element of area has a height y and a base dx. Thus

$$dA = ydx = e^x dx.$$

Summing up these infinitesimal elements in $\ln\left(\dfrac{1}{2}\right) = -\ln 2 < x < 0$ leads to the integral

$$\begin{aligned} A &= \int_{-\ln 2}^0 e^x dx = e^x\Big|_{-\ln 2}^0 \\ &= 1 - e^{-\ln 2} = 1 - e^{\ln(1/2)} \\ &= \dfrac{1}{2} \text{ square units} \end{aligned}$$

29. The region is split into two subregions on either side of the vertical line $x = 1$.

$$\begin{aligned} A &= \int_0^1 \left(x - \dfrac{x}{8}\right)dx \\ &\quad + \int_1^2 \left(\dfrac{1}{x^2} - \dfrac{x}{8}\right)dx \\ &= \dfrac{7x^2}{16}\Big|_0^1 - \left(\dfrac{1}{x} + \dfrac{x^2}{16}\right)\Big|_1^2 \\ &= \dfrac{3}{4} \text{ square units.} \end{aligned}$$

31. The element of area has a height $9x - x^3$ and a base dx. Thus

$$dA = (9x - x^3)dx$$

$y = x^3$ intersects $y = 9x$ at $(0,0)$ and $(3, 27)$.

$$\begin{aligned} A &= \int_0^3 (9x - x^3)dx \\ &= \left(\dfrac{9x^2}{2} - \dfrac{x^4}{4}\right)\Big|_0^3 \\ &= \dfrac{81}{4} \text{ square units} \end{aligned}$$

6.2. APPLICATIONS TO BUSINESS AND ECONOMICS

33. (a)
$$I = \int_0^1 \sqrt{1-x^2}\,dx$$
is the area under the
$$y = \sqrt{1-x^2} \geq 0$$
on $0 \leq x \leq 1$, that is one-quarter of the area of a circle of radius 1. Thus $I = \dfrac{\pi}{4}$.

(b)
$$I = \int_1^2 \sqrt{2x-x^2}\,dx$$
is the area under the curve
$$y = \sqrt{-x^2+2x} = \sqrt{1-(x-1)^2}$$
on $1 \leq x \leq 2$, that is one-quarter of the area of a circle of radius 1. Thus $I = \dfrac{\pi}{4}$.

35.
$$\begin{aligned}
A &= 2\int_{2.34}^{4.2} \sqrt{\frac{2x^2}{5}-2}\,dx \\
&+ \int_{2.6}^{2.97}\left(\sqrt{\frac{2x^2}{5}-2}\right. \\
&\left. -(x^3-8.9x^26.7x-27)\right) = 2.037
\end{aligned}$$

6.2 Applications to Business and Economics

1. Let $V(x)$ denote the value of the machine after x years. Then
$$\frac{dV}{dx} = 220(x-10),$$
and the amount by which the machine depreciates during the 2nd year is
$$\begin{aligned}
V(2)-V(1) &= \int_1^2 220(x-10)\,dx \\
&= 220\left(\frac{x^2}{2}-10x\right)\Big|_1^2 = -\$1,870
\end{aligned}$$
where the negative sign indicates that the value of the machine has decreased.

3. Let $C(q)$ denote the total cost of producing q units. Then the marginal cost is
$$\frac{dC}{dq} = 6(q-5)^2,$$
and the increase in cost is
$$\begin{aligned}
C(13)-C(10) &= \int_{10}^{13} 6(q-5)^2\,dq \\
&= 2(q-5)^3\Big|_{10}^{13} = \$774.
\end{aligned}$$

5. Let $N(t)$ denote the number of bushels that are produced over the next t days. Then
$$\frac{dN}{dt} = 0.3t^2 + 0.6t + 1,$$
and the increase in the crop over the next five days is
$$\begin{aligned}
N(5)-N(0) &= \int_0^5 (0.3t^2+0.6t+1)\,dt \\
dx \\
&= (0.1t^3+0.3t+t)\Big|_0^5 = 25
\end{aligned}$$
bushels. If the price remains fixed at \$3 per bushel, the corresponding increase in the value of the crop is \$75.

7. Let $Q(t)$ denote the production after t hours. Then $t = 0$ at 8:00 a.m., $t = 2$ at 10:0 a.m., and $t = 4$ at noon. Thus,
$$Q(t) = 100\int_2^4 te^{-0.5t}\,dt$$

Note: This problem can be solved by integration by parts, discussed in section 5, or you can use your graphing utility.

$f(t) = t$	$g(t) = e^{-0.5t}$
$f'(t) = 1$	$G(t) = -2e^{-0.5t}$

$$\begin{aligned}
Q(t) &= 100\left[-2te^{-0.5t}\Big|_2^4 + 2\int_0^4 e^{-0.5t}\,dt\right] \\
&= 100\left[-8e^{-2}+4e^{-1}-4e^{-0.5t}\Big|_2^4\right] \\
&= 131.90 \text{ units}
\end{aligned}$$

9. (a) The first plan generates profit at the rate of
$$P_1'(t) = 130 + t^2$$
hundred dollars per year and the second generates profit at the rate of
$$P_2'(t) = 306 + 5t$$
hundred dollars per year. The second plan will be the more profitable until
$$P_1'(t) = P_2'(t),$$
that is, until $130 + t^2 = 306 + 5t$ or $t = 16$ years.

(b) For $0 \leq t \leq 16$, the rate at which the profit generated by the second plan exceeds that of the first plan is
$$P_2(t) - P_1(t).$$
Hence the net excess profit generated by the second plan over the 16-year period is the definite integral
$$\int_0^{16} [P_2(t) - P_1(t)] dt$$
$$= \int_0^{16} [306 + 5t - (130 + t^2)] dt$$
$$= \left(176t + \frac{5t^2}{2} - \frac{t^3}{3}\right)\Big|_0^{16}$$
$$= 2,090.67 \text{ hundred dollars.}$$

(c) In geometric terms, the net excess profit generated by the second plan is the area of the region between the curves $y = P_2(t)$ and $y = P_1(t)$ from $t = 0$ to $t = 16$.

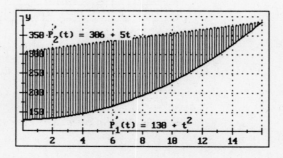

11. (a) The first plan generates profit at the rate of
$$P_1'(t) = 90e^{0.1t}$$
thousand dollars per year and the second generates profit at the rate of
$$P_2'(t) = 140e^{0.07t}$$
thousand dollars per year. The second plan will be the more profitable until
$$P_1'(t) = P_2'(t),$$
that is, until $90e^{0.1t} = 140e^{0.07t}$ or $t = 14.73$ years.

(b) For $0 \leq t \leq 14.73$, the rate at which the profit generated by the second plan exceeds that of the first plan is
$$P_2(t) - P_1(t).$$
Hence the net excess profit generated by the second plan over the 14.73-year period is the definite integral
$$\int_0^{14.73} [P_2(t) - P_1(t)] dt$$
$$= \int_0^{14.73} [140e^{0.07t} - (90e^{0.1t})] dt$$
$$= \left(2000e^{0.07t} - 900e^{0.1t}\right)\Big|_0^{14.73}$$
$$= 582.23 \text{ thousand dollars.}$$

(c) In geometric terms, the net excess profit generated by the second plan is the area of the region between the curves $y = P_2(t)$ and $y = P_1(t)$ from $t = 0$ to $t = 14.73$.

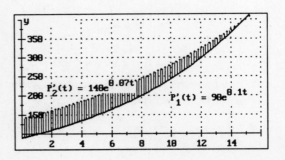

6.2. APPLICATIONS TO BUSINESS AND ECONOMICS

13. **(a)** The machine generates revenue at the rate of

$$R'(t) = 7,250 - 18t^2$$

dollars per year and results in costs that accumulate at the rate of

$$C'(t) = 3,620 + 12t^2$$

dollars per year.
The use of the machine will be profitable as long as the rate at which revenue is generated is greater than the rate at which costs accumulate, that is, until $R'(t) = C'(t)$,

$$7,250 - 18t^2 = 3,620 + 12t^2 \text{ or } t = 11 \text{ years}$$

(b) The difference $R(x) - C(x)$ represents the rate of change of the net earnings generated by the machine.
Hence, the net earnings over the next 11 years is the definite integral

$$\int_0^{11} [R(t) - C(t)] dt$$
$$= \int_0^{11} [(7,250 - 18t^2) - (3,620 + 12t^2)] dx$$
$$= \int_0^{11} (3,530 - 30t^2) dt$$
$$= (3,630t - 10t^3)\Big|_0^{11} = \$26,620$$

(c) In geometric terms, the net earnings in part **(b)** is the area of the region between the curves $y = R'(t)$ and $y = C'(t)$ from $t = 0$ to $t = 11$.

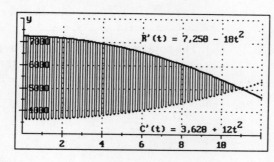

15. **(a)** The campaign generates revenue at the rate of

$$R(t) = 5,000e^{-0.2t}$$

dollars per week and accumulates expenses at the rate of \$676 per week. The campaign will be profitable as long as $R(t)$ is greater than 676, that is, until

$$5,000e^{-0.2t} = 676$$

$$t = -\frac{\ln(676/5,000)}{0.2} \approx 10 \text{ weeks}$$

(b) For $0 \leq t \leq 10$, the difference $R(t) - 676$ is the rate of change with respect to time of the net earnings generated by the campaign. Hence, the net earnings during the 10 week period is the definite integral

$$\int_0^{10} [R(t) - 676] dt$$
$$= \int_0^{10} [5,000e^{-0.2t} - 676] dt$$
$$= -25,000e^{-0.2t}\Big|_0^{10} - 676t\Big|_0^{10} = \$14,857$$

(c) In geometric terms, the net earnings in part **(b)** is the area between the curve $y = R(t)$ and the horizontal line $y = 676$ from $t = 0$ to $t = 10$.

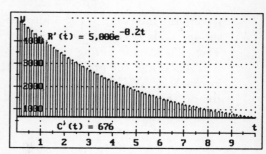

17. Recall that P dollars invested at an annual interest rate of 6 % compounded continuously will be worth

$$Pe^{0.06t}$$

dollars t years later.
To approximate the future value of the income stream, divide the 5-year time interval

$0 \leq t \leq 5$ into n equal sub-intervals of length Δt years and let t_j denote the beginning of the jth sub-interval.

Then, the money deposited during the jth sub-interval is $2,400\Delta t$.

This money will remain in the account approximately $5 - t_j$ years hence. The future value of the money deposited during the jth sub-interval is

$$2,400e^{0.06(5-t_j)}\Delta t.$$

The future value of the income stream is

$$\lim_{n \to \infty} \sum_{j=1}^{n} 2,400e^{0.06(5-t_j)}\Delta t$$

$$= \int_0^5 2,400e^{0.06(5-t)}dt$$

$$= 2,400e^{0.3} \int_0^5 e^{-0.06t}dt$$

$$= \frac{2,400}{-0.06}e^{0.3}e^{-0.06t}\Big|_0^5 = \$13,994.35.$$

19. Recall that the present value of B dollars payable t years from now with an annual interest rate of 12 % compounded continuously is

$$Be^{-0.06t}.$$

Divide the interval $0 \leq t \leq 5$ into n equal sub-intervals of length Δt years.

Then, the income during the jth sub-interval is $1,200\Delta t$ and the present value of this income is $1,200e^{-0.12t_j}\Delta t$.

Hence, the present value of the investment is

$$\lim_{n \to \infty} \sum_{j=1}^{n} 1,200e^{-0.12t_j}\Delta t$$

$$= 1,200 \int_0^5 e^{-0.12t}dt$$

$$= -\frac{1,200e^{-0.12t}}{-0.12}\Big|_0^5 = \$4,511.88.$$

21. $$P(t) = \int_0^{10}(10,000 + 500t)e^{-0.1t}dt.$$

$f(t) = 10,000 + 500t$	$g(t) = e^{-0.1t}$
$f'(t) = 500$	$G(t) = -10e^{-0.1t}$

$$P(t) = -10e^{-0.1t}(10,000 + 500t)\Big|_0^{10}$$

$$+5,000 \int_0^{10} e^{-0.1t}dt$$

$$= -10[15,000e^{-1} - 10,000]$$

$$-50,000e^{-0.1t}\Big|_0^{10}$$

$$= \$76,424.11.$$

23. (a) If the consumers' demand function is

$$D(q) = \frac{300}{(0.1q + 1)^2}$$

dollars per unit, the total amount that consumers are willing to spend to get 5 units is the definite integral

$$\int_0^5 D(q)dq$$

$$= 300 \int_0^5 (0.1q + 1)^{-2}dq$$

$$= -3,000(0.1q + 1)^{-1}\Big|_0^5$$

$$= -3,000\left(\frac{1}{1.5} - 1\right) = \$1,000.$$

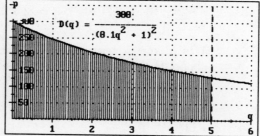

(b) The total willingness to spend in part (a) is the area of the region under the demand curve from $q = 0$ to $q = 5$.

25. (a) If the consumers' demand function is

$$D(q) = \frac{300}{4q + 3}$$

dollars per unit, the total amount that consumers are willing to spend to get 10 units is the definite integral

$$\int_0^{10} D(q)\,dq$$

$$= 300\int_0^{10}(4q+3)^{-1}dq$$

$$= \frac{300}{4}\ln|4q+3|\Big|_0^{10} = 199.70$$

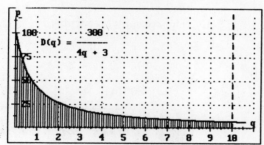

(b) The total willingness to spend in part (a) is the area of the region under the demand curve from $q = 0$ to $q = 10$.

27. (a) If the consumers' demand function is

$$D(q) = 50e^{-0.04q}$$

dollars per unit, the total amount that consumers are willing to spend to get 15 units is the definite integral

$$\int_0^{15} D(q)\,dq$$

$$= 50\int_0^{15} e^{-0.04q}\,dq$$

$$= -\frac{50}{0.04}e^{-0.04q}\Big|_0^{15}$$

$$= 1,250(1 - 0.5488) = \$563.99.$$

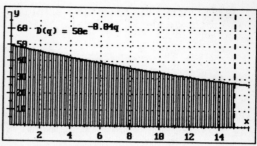

(b) The total willingness to spend in part (a) is the area of the region under the demand curve from $q = 0$ to $q = 15$.

29. (a) The consumers' demand function is

$$D(q) = 150 - 2q - 3q^2$$

dollars per unit. For the market price of 6 units

$$p_0 = 150 - 12 - 108 = 30$$

Thus the consumer's surplus is

$$S(q) = \int_0^6 (150 - 2q - 3q^2)\,dq - (30)(6)$$

$$= 150(6) - 6^2 - 6^3 - 180 = \$468.$$

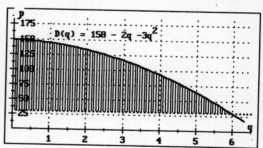

(b) The consumer's surplus in part (a) is the area of the region under the demand curve from $q = 0$ to $q = 6$ from which the actual spending is subtracted.

31. (a) The consumers' demand function is $D(q) = 75e^{-0.04q}$ dollars per unit. The market price for 3 units is

$$D(3) = 75e^{-0.12} = \$66.52$$

Thus the consumer's surplus is

$$S(q) = 75\int_0^3 e^{-0.04q}dq - 3(66.52)$$
$$= -\frac{75}{0.04}e^{-0.04q}\Big|_0^3 - 199.56 = \$12.46.$$

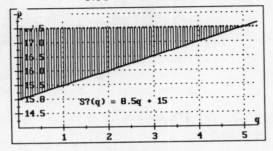

(b) The consumer's surplus in part **(a)** is the area of the region under the demand curve from $q = 0$ to $q = 3$.

33. The producer's supply function is $S(q) = 0.5q + 15$ dollars per unit.

$$p_0 = S(5) = 2.5 + 15 = 17.5$$

The producer's surplus for $q_0 = 5$ is

$$PS = (5)(17.5) - \int_0^5 (0.5q + 15)dq = \$6.25$$

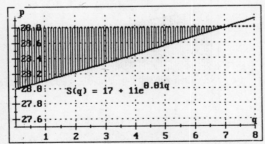

35. The producer's supply function is $S(q) = 17 + 11e^{0.01q}$ dollars per unit.

$$p_0 = S(7) = 17 + 11e^{0.07} = 28.80$$

The producer's surplus for $q_0 = 7$ is is

$$PS = (7)(28.80) - \int_0^7 (17 + 11e^{0.01q})dq = \$2.84$$

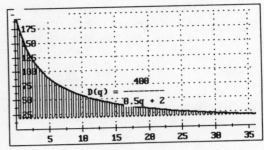

37. (a) $D(q) = p_0$ if $\frac{400}{0.5q + 2} = 20$, $20 = 0.5q + 2$ or $q = 36$.

(b) The consumer's surplus is

$$I = \int_0^{36} \frac{400}{0.5q + 2}dq - 36 \times 20$$
$$= 800\ln|0.5q + 2|\Big|_0^{36} - 720 = \$1,122.07.$$

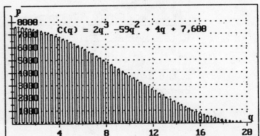

39. (a) The dollar price per unit is $p = 124 - 2q$ and the cost function is

$$C(q) = 2q^3 - 59q^2 + 4q + 7,600.$$

The profit function is

$$P(q) = (124 - 2q)q$$
$$\quad -(2q^3 - 59q^2 + 4q + 7,600)$$
$$= -2q^3 + 57q^2 + 120q - 7,600.$$
$$P'(q) = -(6q^2 - 114q - 120) = 0$$

(b) Profit is maximized when

$$\text{if } q = \frac{57 \pm \sqrt{57^2 + 6(120)}}{6} = 20$$

(since $q > 0$).

$R''(q) = -12q + 114$ and $R(20) < 0$,

so $q = 20$ produces a maximum.

6.2. APPLICATIONS TO BUSINESS AND ECONOMICS

(c) The corresponding consumer's surplus is

$$S(q) = \int_0^{20}(124 - 2q)dq - 20(124 - 40)$$
$$= (124q - q^2)\Big|_0^{20} - (20)(84) = \$400.$$

41. The supply function for a certain commodity is

$$S(p) = \frac{q+1}{3}$$

and the demand function is

$$D(q) = \frac{16}{q+2} - 3.$$

(a) The supply equals the demand if

$$\frac{16}{q+2} - 3 = \frac{q+1}{3},$$
$$48 - 9q - 18 = q^2 + 3q + 2,$$
$$q^2 + 12q - 28 = 0,$$

or $q = -6 \pm \sqrt{36 + 28} = 2$ (since $q > 0$.)

(b) The corresponding consumer's surplus is

$$CS = \int_0^2 \left(\frac{16}{q+2} - 3\right)dq - 2(1)$$
$$= (16\ln|q+2| - 3q)\Big|_0^2 - 2 = \$3.09.$$

$$PS = 2(1) \int_0^2 \frac{q+1}{3}dq = \$0.67$$

(c)

43. (a) Let t denote the time from now in months and $R(t)$ the total revenue generated. The price is

$$P(t) = 18 + 0.3\sqrt{t}$$

dollars per barrel and the revenue is generated at a rate of

$$\frac{dR}{dt} = 300(18 + 0.3\sqrt{t}). \text{ Then}$$

$$R(t) = \int 300(18 + 0.3\sqrt{t})dt$$
$$= 300(18t + 0.2t^{3/2}) + C$$

where $R(0) = 0 = 300(0) + C$ implies $C = 0$. Then

$$R(36) = 300[18(36) + 0.2(36)^{3/2}] = \$207,360$$

(b) Divide the interval $0 \le t \le 36$ into n equal subintervals of length Δt years. Then, the quantity of oil during the jth subinterval is $300\Delta t$ and the revenue generated by that much oil is

$$(18 + 0.3\sqrt{t_j})300\Delta t$$

Hence, the total revenue is

$$\lim_{n \to \infty} \sum_{j=1}^n (18 + 0.3\sqrt{t_j})300\Delta t$$
$$= \int_0^{36}(18 + 0.3\sqrt{t})300dt$$
$$= \$207,360$$

(computed in part (a) above.)

(c) Writing exercise— Answers will vary.

45. During a time interval dt the revenue generated is $f(t)dt$ dollars which has a present value of $f(t)e^{-rt}dt$ at the interest rate r. The present value corresponding to N years is

$$P(t) = \int_0^N f(t)e^{-rt}dt$$
$$= \int_0^{10} 1{,}750e^{-0.095t}dt = 11{,}296.88$$

6.3 Additional Applications of Definite Integration

1. The average value is

$$f_{av} = \frac{1}{4-0}\int_0^4 x\,dx = \frac{1}{4}\frac{x^2}{2}\Big|_0^4 = 2$$

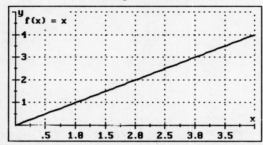

3. The average value is

$$\begin{aligned} f_{av} &= \frac{1}{0-(-4)}\int_{-4}^0 (x+2)^2 dx \\ &= \frac{(x+2)^3}{12}\Big|_{-4}^0 = \frac{4}{3}. \end{aligned}$$

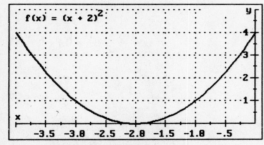

5. The average value is

$$\begin{aligned} f_{av} &= \frac{1}{2-(-1)}\int_{-1}^2 e^{-2t} dt \\ &= -\frac{e^{-2t}}{6}\Big|_{-1}^2 = 1.2285. \end{aligned}$$

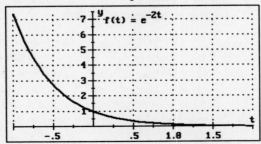

7. The average temperature between 9:00 a.m. and noon is

$$\begin{aligned} f_{av} &= \frac{1}{12-9}\int_9^{12}(-0.3t^2+4t+10)dt \\ &= \frac{1}{3}(-0.1t^3+2t^2+10t)\Big|_9^{12} = 18.7°\text{ C} \end{aligned}$$

9. The average rate during the first 3 months is

$$\begin{aligned} Q_{av} &= \frac{1}{3-0}\int_0^3(700-400e^{-0.5t})dt \\ &= \frac{1}{3}(700t+800e^{-0.5t})\Big|_0^3 \\ &= 492.83 \approx 493 \text{ letters per hour.} \end{aligned}$$

11. $$f(t) = e^{-0.2t}$$

Of the 200 present members, $200f(8)$ will still be in the club in 8 months.
Of the 10 new members picked up t months from now, $10f(8-t)$ will still be members in 8 months. Thus

$$\begin{aligned} P(t) &= 200e^{-1.6} + \int_0^8 10e^{-0.2(8-t)}dt \\ &= 200e^{-1.6} + 10e^{-1.6}\int_0^8 e^{0.2t}dt \\ &= 40.379 + 50e^{-1.6}e^{0.2t}\Big|_0^8 = 80.28 \end{aligned}$$

The club will consist of approximately 80 members in 8 months.

13. Since $f(t) = e^{-t/10}$

is the fraction of members active after t months, and since there were 8,000 charter members, the number of charter members still active at the end of 10 months is

$$8,000f(10) = 8,000e^{-1}$$

Now, divide the interval $0 \le t \le 10$ into n equal subintervals of length Δt months and let t_j denote the beginning of the jth subinterval. During this j^{th} subinterval, $200\Delta t$ new members join, and at the end of the 10 months

6.3. ADDITIONAL APLICATIONS OF DEFINITE INTEGRATION

$(10 - t_j$ months later), the number of these retaining membership is

$$200 f(10 - t_j) = 200 e^{-(10-t_j)/10} \Delta t$$

Hence the number of new members still active 10 months from now is approximately

$$\lim_{n \to \infty} \sum_{j=1}^{n} 200 e^{-(10-t_j)/10} \Delta t$$

$$= \int_0^{10} 200 e^{-(10-t)/10} dt$$

Hence, the total number N of active members 10 months from now is

$$\begin{aligned}
N &= 8,000 e^{-1} + \int_0^{10} 200 e^{-(10-t)/10} dt \\
&= 2,943 + 200 e^{-1} \int_0^{10} e^{t/10} dt \\
&= 2,943 + 200 e^{-1} (10 e^{t/10}) \Big|_0^{10} = 4,207
\end{aligned}$$

15. Divide the interval $0 \le r \le 3$ into n equal subintervals of length Δr, and let r_j denote the beginning of the jth subinterval.
This divides the circular disc of radius 3 into n concentric circles, as shown in the figure.
If Δr is small, the area of the jth ring is

$$2\pi r_j \Delta r$$

where $2\pi r_j$ is the circumference of the circle of radius r_j that forms the inner boundary of the ring and Δr is the width of the ring.
Then, since $D(r) = 5,000 e^{-0.1r}$ is the population density (people per square mile) r miles from the center, it follows that the number of people in the jth ring is

$$\begin{aligned}
&D(r_j)(\text{area of the } j^{th} \text{ ring}) \\
&= 5,000 e^{-0.1 r_j} (2\pi r_j \Delta r)
\end{aligned}$$

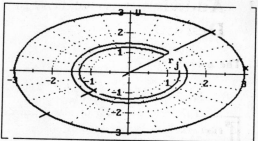

Hence, if N is the total number of people within 3 miles of the center of the city,

$$\begin{aligned}
N &= \lim_{n \to \infty} \sum_{j=1}^{n} 5,000 e^{-0.1 r_j} (2\pi r_j \Delta r) \\
&= \int_0^3 5,000 (2\pi) r e^{-0.1 r} dr \\
&= 10,000 \pi \int_0^3 r e^{-0.1 r} dr
\end{aligned}$$

Applying integration by parts,

$$\begin{aligned}
&\int_0^3 r e^{-0.1 r} dr \\
&= -10 r e^{-0.1 r} \Big|_0^3 - \int_0^3 (-10) e^{-0.1 r} dr \\
&= (-10 r e^{-0.1 r} - 100 e^{-0.1 r}) \Big|_0^3 \\
&= (-30 e^{-0.3} - 100 e^{-0.3}) - (-100) = 3.6936
\end{aligned}$$

Hence, the total number of people within 10 miles of the center of the city is

$$\begin{aligned}
N &= 10,000 \pi \int_0^3 r e^{-0.1 r} dr \\
&= 116,038
\end{aligned}$$

(Your answer may differ slightly due to round-off errors.)

17. Let $S(r) = k(R^2 - r^2)$ denote the speed of the blood in centimeters per second at a distance r from the central axis of the artery of (fixed) radius R.
The area of a small circular ring at a distance r_j is (approximately) $2\pi r_j \Delta r$ square

centimeters, so the amount of blood passing through the ring is

$$V(r) = 2\pi r_j \Delta r[k(R^2 - r_j^2)]$$
$$= 2\pi k(R^2 r_j - r_j^3)\Delta r$$

cubic centimeters per second.
Hence, the total quantity of blood flowing through the artery per second is

$$\lim_{n\to\infty} \sum_{j=1}^{n} 2\pi k(R^2 r_j - r_j^3)\Delta r$$
$$= 2\pi k \int_0^R (rR^2 - r^3)dr$$
$$= 2\pi k \left(\frac{R^2 r^2}{2} - \frac{r^4}{4}\right)\Big|_0^R = \frac{\pi k R^4}{2}$$

The area of the artery is πR^2 and the average velocity of the blood through the artery is

$$V_{ave} = \frac{\pi k R^4 / 2}{\pi R^2} = \frac{kR^2}{2}$$

The maximum speed for the blood occurs at $r = 0$, so $S(0) = kR^2$. Thus

$$V_{ave} = \frac{1}{2}S(0)$$

19. Let $s(t)$ denote the distance traveled after t minutes.

$$\frac{ds}{dt} = 3t^2 + 2t + 5$$
$$s(t) = t^3 + t^2 + 5t + s_0$$
$$s(2) = 8 + 4 + 10 + s_0$$
$$s(1) = 1 + 1 + 5 + s_0$$
$$s(2) - s(1) = 15 \text{ meters}$$

21. (a) The average speed is

$$S_{av} = \frac{1}{N}\int_0^N S(t)dt.$$

(b) The total distance is

$$d = \int_0^N S(t)dt.$$

(c) The average speed is

$$\frac{\text{total distance}}{\text{number of hours}}.$$

23. Radioactive material decays exponentially so that if $A(t)$ denotes the amount of radioactive material present after t years, $A(t) = A_0 e^{-kt}$, where A_0 is the amount present initially and k is a positive constant.
Since the half-life is 28 years,

$$\frac{A_0}{2} = A(28) = A_0 e^{-28k},$$
$$-28k = \ln\frac{1}{2} = -\ln 2 \text{ or } k = \frac{\ln 2}{28}$$

Now divide the interval $0 \le t \le 140$ into n equal sub-intervals of length Δt years and let t_j denote the beginning of the j_{th} sub-interval. During the j_{th} sub-interval, $500\Delta t$ pounds of radioactive material is produced, and the amount left 140 years from now (i.e., after $140 - t_j$ years) is approximately

$$(500\Delta t)A(140 - t_j) = 500A(140 - t_j)\Delta t.$$

Hence, the total waste after 140 years is

$$N = \lim_{n\to\infty}\sum_{j=1}^{n} 500A(140 - t_j)\Delta t$$
$$= \int_0^{140} 500A(140 - t)dt$$
$$= 500\int_0^{140} e^{-(140-t)k}dt$$
$$= 500e^{-140k}\int_0^{140} e^{kt}dt$$
$$= \frac{500}{k}e^{-140k+kt}\Big|_0^{140}$$
$$= \frac{500}{k}(1 - e^{-140k}).$$

Since $k = \frac{\ln 2}{28}$ it follows that

$$N = \frac{28(500)}{\ln 2}[1 - e^{-(140\ln 2)/28}]$$
$$= \frac{28(500)}{\ln 2}[1 - e^{-5\ln 2}]$$
$$= \frac{28(500)}{\ln 2}[1 - 2^{-5}] = 19,567 \text{ pounds}.$$

6.3. ADDITIONAL APLICATIONS OF DEFINITE INTEGRATION

25.
$$C(t) = \frac{c}{b-a}(e^{-at} - e^{-bt})$$
$$C'(t) = \frac{c}{b-a}(-ae^{-at} + be^{-bt})$$
$$C''(t) = \frac{c}{b-a}(a^2 e^{-at} - b^2 e^{-bt})$$
$$C''(t) = 0 \text{ when } a^2 e^{-at} = b^2 e^{-bt}$$
$$\frac{a^2}{b^2} = e^{(a-b)t} \text{ or } t = \frac{2}{a-b}\ln\left(\frac{a}{b}\right)$$

Let $b = 2a$ then $t = \frac{2}{a}\ln 2$

$$\frac{a}{2\ln 2}\int_0^{(2/a)\ln 2} \frac{c}{a}(e^{-at} - e^{-2at})dt$$
$$= \frac{c}{4a\ln 2}(-2e^{-at} + e^{-2at})\Big|_0^{(2/a)\ln 2}$$
$$= \frac{c}{2\ln 2}\left[-\frac{1}{4a} + \frac{1}{32a} - \left(-\frac{1}{a} + \frac{1}{2a}\right)\right]$$
$$= \frac{9c}{64a\ln 2}$$

27. Let $f(t)$ denote the fraction of the membership of the group that will remain active for at least t years, P_0 the initial membership, and $r(t)$ the rate per year at which additional members are added to the group.

Then, the size of the group N years from now is the number of initial members still active plus the number of new members still active. Of the P_0 initial members, $f(N)$ is the fraction remaining active for N years. Hence, the number of initial members still active after N years is

$$P_0 f(N).$$

To find the number of new members still active after N years, divide the interval $0 \leq t \leq N$ into n equal sub-intervals of length Δt years and let t_j denote the beginning of the jth sub-interval. During the jth sub-interval, approximately

$$r(t_j)\Delta t$$

new members joined the group. Of these, the fraction still active $t = N$ (that is $N - t_j$ years later) is $f(N - t_j)$, and so the number of these still active after N years is

$$\lim_{n\to\infty}\sum_{j=1}^{n} r(t_j)f(N-t_j)\Delta t$$
$$= \int_0^N r(t)f(N-t)dt.$$

Putting it all together, the total number of active members N years from now is

$$P_0 f(N) + \int_0^N r(t)f(N-t)dt.$$

29.
$$L(x) = x^2$$
$$G = 2\int_0^1 [x - L(x)]dx$$
$$= 2\int_0^1 [x - x^2]dx$$
$$= 2\left(\frac{x^2}{2} - \frac{x^3}{3}\right)\Big|_0^1 = \frac{1}{3}$$

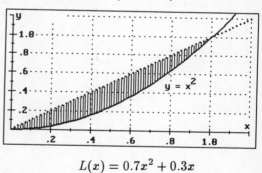

31.
$$L(x) = 0.7x^2 + 0.3x$$
$$G = 2\int_0^1 [x - L(x)]dx$$
$$= 2\int_0^1 (0.7x - 0.7x^2)dx$$
$$= 1.4\left[\frac{x^2}{2} - \frac{x^3}{3}\right]\Big|_0^1 = \frac{0.7}{3}$$

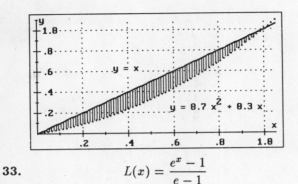

33. $L(x) = \dfrac{e^x - 1}{e - 1}$

$$G = 2\int_0^1 [x - L(x)]dx$$
$$= 2\int_0^1 \left[x - \dfrac{e^x - 1}{e - 1}\right]dx$$
$$= 2\left[\dfrac{x^2}{2} - \dfrac{1}{e-1}(e^x - x)\right]\Big|_0^1 = 1 - \dfrac{2(e-2)}{e-1}$$

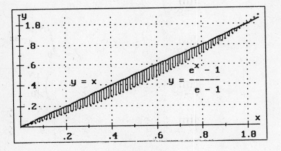

35. (a) Writing exercise —
  Answers will vary.

  (b) If $L(x) = x$ then the area between $L(x)$ and $y = x$ is 0.
  If $L(x) = 0$ then the area between $L(x)$ and $y = 0$ is 0.5 (from 0 and 1.

  (c) Writing exercise —
  Answers will vary.

37. The Gini index for computer engineers is

$$G_1 = 2\int_0^1 [x - x^{1.8}]dx$$
$$= 2\left[\dfrac{x^2}{2} - \dfrac{x^{2.8}}{2.8}\right]\Big|_0^1$$
$$= 0.2857$$

The Gini index for stock brokers is

$$G_2 = 2\int_0^1 [0.75x - 0.75x^2]dx$$
$$= 1.5\left[\dfrac{x^2}{2} - \dfrac{x^3}{3}\right]\Big|_0^1$$
$$= 0.25$$

The distribution of incomes amongst stock brokers is more fairly distributed.

39.

land	owners	land	owners
0	0	0.1	0.025
0.2	0.05	0.3	0.075
0.4	0.1	0.5	0.13
0.6	0.18	0.7	0.22
0.8	0.28	0.9	0.42
1.0	1.00		

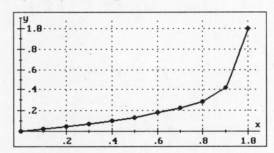

41.

land	owners	land	owners
0	0	0.1	0.001
0.2	0.002	0.3	0.005
0.4	0.1	0.5	0.18
0.6	0.28	0.7	0.58
0.8	0.11	0.9	0.21
1.0	1.00		

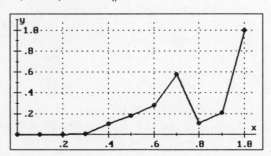

6.4 Improper Integrals

1. $$\int_1^\infty \frac{1}{x^3}dx = \lim_{N\to\infty}\int_1^N \frac{1}{x^3}dx$$
$$= \lim_{N\to\infty}\left.\frac{-1}{2x^2}\right|_1^N = \frac{1}{2}.$$

3. $$\int_1^\infty \frac{1}{\sqrt{x}}dx = \lim_{N\to\infty}\int_1^N x^{-1/2}dx$$
$$= \lim_{N\to\infty}\left. 2x^{1/2}\right|_1^N = \infty.$$

5. $$\int_3^\infty \frac{1}{2x-1}dx$$
$$= \lim_{N\to\infty}\int_3^N \frac{1}{2x-1}dx$$
$$= \frac{1}{2}\lim_{N\to\infty}\left.\ln|2x-1|\right|_3^N = \infty.$$

7. $$\int_3^\infty \frac{1}{(2x-1)^2}dx$$
$$= \lim_{N\to\infty}\int_3^N (2x-1)^{-2}dx$$
$$= \frac{1}{2}\lim_{N\to\infty}\left.\frac{-1}{2x-1}\right|_3^N = \frac{1}{10}.$$

9. $$\int_0^\infty 5e^{-2x}dx$$
$$= 5\lim_{N\to\infty}\int_0^N e^{-2x}dx$$
$$= -\frac{5}{2}\lim_{N\to\infty}\left. e^{-2x}\right|_0^N = \frac{5}{2}.$$

11. $$\int_1^\infty \frac{x^2}{(x^3+2)^2}dx$$
$$= \lim_{N\to\infty}\frac{1}{3}\int_1^N 3x^2(x^3+2)^{-2}dx$$
$$= \frac{1}{3}\lim_{N\to\infty}\left.\frac{-1}{x^3+2}\right|_1^N = \frac{1}{9}.$$

13. $$\int_1^\infty \frac{x^2}{\sqrt{x^3+2}}dx$$
$$= \lim_{N\to\infty}\frac{1}{3}\int_1^N 3x^2(x^3+2)^{-1/2}dx$$
$$= \frac{1}{3}\lim_{N\to\infty}\left. 2(x^3+2)^{1/2}\right|_1^N = \infty.$$

15. $$\int_1^\infty \frac{e^{-\sqrt{x}}}{\sqrt{x}}dx$$
$$= \lim_{N\to\infty} 2\int_1^N e^{-\sqrt{x}}\left(\frac{1}{2\sqrt{x}}\right)dx$$
$$= -2\lim_{N\to\infty}\left. e^{-\sqrt{x}}\right|_1^N = \frac{2}{e}.$$

17. $$\int_0^\infty 2xe^{-3x}dx$$
$$= \lim_{N\to\infty}\int_0^N 2xe^{-3x}dx$$
$$= \lim_{N\to\infty}\left(-\frac{2}{3}xe^{-3x}\Big|_0^N + \frac{2}{3}\int_0^N e^{-3x}dx\right)$$
$$= \lim_{N\to\infty}\left.\left(-\frac{2}{3}xe^{-3x} - \frac{2}{9}e^{-3x}\right)\right|_0^N$$
$$= \lim_{N\to\infty}\left.\left[-\frac{2}{3}e^{-3x}\left(x+\frac{1}{3}\right)\right]\right|_0^N = \frac{2}{9}$$

19. $$\int_0^\infty 5xe^{10-x}dx$$
$$= 5e^{10}\lim_{N\to\infty}\int_0^N xe^{-x}dx$$
$$= 5e^{10}\lim_{N\to\infty}(-xe^{-x})\Big|_0^N + \int_0^N e^{-x}dx$$
$$= 5e^{10}\lim_{N\to\infty}\left.(-xe^{-x} - e^{-x})\right|_0^N = 5e^{10}$$

21. $$\int_2^\infty \frac{1}{x\ln x}dx$$
$$= \lim_{N\to\infty}\int_2^N \frac{1}{\ln x}\left(\frac{1}{x}\right)dx$$
$$= \lim_{N\to\infty}\left.\ln|\ln x|\right|_2^N = \infty$$

23.
$$\int_0^\infty x^2 e^{-x}\,dx$$
$$= \lim_{N\to\infty} \int_0^N x^2 e^{-x}\,dx$$
$$= \lim_{N\to\infty} \left(-x^2 e^{-x}\Big|_0^N + 2\int_0^N x e^{-x}\,dx \right)$$
$$= \lim_{N\to\infty} \left[(-x^2 e^{-x} - 2xe^{-x})\Big|_0^N + \int_0^N 2e^{-x}\,dx \right]$$
$$= \lim_{N\to\infty} [(-x^2 - 2x - 2)e^{-x}\Big|_0^N = 2.$$

25. To find the present value of the investment of $2,400 per year for N years, divide the N-year interval $0 \le t \le N$ into n equal sub-intervals of length Δt years, and let t_j denote the beginning of the jth sub-interval. Then, during the jth sub-interval, the amount generated is approximately $2,400\Delta t$ and the present value is

$$2,400 e^{-0.12 t_j}\Delta t.$$

Hence, the present value of an N-year investment is

$$\lim_{n\to\infty} \sum_{j=1}^n 2,400 e^{-0.12 t_j}\Delta t$$
$$= \int_0^N 2,400 e^{-0.12t}\,dt.$$

To find the present value P of the total investment, let $N \to \infty$ to get

$$P = \lim_{N\to\infty} \int_0^N 2,400 e^{-0.12t}\,dt$$
$$= -\frac{2,400}{0.12} \lim_{N\to\infty} e^{-0.12t}\Big|_0^N$$
$$= -\frac{2,400}{0.12} \lim_{N\to\infty} (e^{-0.12N} - 1) = \$20,000.$$

27. To find the present value of an apartment complex generating

$$f(t) = 10,000 + 500t$$

dollars per year for N years, divide the N-year interval $0 \le t \le N$ into n equal sub-intervals of length Δt years, and let t_j denote the beginning of the jth sub-interval. Then, during the jth sub-interval, the amount generated is approximately $f(t_j)\Delta t$ and at the interest rate of 10 %, the present value is

$$f(t_j)e^{-0.1 t_j}\Delta t.$$

Hence, the present value of the apartment complex over an N-year period is

$$\lim_{n\to\infty} \sum_{j=1}^n f(t_j)e^{-0.1 t_j}\Delta t$$
$$= \int_0^N f(t)e^{-0.1t}\,dt.$$

To find the present value P of the total income, let $N \to \infty$ to get

$$P = \lim_{N\to\infty} \int_0^N (10,000 + 500t)e^{-0.1t}\,dt$$
$$= -\lim_{N\to\infty} 10(10,000 + 500t)e^{-0.1t}\Big|_0^N$$
$$\quad + \lim_{N\to\infty} \int_0^N (5,000)e^{-0.1t}\,dt$$
$$= \lim_{N\to\infty} (-100,000 - 5,000t - 50,000)$$
$$e^{-0.1t}\Big|_0^N$$
$$= \lim_{N\to\infty} (-150,000 - 5,000t)e^{-0.1t}\Big|_0^N = \$150,000.$$

29. To find the present value of an investment generating $f(t) = A + Bt$ dollars per year for N years, divide the N-year time interval $0 \le t \le N$ into n equal sub-intervals of length Δt years, and let t_j denote the beginning of the jth sub-interval.
Then, during the jth sub-interval, the amount generated is approximately $f(t_j)\Delta t$. Hence the present value of an N-year investment is

$$\lim_{n\to\infty} \sum_{j=1}^n f(t_j)e^{-rt_j}\Delta t$$
$$= \int_0^N (A + Bt)e^{-rt}\,dt.$$

6.4. IMPROPER INTEGRALS

To find the present value P of the total income, let $N \to \infty$ to get

$$\begin{aligned} P &= \lim_{N \to \infty} \int_0^N (A + Bt)e^{-rt} dt \\ &= \lim_{N \to \infty} \left[-\frac{1}{r}(A+Bt)e^{-rt}\Big|_0^N + \frac{B}{r}\int_0^N e^{-rt} dt \right] \\ &= \lim_{N \to \infty} \left[-\frac{1}{r}(A+Bt)e^{-rt} - \frac{B}{r^2}e^{-rt} \right]\Big|_0^N \\ &= -\frac{1}{r}\lim_{N \to \infty} \left[(A+Bt) + \frac{B}{r} \right] e^{-rt}\Big|_0^N \\ &= \frac{1}{r}\left(A + \frac{B}{r} \right) = \frac{A}{r} + \frac{B}{r^2}. \end{aligned}$$

31. To find the number of patients after N months, divide the N-month time interval $0 \le t \le N$ into n equal sub-intervals of length Δt months, and let t_j denote the beginning of the jth sub-interval.
Then, the number of people starting treatment during the jth sub-interval is approximately $10\Delta t$. Of these, the number still receiving treatment at time $t = N$ (that is, $N - t_j$ months later) is approximately $10f(N - t_j)\Delta t$. Hence, the number of patients receiving treatment at time $t = N$ is

$$\lim_{n \to \infty} \sum_{j=1}^n 10f(N - t_j)\Delta t$$
$$= \int_0^N 10f(N - t) dt$$

and the number of patients receiving treatment in the long run is

$$\begin{aligned} P &= \lim_{N \to \infty} 10 \int_0^N e^{-(N-t)/20} dt \\ &= \lim_{N \to \infty} 10 e^{-N/20} \int_0^N e^{t/20} dt \\ &= \lim_{N \to \infty} 200 e^{-N/20} e^{t/20}\Big|_0^N = 200 \text{ patients.} \end{aligned}$$

33. To find the number of units of the drug in the patient's body after N hours, divide the N-hour time interval $0 \le t \le N$ into n equal sub-intervals of length Δt hours, and let t_j denote the beginning of the jth sub-interval. Then, during the jth sub-interval, approximately $5\Delta t$ units of the drug are received. Of these, the number remaining at time $t = N$ (that is, $N - t_j$ hours later) is approximately $5f(N - t_j)\Delta t$.
Hence, the number of units of the drug in the patient's body at time $t = N$ is

$$\lim_{n \to \infty} \sum_{j=1}^n 5f(N - t_j)\Delta t$$
$$= \int_0^N 5f(N - t) dt$$

and the number Q of units in the patient's body in the long run is

$$\begin{aligned} Q &= \lim_{N \to \infty} 5 \int_0^N f(N - t) dt \\ &= \lim_{N \to \infty} 5 \int_0^N e^{-(N-t)/10} dt \\ &= \lim_{N \to \infty} 5 e^{-N/10} \int_0^N e^{t/10} dt \\ &= \lim_{N \to \infty} 50 e^{-N/10} e^{t/10}\Big|_0^N = 50 \text{ units.} \end{aligned}$$

35. (a) $\quad I = \int_0^2 \frac{x}{2} dx = \frac{x^2}{4}\Big|_0^2 = 1$

(b)
$$\begin{aligned} I &= \int_1^2 \frac{x}{2} dx \\ &= \frac{x^2}{4}\Big|_1^2 = \frac{3}{4} \end{aligned}$$

(c) $\quad I = \int_0^1 \frac{x}{2} dx = \frac{x^2}{4}\Big|_0^1 = \frac{1}{4}$

37. (a)
$$\begin{aligned} I &= \frac{3}{32} \int_0^4 (4x - x^2) dx \\ &= \frac{3}{32}\left(2x^2 - \frac{x^3}{3} \right)\Big|_0^4 = 1 \end{aligned}$$

(b)
$$\begin{aligned} I &= \frac{3}{32} \int_1^2 (4x - x^2) dx \\ &= \frac{3}{32}\left(2x^2 - \frac{x^3}{3} \right)\Big|_1^2 = \frac{11}{32} \end{aligned}$$

(c)
$$I = \frac{3}{32}\int_0^1 (4x - x^2)dx$$
$$= \frac{3}{32}\left(2x^2 - \frac{x^3}{3}\right)\bigg|_0^1 = \frac{5}{32}$$

39. (a)
$$I = \frac{1}{10}\int_0^\infty e^{-x/10}dx$$
$$= \frac{1}{10}\lim_{N\to\infty}\int_0^N e^{-x/10}dx$$
$$= -\lim_{N\to\infty} e^{-x/10}\bigg|_0^N = 1$$

(b)
$$I = \frac{1}{10}\int_0^2 e^{-x/10}dx$$
$$= -e^{-x/10}\bigg|_0^2 = 0.1813$$

(c)
$$I = \frac{1}{10}\int_5^\infty e^{-x/10}dx$$
$$= \frac{1}{10}\lim_{N\to\infty}\int_5^N e^{-x/10}dx$$
$$= -\lim_{N\to\infty} e^{-x/10}\bigg|_5^N = 0.6065$$

41. (a)
$$I = \frac{1}{4}\int_0^\infty xe^{-x/2}dx$$
$$= \frac{1}{4}\lim_{N\to\infty}\int_0^N xe^{-x/2}dx$$

$f(x) = x$	$g(x) = e^{-x/2}$
$f'(x) = 1$	$G(x) = -2e^{-x/2}$

$$I = \frac{1}{4}\lim_{N\to\infty}\left[-\frac{2x}{e^{x/2}}\bigg|_0^N + 2\int_0^N e^{-x/2}dx\right]$$
$$= \frac{1}{4}\lim_{N\to\infty}\left[-\frac{2}{\frac{1}{2}e^{N/2}} - 4e^{-x/2}\bigg|_0^N\right]$$
$$= \frac{1}{4}(4+4) = 2$$

(b)
$$I = \frac{1}{4}\int_2^4 xe^{-x/2}dx$$
$$= \frac{1}{4}\int_2^4\left(-\frac{2x}{e^{x/2}}\bigg|_2^4 + 2\int_2^4 e^{-x/2}dx\right)$$
$$= \frac{1}{4}\left[-2\left(\frac{4}{e^2} - \frac{2}{e}\right) - 4e^{-x/2}\bigg|_2^4\right] = 0.3298$$

(c)
$$I = \frac{1}{4}\int_6^\infty xe^{-x/2}dx$$
$$= \frac{1}{4}\lim_{N\to\infty}\int_6^N xe^{-x/2}dx$$
$$= \frac{1}{4}\lim_{N\to\infty}\left[-\frac{2x}{e^{x/2}}\bigg|_6^N + 2\int_6^N e^{-x/2}dx\right]$$
$$= \frac{1}{4}\lim_{N\to\infty}\left[-\frac{4}{e^{N/2}} + \frac{12}{e^3} - 4e^{-x/2}\bigg|_6^N\right]$$
$$= 0.1991$$

43. (a)
$$f(x) = 0.2e^{-0.2x}$$
$$P(10 < x < 15) = 0.2\int_{10}^{15} e^{-0.2x}dx$$
$$= -e^{-0.2x}\bigg|_{10}^{15} = 0.0855$$

(b)
$$P(x < 8) = 0.2\int_0^8 e^{-0.2x}dx$$
$$= -e^{-0.2x}\bigg|_0^8 = 0.7981$$

(c)
$$P(12 < x) = 0.2\int_{12}^\infty e^{-0.2x}dx$$
$$= -e^{-0.2x}\bigg|_{12}^\infty = 0.0907$$

45. (a)
$$f(x) = 0.2e^{-0.2x}$$
$$P(0 < x < 5) = 0.2\int_0^5 e^{-0.2x}dx$$
$$= -e^{-0.2x}\bigg|_0^5 = 0.6321$$

(b)
$$P(6 \leq x) = 1 - P(0 \leq x \leq 6)$$
$$= 1 - 0.2 \int_0^6 e^{-0.2x} dx$$
$$= 1 + e^{-0.2x}\Big|_0^6 = 0.3012$$

47. The waiting period is represented by the function
$$f(t) = \frac{1}{20} \text{ for } 0 \leq t \leq 20$$

$$I = \frac{1}{20} \int_8^{20} dt = 0.6$$

49.
$$I = \frac{1}{3} \int_3^\infty e^{-x/3} dx$$
$$= \frac{1}{3} \lim_{N \to \infty} \int_3^N e^{-x/3} dx$$
$$= -\lim_{N \to \infty} e^{-x/3}\Big|_3^N$$
$$= -\lim_{N \to \infty} (e^{-N/3} - e^{-1}) = 0.3679$$

51. $P(12) = 0.08 \int_0^{12} e^{-0.08t} dt = 0.6171$

The probability that the grenade is defective after one year and the spy will expire is $1 - 0.6171 = 0.3829$.

53. (a) The capitalized cost for the first machine is

$$M_1 = 10,000 + \int_0^\infty 1,000(1+0.06t)e^{-0.09t} dt$$

$f(t) = 1,000(1+.06t)$	$g(x) = e^{-.09t}$
$f'(t) = 60$	$G(t) = -11.11e^{-.09t}$

$$M_1 = 10^4 - \lim_{N \to \infty} 11,111.11(1+.06t)e^{-.09t}\Big|_0^N$$
$$+ \int_0^N 666.67 e^{-.09t} dt = 29,185.19$$

$$M_2 = 8,000 + \int_0^\infty 1,100 e^{-0.09t} dt$$
$$= 8,000 - \lim_{N \to \infty} 12,222.22 e^{-.09t}\Big|_0^\infty$$
$$= 20,222.22$$

Thus the second machine ought to be purchased.

(b) Continuing as above Writing exercise — Answers will vary.

Review Problems

1.
$$\int_0^1 (5x^4 - 8x^3 + 1) dx$$
$$= (x^5 - 2x^4 + x)\Big|_0^1 = (1 - 2 + 1) - 0 = 0$$

2.
$$\int_1^4 (\sqrt{x} + x^{-3/2}) dx$$
$$= \frac{2x^{3/2}}{3} - 2x^{-1/2}\Big|_1^4 = \frac{17}{3}.$$

3. Let $u = 5x - 2$. Then $du = 5dx$ or $dx = \frac{1}{5} du$. When $x = -1$, $u = -7$, and when $x = 2$, $u = 8$. Hence,

$$\int_{-1}^2 30(5x - 2)^2 dx$$
$$= 6\int_{-7}^8 u^2 du = 2u^3\Big|_{-7}^8 = 1,710.$$

4. Let $u = x^2 - 1$. Then $du = 2x dx$. When $x = 0$, $u = -1$, and when $x = 1$, $u = 0$.

Hence $\int_0^1 2x e^{x^2 - 1} dx = \int_{-1}^0 e^u du$
$$= e^u\Big|_{-1}^0 = e^0 - e^{-1} = 0.6321.$$

5. Let $u = x^2 - 6x + 2$. Then $du = (2x - 6)dx$ or $(x-3)dx = \frac{1}{2}du$.
   When $x = 0$, $u = 2$, and when $x = 1$, $u = -3$.

   Hence $\int_0^1 (x-3)(x^2 - 6x + 2)^3 dx$
   $= \frac{1}{2}\int_2^{-3} u^3 du = \frac{u^4}{8}\Big|_2^{-3} = \frac{65}{8}$

6. Let $u = x^2 + 4x + 5$. Then $du = (2x + 4)dx$ or $(3x + 6)dx = \frac{3}{2}du$.
   When $x = -1$, $u = 2$, and when $x = 1$, $u = 10$.

   Hence $\int_{-1}^1 \frac{3x+6}{(x^2+4x+5)^2}dx$
   $= \frac{3}{2}\int_2^{10} u^{-2} du$
   $= -\frac{3}{2u}\Big|_2^{10} = \frac{3}{5}$

7. Let $g(x) = e^x$ and $f(x) = x$.
   Then, $G(x) = e^x$ and $f'(x) = 1$. Thus

   $\int_{-1}^1 xe^x dx = xe^x\Big|_{-1}^1 - \int_{-1}^1 e^x dx$
   $= (xe^x - e^x)\Big|_{-1}^1 = 2e^{-1} = 0.7358$

8. Let $u = \ln x$. Then $du = \frac{1}{x}dx$. When $x = e$, $u = 1$, and when $x = e^2$, $u = 2$. Hence,

   $\int_e^{e^2} \frac{1}{x(\ln x)^2}dx = \int_1^2 u^{-2}du = -\frac{1}{u}\Big|_1^2 = \frac{1}{2}$

9. Let $g(x) = x^2$ and $f(x) = \ln x$. Then, $G(x) = \frac{x^3}{3}$ and $f'(x) = \frac{1}{x}$. Thus

   $\int_1^e x^2 \ln x\, dx$
   $= \frac{x^3}{3}\ln x\Big|_1^e - \frac{1}{3}\int_1^e x^3 \frac{1}{x}dx$
   $= \frac{x^3}{3}\ln x\Big|_1^e - \frac{1}{3}\int_1^e x^2 dx$
   $= \left(\frac{x^3}{3}\ln x - \frac{x^3}{9}\right)\Big|_1^e$
   $= \frac{x^3}{3}\left(\ln x - \frac{1}{3}\right)\Big|_1^e = \frac{2e^3 + 1}{9}$

10. Let $g(x) = e^{0.2x}$ and $f(x) = 2x + 1$. Then, $G(x) = 5e^{0.2x}$ and $f'(x) = 2$. Thus

    $\int_0^{10} (2x+1)e^{0.2x}dx$
    $= 5(2x+1)e^{0.2x}\Big|_0^{10} - 10\int_0^{10} e^{0.2x}dx$
    $= [5(2x+1)e^{0.2x} - 50e^{0.2x}]\Big|_0^{10} = 55e^2 + 45$

11. $\int_0^\infty \frac{1}{\sqrt[3]{1+2x}}dx$
    $= \lim_{N\to\infty}\int_0^N (1+2x)^{-1/3}dx$
    $= \lim_{N\to\infty} \frac{3}{4}(1+2x)^{2/3}\Big|_0^N = \infty$.

12. $\int_0^\infty (1+2x)^{-3/2}dx$
    $= \lim_{N\to\infty}\int_0^N (1+2x)^{-3/2}dx$
    $= -\lim_{N\to\infty}(1+2x)^{-1/2}\Big|_0^N = 1$.

13. $\int_0^\infty \frac{3x}{x^2+1}dx$
    $= 3\lim_{N\to\infty}\int_0^N x^{-1}(x^2+1)^{-1}dx$
    $= 3\lim_{N\to\infty}\frac{1}{2}\ln(x^2+1)\Big|_0^N = \infty$.

14. $\int_0^\infty 3e^{-5x}dx = 3\lim_{N\to\infty}\int_0^N e^{-5x}dx$
    $= -\frac{3}{5}\lim_{N\to\infty} e^{-5x}\Big|_0^N = \frac{3}{5}$.

15.
$$\int_0^\infty xe^{-2x}dx$$
$$= \lim_{N\to\infty} \int_0^N xe^{-2x}dx$$
$$= \lim_{N\to\infty} \left(-\frac{1}{2}xe^{-2x}\right)\Big|_0^N + \frac{1}{2}\int_0^N e^{-2x}dx$$
$$= \lim_{N\to\infty} \left(-\frac{1}{2}xe^{-2x} - \frac{1}{4}e^{-2x}\right)\Big|_0^N = \frac{1}{4}$$

16.
$$\int_0^\infty 2x^2 e^{-x^3}dx$$
$$= 2\lim_{N\to\infty} \int_0^N x^2 e^{-x^3}dx$$
$$= 2\lim_{N\to\infty} \left(-\frac{1}{3}\right)e^{-x^3}\Big|_0^N = \frac{2}{3}.$$

17.
$$\int_0^\infty x^2 e^{-2x}dx = \lim_{N\to\infty}\int_0^N x^2 e^{-2x}dx$$
$$= -\lim_{N\to\infty} \frac{1}{2}x^2 e^{-2x}\Big|_0^N$$
$$+ \lim_{N\to\infty}\int_0^N xe^{-2x}dx$$
$$= -\lim_{N\to\infty}\frac{1}{2}x^2 e^{-2x}\Big|_0^N$$
$$-\lim_{N\to\infty}\frac{1}{2}xe^{-2x}\Big|_0^N$$
$$-\lim_{N\to\infty}\frac{1}{4}e^{-2x}\Big|_0^N = \frac{1}{4}$$

18.
$$\int_2^\infty \frac{1}{x(\ln x)^2}dx$$
$$= \lim_{N\to\infty}\int_2^N \frac{1}{(\ln x)^2}\frac{dx}{x}$$
$$= \lim_{N\to\infty}\left(-\frac{1}{\ln x}\right)\Big|_2^N = \frac{1}{\ln 2}$$

19.
$$\int_0^\infty \frac{x-1}{x+2}dx$$
$$= \lim_{N\to\infty}\int_0^N \left(1 - \frac{3}{x+2}\right)dx$$
$$= \lim_{N\to\infty}(x - 3\ln|x+2|)\Big|_0^N = \infty.$$

(Note: x grows much more quickly than $\ln x$.)

20.
$$\int_0^\infty x^5 e^{-x^3}dx = \lim_{N\to\infty}\int_0^N x^3(x^2 e^{-x^3})dx$$
$$= \lim_{N\to\infty}\left[\left(-\frac{1}{3}x^3 e^{-x^3}\right)\Big|_0^N + \int_0^N x^2 e^{-x^3}dx\right]$$
$$= \lim_{N\to\infty}\left(-\frac{1}{3}x^3 e^{-x^3} - \frac{1}{3}e^{-x^3}\right)\Big|_0^N = \frac{1}{3}$$

21. (a) $P(1 \leq X \leq 4) = \int_1^4 f(x)dx$
$$= \int_1^4 \frac{1}{3}dx = \frac{x}{3}\Big|_1^4 = 1.$$

(b) $P(2 \leq X \leq 3) = \int_2^3 f(x)dx$
$$= \int_2^3 \frac{1}{3}dx = \frac{x}{3}\Big|_2^3 = \frac{1}{3}.$$

(c) $P(X \leq 2) = \int_{-\infty}^2 f(x)dx$
$$= \int_1^2 \frac{1}{3}dx = \frac{x}{3}\Big|_1^2 = \frac{1}{3}.$$

22. (a) $P(0 \leq X \leq 3) = \int_0^3 f(x)dx$
$$= \int_0^3 \frac{2(3-x)}{9}dx$$
$$= \frac{2}{9}\left(3x - \frac{x^2}{2}\right)\Big|_0^3$$
$$= \frac{2}{9}\left(9 - \frac{9}{2} - 0\right) = 1.$$

(b)
$$P(1 \leq X \leq 2)$$
$$= \int_1^2 f(x)dx$$
$$= \int_1^2 \frac{2(3-x)}{9}dx$$
$$= \frac{2}{9}\left(3x - \frac{x^2}{2}\right)\Big|_1^2$$
$$= \frac{2}{9}\left[\left(6 - \frac{4}{2}\right) - \left(3 - \frac{1}{2}\right)\right] = \frac{1}{3}.$$

23. (a)
$$P(0 \leq X) = \int_0^\infty f(x)dx$$
$$= \lim_{N \to \infty} \int_0^N 0.2e^{-0.2x}dx$$
$$= \lim_{N \to \infty} (-e^{-0.2x})\Big|_0^N$$
$$= \lim_{N \to \infty} (-e^{-0.2N} + 1) = 1.$$

(b)
$$P(1 \leq X \leq 4)$$
$$= \int_1^4 f(x)dx$$
$$= \int_1^4 0.2e^{-0.2x}dx$$
$$= -e^{-0.2x}\Big|_1^4$$
$$= -e^{-0.8} + e^{-0.2} = 0.3694.$$

(c)
$$P(5 \leq x)$$
$$= \int_5^\infty f(x)dx$$
$$= \lim_{N \to \infty} \int_5^N 0.2e^{-0.2x}dx$$
$$= -\lim_{N \to \infty} e^{-0.2x}\Big|_5^N$$
$$= \lim_{N \to \infty} [-e^{-0.2N} + e^{-1}] = 0.3679.$$

24. (a)
$$P(0 < X) = \lim_{N \to \infty} \int_0^N \frac{5}{(x+5)^2}dx$$
$$= -5\lim_{N \to \infty} \frac{1}{x+5}\Big|_0^N = 1$$

(b)
$$P(1 \leq X \leq 9) = \int_1^9 \frac{5}{(x+5)^2}dx$$
$$= -\frac{5}{x+5}\Big|_1^9 = \frac{5}{6} - \frac{5}{14} = 0.4762$$

(c)
$$P(3 \leq X)$$
$$= \lim_{N \to \infty} \int_3^N \frac{5}{(x+5)^2}dx$$
$$= -\lim_{N \to \infty} \frac{5}{x+5}\Big|_3^N$$
$$= \frac{5}{8} = 0.625.$$

25.
$$y = f(x) = x^3 - 3x + \sqrt{2}x^{1/2}$$
$$f_{av} = \frac{1}{7}\int_1^8 (x^3 - 3x + \sqrt{2}x^{1/2})dx$$
$$= \frac{1}{7}\left[\frac{x^4}{4} - \frac{3x^2}{2} + \frac{2\sqrt{2}x^{3/2}}{3}\right]\Big|_1^8 = 135.6629$$

26.
$$f(t) = t\sqrt[3]{10 - 2t^2}$$
$$f_{av} = \int_0^1 t(10 - 2t^2)^{1/3}dt$$
$$= -\frac{3}{16}(10 - 2t^2)^{4/3}\Big|_0^1 = 1.0396$$

27.
$$g(u) = e^{-u}(u + 3)$$
$$g_{av} = \frac{1}{2}\int_0^2 e^{-u}(u+3)du$$

$g(u) = u + 3$	$h(u) = e^{-u}$
$g'(u) = 1$	$H(u) = -e^{-u}$

$$g_{av} = \frac{1}{2}[-(u+3)e^{-u} - e^{-u}]\Big|_0^2 = 1.594$$

28.
$$f(x) = x\ln\sqrt{x} = \frac{1}{2}x\ln x$$
$$f_{av} = \frac{1}{8}\int_1^5 x\ln x\,dx$$

$f(x) = \ln x$	$g(x) = x$
$f'(x) = \frac{1}{x}$	$G(x) = \frac{x^2}{2}$

$$f_{av} = \frac{1}{8}\left(\frac{x^2}{2}\ln x - \frac{x^2}{4}\right)\Big|_1^5 = 1.7647$$

REVIEW PROBLEMS

29. Recall that P dollars invested at an annual interest rate of 8 percent compounded continuously will be worth

$$Pe^{0.08t}$$

dollars after t years.

Let t_j denote the time (in years) of the jth deposit of \$1,200. This deposit will remain in the account for $5 - t_j$ years and hence will grow to

$$1,200e^{0.08(5-t_j)}$$

dollars. At the end of 5 years the amount in the account is

$$\sum_{j=1}^{n} 1,200e^{0.08(5-t_j)}.$$

Rewrite the sum as

$$\sum_{j=1}^{n} 1,200e^{0.08(5-t_j)}\Delta t.$$

This sum can be approximated by the definite integral

$$\int_0^5 1,200e^{0.08(5-t)}dt$$

$$= 1,200e^{0.4}\int_0^5 e^{-0.08t}dt$$

$$= \frac{1,200e^{0.4}}{-0.08}(e^{-0.08t})\Big|_0^5 = \$7,377.37.$$

30. Recall that the present value of B dollars payable t years from now with an annual interest rate of 7 percent compounded continuously is

$$Be^{-0.07t}.$$

Divide the interval $0 \leq t \leq 10$ into n equal sub-intervals of length Δt years, and let t_j denote the beginning of the jth sub-interval. During the jth sub-interval, the income will equal (dollars per year)(number of years)$= 1,000\Delta t$ and the present value of the income will

$$= 1,000(\Delta t)e^{-0.07t_j} = 1,000e^{-0.07t_j}\Delta t.$$

Hence, over the entire 10 years, the present value of the investment

$$= \sum_{j=1}^{n} 1,000e^{-0.07t_j}\Delta t.$$

Now as n increases without bound, this approximation improves and the sum approaches the corresponding integral. Hence, the present value is

$$\int_0^{10} 1,000e^{-0.07t}dt$$

$$= \frac{1,000}{-0.07}e^{-0.07t}\Big|_0^{10}$$

$$= -\frac{1,000}{0.07}(e^{-0.7} - 1) = \$7,191.64.$$

31. Let $f(t) = e^{-0.2t}$ denote the fraction of the homes that will remain unsold for t weeks. Of the 200 homes currently on the market, the number that will still be on the market 10 weeks from now is

$$200f(10) = 200e^{-2}.$$

To find the number of additional homes on the market 10 weeks from now, divide the interval $0 \leq t \leq 10$ into n equal sub-intervals of length Δt weeks and let t_j denote the beginning of the jth sub-interval.

During the jth sub-interval, $8\Delta t$ additional homes are placed on the market, and 10 weeks from now (that is $10 - t_j$ weeks later) the number of these still on the market will be

$$8(\Delta t)f(10 - t_j).$$

Hence, the total number N of homes on the market 10 weeks from now will be approximately

$$N = 200f(10) + \sum_{j=1}^{n} 8f(10 - t_j)\Delta t.$$

Now, as n increases without bound, this approximation improves while the sum approaches the corresponding integral. Hence,

$$\begin{aligned} N &= 200f(10) + \int_0^{10} 8f(10-t)dt \\ &= 200e^{-2} + 8\int_0^{10} e^{-0.2(10-t)}dt \\ &= 200e^{-2} + 8e^{-2}\int_0^{10} e^{0.2t}dt \\ &= 200e^{-2} + \frac{8e^{-2}}{0.2}e^{0.2t}\Big|_0^{10} = 61.65 \end{aligned}$$

or about 62 homes.

32. Let's focus on one bicycle at a time. The price per month is

$$P(x) = 80 + 3\sqrt{x}$$

so the monthly revenue for one bicycle is

$$R(x) = \int (80 + 3x^{1/2})dx = 80x + 3\frac{x^{3/2}}{3/2} + C$$

The initial revenue, when $x = 0$, will be $R(0)$. Thus

$$R(x) = 80x + 3\frac{x^{3/2}}{3/2} + R(0)$$

Over 16 months, that is when $x = 16$,

$$R(16) = 1,280 + 128 + R(0)$$

The revenue per bicycle is $R(16) - R(0)$ and for 5,000 bicycles it is
$\$5,000 \times 1,408 = 7,040,000$

33. The density at a point r miles from the center is

$$D(r) = 25,000e^{-0.05r}$$

people per square mile.
The area of the ring of thickness dr at r miles from the center is $2\pi r dr$, so

$$50,000\pi r e^{-0.05r} dr$$

people live in that ring.
The total number of people is

$$N(r) = 50,000\pi \int_1^2 re^{-0.05r} dr.$$

$f(r) = r$	$g(r) = e^{-0.05r}$
$f'(r) = 1$	$G(r) = -20e^{-0.05r}$

$$\begin{aligned} N(r) &= 50,000\pi[-20re^{-0.05r}\Big|_1^2 \\ &\quad + 20\int_1^2 e^{-0.05r} dr] \\ &= 50,000\pi[-20(1.8097 - 0.95123) \\ &\quad -400(-0.04639)] \\ &= 218,010 \text{ people.} \end{aligned}$$

34. Divide the interval $0 \le r \le 3$ into n equal subintervals of length Δr, and let r_j denote the beginning of the jth subinterval.
This divides the circular disc of radius 3 into n concentric circles, as shown in the figure.
If Δr is small, the area of the jth ring is

$$2\pi r_j \Delta r$$

where $2\pi r_j$ is the circumference of the circle of radius r_j that forms the inner boundary of the ring and Δr is the width of the ring.
Then, since $D(r) = 5,000e^{-0.1r}$ is the population density (people per square mile) r miles from the center, it follows that the number of people in the jth ring is

$$\begin{aligned} &D(r_j)(\text{area of the } j^{th} \text{ ring}) \\ &= 5,000e^{-0.1r_j}(2\pi r_j \Delta r) \end{aligned}$$

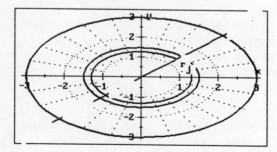

Hence, if N is the total number of people within 3 miles of the center of the city,

$$N = \lim_{n\to\infty} \sum_{j=1}^n 5,000e^{-0.1r_j}(2\pi r_j \Delta r)$$

REVIEW PROBLEMS

$$= \int_0^3 5,000(2\pi)re^{-0.1r}dr$$

$$= 10,000\pi \int_0^3 re^{-0.1r}dr$$

Applying integration by parts,

$$\int_0^3 re^{-0.1r}dr$$

$$= -10re^{-0.1r}\Big|_0^3 - \int_0^3 (-10)e^{-0.1r}dr$$

$$= (-10re^{-0.1r} - 100e^{-0.1r})\Big|_0^3$$

$$= (-30e^{-0.3} - 100e^{-0.3}) - (-100) = 3.6936$$

Hence, the total number of people within 10 miles of the center of the city is

$$N = 10,000\pi \int_0^3 re^{-0.1r}dr$$

$$= 116,038$$

(Your answer may differ slightly due to round-off errors.)

35. Let $D(t)$ denote the demand for oil after t years. Then

$$D'(t) = D_0 e^{0.1t} \text{ and } D_0 = 30$$

$$D(t) = 30 \int_0^{10} e^{0.1t}dt = 30 \times 10e^{0.1t}\Big|_0^{10}$$

$$= 300(e-1) = 515.48 \text{ billion barrels}$$

36. Let $D(t)$ denote the demand for the product. Since the current demand is 5,000 and the demand increases exponentially,

$$D(t) = 5,000e^{0.02t}$$

units per year.
Let $R(t)$ denote the total revenue t years from now. Then the rate of change of revenue is

$$\frac{dR}{dt} = \frac{\text{dollars}}{\text{year}}$$

$$= \frac{\text{dollars}}{\text{unit}} \frac{\text{units}}{\text{year}}$$

$$= 400D(t) = 400(5,000e^{0.02t})$$

$$= 2,000,000e^{0.02t}.$$

The increase in revenue over the next 2 years is

$$R(2) - R(0) = \int_0^2 2,000,000e^{0.02t}dt$$

$$= 100,000,000e^{0.02t}\Big|_0^2$$

$$= \$4,081,077.$$

37. Let $Q(t)$ denote the production after t hours. Then $t = 0$ at 8:00 a.m., $t = 2$ at 10:0 a.m., and $t = 4$ at noon. Thus,

$$Q(t) = 100 \int_2^4 te^{-0.5t}dt$$

Note: This problem can be solved by integration by parts, discussed in section 5, or you can use your graphing utility.

$f(t) = t$	$g(t) = e^{-0.5t}$
$f'(t) = 1$	$G(t) = -2e^{-0.5t}$

$$Q(t) = 100\left[-2te^{-0.5t}\Big|_2^4 + 2\int_0^4 e^{-0.5t}dt\right]$$

$$= 100\left[-8e^{-2} + 4e^{-1} - 4e^{-0.5t}\Big|_2^4\right]$$

$$= 131.90 \text{ units}$$

38. Let $R(t)$ denote the total revenue generated during the next t months and $P(t)$ the price of oil t months from now.
Then, $P(t) = 16 + 0.08t$ and

$$\frac{dR}{dt} = \frac{\text{dollars}}{\text{month}}$$

$$= \frac{\text{dollars}}{\text{barrel}}\frac{\text{barrels}}{\text{month}} = P(t)(900)$$

$$= 900(16 + 0.08t).$$

$$R(t) = 900\int (16 + 0.08t)dt$$

$$= 900(16t + 0.04t^2) + C.$$

Since $R(0) = 0$ it follows that $C = 0$ and the appropriate particular solution is

$$R(t) = 900(16t + 0.04t^2).$$

Since the well will run dry in 36 months, the total future revenue will be

$$R(36) = 900[16(36) + 0.04(36)^2] = \$565,056.$$

39.
$$\begin{aligned} A &= \int_{-1}^{3}(3x^2+2)dx \\ &= (x^3+2x)\Big|_{-1}^{3} = 36 \end{aligned}$$

40. $A = \int_{1}^{4} \dfrac{1}{x^2}dx = -\dfrac{1}{x}\Big|_{1}^{4} = \dfrac{3}{4}$

41.
$$\begin{aligned} A &= \int_{-1}^{2}(2+x-x^2)dx \\ &= \left(2x+\dfrac{x^2}{2}-\dfrac{x^3}{3}\right)\Big|_{-1}^{2} = \dfrac{9}{2} \end{aligned}$$

42. Break R into two subregions R_1 and R_2 as in the accompanying figure.
The area of $R =$ that of R_1+ that of
$$\begin{aligned} R_2 &= \int_{0}^{4}\sqrt{x}\,dx + \int_{4}^{8}\dfrac{8}{x}dx \\ &= \dfrac{2}{3}x^{3/2}\Big|_{0}^{4} + 8\ln|x|\Big|_{4}^{8} = \dfrac{16}{3} + 8\ln 2. \end{aligned}$$

43. $\int_{0}^{1}(x - x^4)dx = \left(\dfrac{x^2}{2} - \dfrac{x^5}{5}\right)\Big|_{0}^{1} = \dfrac{3}{10}$

44. Break R into two subregions R_1 and R_2 as in the accompanying figure.
Area of $R =$ area of R_1+ area of R_2
$$\begin{aligned} &= \int_{0}^{1}(7x-x^2)dx + \int_{1}^{2}[(8-x^2)-x^2]dx \\ &= \int_{0}^{1}(7x-x^2)dx + \int_{1}^{2}(8-2x^2)dx \\ &= \left(\dfrac{7x^2}{2}-\dfrac{x^3}{3}\right)\Big|_{0}^{1} + \left(8x-\dfrac{2x^3}{3}\right)\Big|_{1}^{2} = \dfrac{13}{2}. \end{aligned}$$

45. (a) The demand function is
$$D(q) = 50 - 3q - q^2$$
dollars per unit.

To find the number of units bought when the price is $p = 32$, solve the equation $32 = D(q)$ for q to get

$$32 = 50 - 3q - q^2,$$
$$(q+6)(q-3) = 0 \text{ or } q = 3 \text{ units}$$

(b) The amount that consumers are willing to spend to get 3 units of the commodity is

$$\begin{aligned} &\int_{0}^{3} D(q)dq \\ &= \int_{0}^{3}(50 - 3q - q^2)dq \\ &= \left(50q - \dfrac{3q^2}{2} - \dfrac{q^3}{3}\right)\Big|_{0}^{3} = \$127.50. \end{aligned}$$

(c) When the market price is \$32 per unit, 3 units will be bought and the consumer's surplus will be

$$\begin{aligned} &\int_{0}^{3} D(q)dq - (32)(3) \\ &= \int_{0}^{3}(50 - 3q - q^2)dq - 96 = \$31.5. \end{aligned}$$

(d) The consumer's willingness to spend in part b) is equal to the area under the demand curve $p = D(q)$ from $q = 0$ to $q = 3$.
The consumers' surplus in part (c) is equal to the area of the region between the demand curve and the horizontal line $p = 32$.

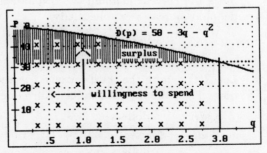

46. Since the price of chicken t months after the beginning of the year is

$$P(t) = 0.06t^2 - 0.2t + 1.2$$

dollars per pound, the average price during the first six months is

$$\frac{1}{6-0}\int_0^6 (0.06t^2 - 0.2t + 1.2)dx$$

$$= \frac{1}{6}(0.02t^3 - 0.1t^2 + 1.2t)\Big|_0^6$$

$$= \frac{1}{6}[0.02(6)^3 - 0.1(6)^2 + 1.2(6)] = \$1.32$$

per pound.

47. In N years the population of the city will be

$$P_0 f(N) + \int_0^N r(t)f(N-t)dt$$

where $P_0 = 100,000$ is the current population,

$$f(t) = e^{-t/20}$$

is the fraction of the residents remaining for at least t years, and

$$r(t) = 100t$$

is the rate of new arrivals. Hence, in the long run, the number of residents will be

$$\lim_{N\to\infty}[100,000e^{-N/20} + \int_0^N 100te^{-(N-t)/20}dt]$$

$$= 0 + \lim_{N\to\infty} 100e^{-N/20}\int_0^N te^{t/20}dt]$$

$$= \lim_{N\to\infty} 100e^{-N/20}[20te^{t/20} - 400e^{t/20}]\Big|_0^N$$

$$= \lim_{N\to\infty} 100(20N - 400 + 400e^{-N/20}) = \infty.$$

Thus the population will increase without bound.

48. The probability density function is

$$f(x) = 0.4e^{-0.4x}.$$

(a)
$$P(1 \leq X \leq 2) = \int_1^2 0.4e^{-0.4x}dx$$

$$= -e^{-0.4x}\Big|_1^2 = 0.2210$$

which corresponds to 22.10 %.

(b)
$$P(X \leq 2) = \int_0^2 0.4e^{-0.4x}dx$$

$$= -e^{-0.4x}\Big|_0^2 = 0.5507$$

which corresponds to 55.07%.

(c)
$$P(X \geq 2) = 1 - P(X \leq 2) = 1 - 0.5507 = 0.4493$$

which corresponds to 44.93%.

49. It was determined in exercise 37 of section 6.1 that the number of subscribers in N years will be

$$P_0 f(N) + \int_0^N r(t)f(N-t)dt$$

where $P_0 = 20,000$ is the current number of subscribers,

$$f(t) = e^{-t/10}$$

is the fraction of subscribers remaining at least t years, and

$$r(t) = 1,000$$

is the rate at which new subscriptions are sold. Hence, the number of subscribers in the long run is

$$\lim_{N\to\infty}[(20,000e^{-N/10})$$

$$+ \int_0^N 1,000e^{-(N-t)/10}dt]$$

$$= \lim_{N\to\infty}[(20,000e^{-N/10})$$

$$+ 1,000e^{-N/10}\int_0^N e^{t/10}dt]$$

$$= 0 + 10,000\lim_{N\to\infty} e^{-N/10}e^{t/10}\Big|_0^N = 10,000.$$

50. To find the present value of the investment in N years, divide the N-year interval $0 \leq t \leq N$ into n equal sub-intervals of length Δt years, and let t_j denote the beginning of the jth sub-interval.

Then, during the jth sub-interval, the amount generated is approximately $f(t_j)\Delta t$ and the present value is

$$f(t_j)e^{-0.1t_j}\Delta t$$

Hence, the present value of an N-year investment is

$$\lim_{n\to\infty}\sum_{j=1}^{n}f(t_j)e^{-0.1t_j}\Delta t$$

$$= \int_0^N f(t)e^{-0.1t}dt.$$

To find the present value P of the total investment, let

$$\begin{aligned}P &= \lim_{N\to\infty}\int_0^N f(t)e^{-0.1t}dt \\ &= \lim_{N\to\infty}\int_0^N (8,000+400t)e^{-0.1t}dt \\ &= \lim_{N\to\infty}[-10(8,000+400t)e^{-0.1t}]\Big|_0^N \\ &\quad +4,000\lim_{N\to\infty}\int_0^N e^{-0.1t}dt \\ &= \lim_{N\to\infty}[-10(8,000+400t)e^{-0.1t} \\ &\quad -40,000e^{-0.1t}]\Big|_0^N \\ &= \lim_{N\to\infty}(-120,000-4,000t)e^{-0.1t}\Big|_0^N = \$120,000.\end{aligned}$$

51. Let x denote the time (in minutes) between your arrival and the next batch of cookies. Then x is uniformly distributed with probability density function

$$f(x) = \begin{cases} \dfrac{1}{45} & \text{if } 0 \leq x \leq 45 \\ 0 & \text{otherwise}\end{cases}$$

Hence, the probability that you arrive within 5 minutes (before or after) the cookies were baked is

$$P(0 \leq X \leq 5) + P(40 \leq X \leq 45)$$
$$= 2P(0 \leq X \leq 5)$$
$$= 2\int_0^5 \frac{1}{45}dx = \frac{2x}{45}\Big|_0^5 = \frac{2}{9}.$$

52. Let x denote the time (in minutes) between the arrivals of successive cars. Then the probability density function is

$$f(x) = \begin{cases} 0.5e^{-0.5x} & \text{if } 0 \leq x \\ 0 & \text{otherwise}\end{cases}$$

The probability that two cars will arrive at least 6 minutes apart is

$$\begin{aligned}&P(6 \leq X < \infty) \\ &= \int_6^\infty f(x)dx \\ &= \lim_{N\to\infty}\int_6^N 0.5e^{-0.5x}dx \\ &= \lim_{N\to\infty}(-e^{-0.5x})\Big|_6^N = 0.0498.\end{aligned}$$

53. (a)
$$\int_5^\infty 0.07e^{-0.07u}du = -\lim_{N\to\infty}e^{-0.07u}\Big|_5^N = 0.7047$$

(b)
$$\int_{10}^{15} 0.07e^{-0.07u}du = 0.1466$$

54. (a)
$$f(x) = 2\sqrt{x} + \frac{1}{x+1}$$

$$I = \int_0^2 f(x)dx$$

x	$f(x)$	x	$f(x)$
0.0	1 ***	0.0	1 ***
0.5	2.08	0.25	1.8
1.0	2.5	0.5	2.08
1.5	2.849	0.75	2.303
2.0	3.162	1.0	2.5
		1.25	2.681
		1.5	2.849
		1.75	3.009
		2.0	3.162
total:	10.591	total:	20.384

(b) $I_4 = 0.5(10.591) = 5.296$

$I_8 = 0.25(20.384) = 5.096$

The leftmost point was not used in the calculation. Since the graph of the function increases, the approximating rectangles, using the right endpoint of each interval, lead to an overestimate. The actual value of the integral, as estimated by the graphing utility, is $I = 4.8698$.

55.
$$f(x) = \frac{\ln x}{x}$$

$$f_{av} = \frac{1}{e^2 - 1} \int_1^{e^2} \frac{\ln x}{x} = 0.313$$

First turn on your graphics utility – that's very important.
For the HP48G:
Enter $\frac{\ln x}{x}$ in equation writer form and store in EXPR.
Now numerically integrate from 1 to 7.389 to get 2 which is divided by $e^2 - 1$ for the result to be 0.313.
For the TI-85:
Press 2nd CALC and F5 for FNINT. Enter $((\ln x)/x, x, 1, e^2)$ ENTER. The result is 1.9999999997 (say 2) and divide by $e^2 - 1$ just like above.

56.
$$I_1 = \int_0^{10} e^{-1.1x} dx = 0.909\,076$$

$$I_2 = \int_0^{50} e^{-1.1x} dx = 0.909\,0909$$

$$I_3 = \int_0^{100} e^{-1.1x} dx = 0.909\,090\,909$$

$$I_4 = \int_0^{1,000} e^{-1.1x} dx = 0.909\,090\,909$$

This integral converges to 0.90909.
Hint: For the HP48G Compose $f(n) = \int_0^n e^{-1.1x} dx$ using the equation writer and store as a function. Then press $f(10)$, $f(50)$, $f(100)$, $f(1,000)$ and numerically evaluate each result.
For the TI-85, use 2nd CALC F5, FNINT(), ENTER, and 2nd ENTRY to reproduce the previous command line.

57.
$$I_1 = \int_0^{10} \frac{1}{x+2} dx = 1.7918$$

$$I_1 = \int_0^{50} \frac{1}{x+2} dx = 3.2581$$

$$I_1 = \int_0^{100} \frac{1}{x+2} dx = 3.9318$$

$$I_1 = \int_0^{1,000} \frac{1}{x+2} dx = 6.2166$$

This integral does not behave as if it were finite. It diverges.

58. $f(x) = 1 - x^2$, $a = 0$, and $b = 1$
$$A = \int_0^1 (1-x^2) dx = \left(x - \frac{x^3}{3}\right)\bigg|_0^1 = \frac{2}{3}$$

$$\bar{x} = \frac{3}{2} \int_0^1 (x - x^3) dx$$
$$= \frac{3}{2}\left(\frac{x^2}{2} - \frac{x^4}{4}\right)\bigg|_0^1 = \frac{3}{8}$$

$$\bar{y} = \frac{3}{4} \int_0^1 (1 - 2x^2 + x^4) dx$$
$$= \frac{3}{4}\left(x - \frac{2x^3}{3} + \frac{x^5}{5}\right)\bigg|_0^1 = \frac{2}{5}$$

59.
$$y = \ln x$$
$$A = \int_1^{e^2} \ln x\, dx$$
$$= (x \ln x - x)\bigg|_1^{e^2} = e^2 + 1$$

For the numerator of $\bar{x}$

$$I_1 = \int_1^{e^2} x \ln x\, dx$$

$f(x) = \ln x$	$g(x) = x\, dx$
$f'(x) = \dfrac{dx}{x}$	$G(x) = \dfrac{x^2}{2}$

$$I_1 = \frac{1}{2} x^2 \ln x \bigg|_1^{e^2} - \int_1^{e^2} \frac{x}{2} dx$$
$$= e^4 - \frac{x^2}{4}\bigg|_1^{e^2} = \frac{3e^4 + 1}{4}$$

Thus $\bar{x} = \dfrac{3e^4+1}{4(e^2+1)} \approx 4.911$

For the numerator of $\bar{y}$

$$\begin{aligned} I_2 &= \int_1^{e^2} (\ln x)^2 dx \\ &= x(\ln x)^2 \Big|_1^{e^2} - 2\int_1^{e^2} \ln x\, dx \\ &= 4e^2 - 2(x\ln x - x)\Big|_1^{e^2} = 2(e^2-1) \end{aligned}$$

Thus $\bar{y} = \dfrac{2(e^2-1)}{2(e^2+1)} \approx 0.762$

60. The Gini index for high school teachers is

$$G_1 = 2\int_0^1 [x - 0.33x^3 - 0,67x^4]dx \approx 0.567$$

The Gini index for real estate brokers is

$$\begin{aligned} G_2 &= 2\int_0^1 [0.72x - 0.72x^2]dx \\ &= 1.44\left[\dfrac{x^2}{2} - \dfrac{x^3}{3}\right]\Big|_0^1 \\ &= 0.24 \end{aligned}$$

The distribution of incomes amongst real estate brokers is more fairly distributed.

61. $y = -x^3 - 2x^2 + 5x - 2$

intersects $y = x\ln x$

at $(0.4062, -0.3660)$ and $(1, 0)$ according to the graphics utility.
Numerically integrate

$$\int_{0.406}^1 (-x^3 - 2x^2 + 5x - 2 - x\ln x)dx$$

to get $A = 0.1692$

62. $y = \dfrac{x-2}{x+1}$

intersects

$y = \sqrt{25-x^2}$

at $(-1.8204, 4.6568)$ and $(-4.6568, 1.8204)$ according to the graphics utility.
To find the area numerically integrate

$$\int_{-4.6568}^{-1.8204} \left[\sqrt{25-x^2} - \dfrac{x-2}{x+1}\right]dx$$

to get $A = 2.9987$

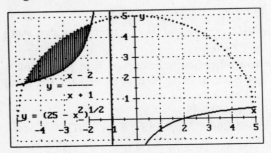

Chapter 7

Calculus of Several Variables

7.1 Functions of Several Variables

1.
$$f(x,y) = (x-1)^2 + 2xy^3.$$

The domain consists of all ordered pairs (x, y) of real numbers. Moreover

$$f(2,-1) = (2-1)^2 + 2(2)(-1)^3 = -3$$
$$f(1,2) = (1-1)^2 + 2(1)(2)^3 = 16.$$

3.
$$g(x,y) = \sqrt{y^2 - x^2}$$

The domain consists of all ordered pairs (x, y) of real numbers for which $y^2 - x^2 \geq 0$, or equivalently, for which $|y| \geq |x|$. Moreover

$$g(4,5) = \sqrt{5^2 - 4^2} = \sqrt{9} = 3$$
$$g(-1,2) = \sqrt{2^2 - (-1)^2} = \sqrt{3} = 1.732.$$

5.
$$f(r,s) = \frac{s}{\ln r}.$$

The domain consists of all ordered pairs (r, s) of real numbers for which $r > 0$ (since $\ln r$ is defined only for positive values of r) and $r \neq 1$ (since $\ln 1 = 0$). Moreover

$$f(e^3, 3) = \frac{3}{\ln e^2} = \frac{3}{2}$$

$$f(\ln 9, e^3) = \frac{e^3}{\ln(\ln 9)} = 25.515$$

7.
$$g(x,y) = \frac{y}{x} + \frac{x}{y}$$
$$g(1,2) = \frac{2}{1} + \frac{1}{2} = \frac{5}{2}$$
$$g(2,-3) = \frac{-3}{2} + \frac{2}{-3} = -\frac{13}{6}$$

9.
$$f(x,y,z) = xyz$$
$$f(1,2,3) = (1)(2)(3) = 6$$
$$f(3,2,1) = (3)(2)(1) = 6$$

11.
$$F(r,s,t) = \frac{\ln(r+t)}{r+s+t}$$
$$f(1,1,1) = \frac{\ln(2)}{3}$$
$$f(0, e^2, 3e^2) = \frac{\ln(3e^2)}{4e^2} = \frac{2 + \ln 3}{4e^2}$$

13.
$$f(x,y) = \frac{5x + 2y}{4x + 3y}$$

The domain of $f(x, y)$ is the set of real numbers excluded from the line $4x + 3y = 0$

15.
$$f(x,y) = \sqrt{x^2 - y}$$
The domain of $f(x, y)$ is consists of $y \leq x^2$

17.
$$f(x,y) = \ln(x + y - 4)$$
The domain of $f(x, y)$ is consists of $x + y > 4$

193

19.

$$f(x,y) = x + 2y.$$

With $C = 1$, $C = 2$, and $C = -3$, the three sketched level curves have equations $x + 2y = 1$, $x + 2y = 2$, and $x + 2y = -3$.

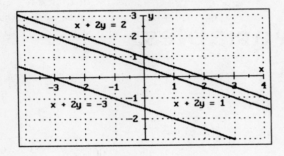

21.

$$f(x,y) = x^2 - 4x - y.$$

With $C = -4$ and $C = 5$, the two sketched level curves have equations $x^2 - 4x - y = -4$ and $x^2 - 4x - y = 5$.

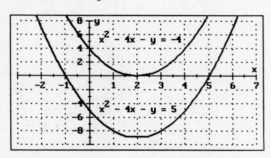

23.

$$f(x,y) = xy.$$

With $C = 1$, $C = -1$, $C = 2$, and $C = -2$, the four sketched level curves have equations $xy = 1$, $xy = -1$, $xy = 2$, and $xy = -2$.

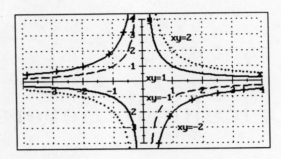

25.

$$f(x,y) = xe^y.$$

With $C = 1$ and $C = e$, the two sketched level curves have equations $xe^y = 1$ and $xe^y = e$.

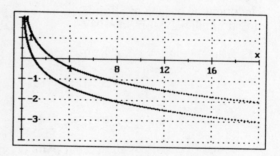

27. (a) $Q(x,y) = 10x^2 y$ and $x = 20$, $y = 40$.

$Q(20, 40) = 10(20)^2(40) = 160,000$ units.

(b) With one more skilled worker, $x = 21$ and the additional output is

$Q(21, 40) - Q(20, 40) = 16,400$ units.

(c) With one more unskilled worker, $y = 41$ and the additional output is

$Q(20, 41) - Q(20, 40) = 4,000$ units.

(d) With one more skilled worker and one more unskilled worker, $x = 21$ and $y = 41$, so the additional output is

$Q(21, 41) - Q(20, 40) = 20,810$ units.

7.1. FUNCTIONS OF SEVERAL VARIABLES

29. (a) Let R denote the total monthly revenue. Then,

$$\begin{aligned} R &= \text{(revenue from the first brand)} \\ &\quad + \text{(revenue from the second brand)} \\ &= x_1 D_1(x_1, x_2) + x_2 D_2(x_1, x_2). \end{aligned}$$

Hence,

$$\begin{aligned} R(x_1, x_2) &= x_1(200 - 10x_1 + 20x_2) \\ &\quad + x_2(100 + 5x_1 - 10x_2) \\ &= 200x_1 - 10x_1^2 + 25x_1 x_2 \\ &\quad + 100x_2 - 10x_2^2. \end{aligned}$$

(b) If $x_1 = 6$ and $x_2 = 5$, then

$$\begin{aligned} R(6,5) &= 200(6) - 10(6)^2 + 25(6)(5) \\ &\quad + 100(5) - 10(5)^2 \\ &= \$1,840. \end{aligned}$$

31.
$$f(x,y) = Ax^a y^b.$$

$$\begin{aligned} f(2x, 2y) &= A(2x)^a (2y)^b = A(2)^a x^a (2)^b y^b \\ &= (2^{a+b}) Ax^a y^b. \end{aligned}$$

$x \geq 0$, $y \geq 0$, and $A > 0$.

(a) If $a + b > 1$, $2^{a+b} > 2$ and f more than doubles.

(b) If $a + b < 1$, $2^{a+b} < 2$ and f increases but does not double.

(c) If $a + b = 1$, $2^{a+b} = 2$ and f doubles (exactly).

33. Let y denote the number of machines sold in foreign markets and x the number of machines sold domestically. The sales price domestically is

$$60 - \frac{x}{5} + \frac{y}{20}$$

and

$$50 - \frac{y}{10} + \frac{x}{20}$$

in foreign markets.

The revenue is

$$\begin{aligned} R(x,y) &= x\left(60 - \frac{x}{5} + \frac{y}{20}\right) + y\left(50 - \frac{y}{10} + \frac{x}{20}\right) \\ &= 60x + 50y - \frac{x^2}{5} - \frac{y^2}{10} + \frac{xy}{10}. \end{aligned}$$

If it were given that the cost of manufacturing, storing, and selling a machine is a constant a for foreign markets and b at home, the profit would be

$$\begin{aligned} P(x,y) &= 60x + 50y - \frac{x^2}{5} - \frac{y^2}{10} \\ &\quad + \frac{xy}{10} - bx - ay. \end{aligned}$$

35. (a)
$$S(W, H) = 0.0072 W^{0.425} H^{0.725}$$

Enter this function in your graphics utility. Then enter the independent variables in the order in which they were set up, that is 15.83 first followed by 87.11.

$$S(15.83, 87.11) = 0.5938$$

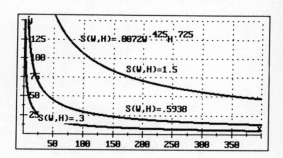

(b)
$$0.648 = 0.0072(18.37)^{0.425} H^{0.725},$$
$$H^{0.725} = 26.121, \; H = 90.05$$

(c)
$$\begin{aligned} S_{old}(W, H) &= 0.0072 W^{0.425} H^{0.725} \\ S_{new}(W, H) &= 0.0072(2W)^{0.425}(3H)^{0.725} \\ &= 2.9775 S_{old} \end{aligned}$$

Thus $100 \dfrac{S_{new} - S_{old}}{S_{old}} = 197.75\%$.

(d) Writing exercise — Answers will vary.

37. (a)
$$Q(10, 20) = 30 + 40 = 70$$

(b)
$$C = 3x + 2y = 70$$

(c)

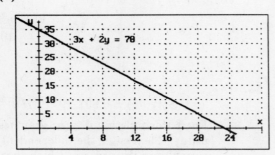

(d)
$$\frac{\partial C}{\partial x} = 0 = 3 + 2\frac{dy}{dx}.$$
$$\text{Thus } \frac{dy}{dx} = -\frac{3}{2}.$$
If $\Delta = +2$ then $\Delta y = -3$.

39. Let $U(x, y) = (x+1)(y+2)$

denote the consumer's utility function. With $x = 25$ and $y = 8$,
$U(25, 8) = (25+1)(8+2) = 260$.

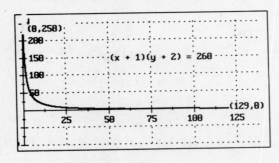

41.
$$V(P, L, r) = \frac{9.3P}{L}(R^2 - r^2)$$

(a)
$$V(P, L, r) = \frac{9.3P}{L}(0.0075^2 - r^2)$$
so $V(3,875, 1.675, 0.004) = 0.866$

(b)
$$V(P, L, 0.5R) = \frac{9.3P}{L}(0.75R^2)$$
With $R = 0.0075$,
$$V(P, L) = \frac{3.9 \times 10^{-4}P}{L}$$
The level curve $C = \dfrac{3.9 \times 10^{-4}P}{L}$ corresponding to $L = \dfrac{3.9 \times 10^{-4}P}{C}$ is a line with slope $\dfrac{3.9 \times 10^{-4}}{C}$.

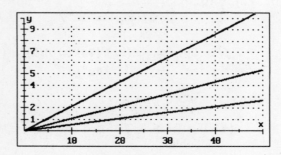

43. (a)

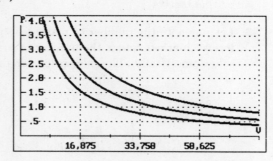

7.2. PARTIAL DERIVATIVES

(b)
$$T(P,V) = 0.0122\left(P + \frac{6.49E6}{V^2}\right)(V-56.2) - 273.15$$

Enter this function in your graphing utility. Then
$$T(1.13, 31, 275) = 159.76$$

45. (a)
$$P(v, A) = abAv^3$$
With $a = 0.5$ $P(v, A) = 0.06Av^3$ and
$$P(22, 225\pi) = 4,515,977 \frac{\text{kg}}{\text{sec}}$$

(b) With $a = \frac{8}{27}$ $P(v, A) = \frac{8(1.2)}{27} Av^3$ and
$$P(11, 900\pi) = 1,338,067 \frac{\text{kg m}^2}{\text{sec}^3}$$

(c) Writing exercise — Answers will vary.

47.
$$Q(K, L) = 57K^{1/4}L^{3/4}$$

K	277	311	493	554	718
L	743	823	1,221	1,486	3,197
Q	33,093	36,780	55,478	66,186	125,448

$$Q(3(277), 3(743)) = 99,279 = 3(33,093)$$
$$Q(.5(277), .5(743)) = 16,546.5 = .5(33,093)$$
$$Q(2(277), .5(743)) = 23,400.3 = \frac{33,093}{\sqrt{2}}$$

For this last Q,
$$Q(2K, .5L) = 2^{-1/2}(57)K^{1/4}L^{3/4} = \frac{1}{\sqrt{2}}Q(K, L)$$

7.2 Partial Derivatives

1.
$$f(x, y) = 2xy^5 + 3x^2y + x^2$$
$$f_x(x, y) = 2y^5 + 6xy + 2x$$
$$f_y(x, y) = 2x(5y^4) + 3x^2 = 10xy^4 + 3x^2$$

3.
$$\begin{aligned}z &= (3x + 2y)^5 \\ \frac{\partial z}{\partial x} &= 5(3x + 2y)^4 \frac{\partial}{\partial x}(3x + 2y) \\ &= 15(3x + 2y)^4 \text{ and} \\ \frac{\partial z}{\partial y} &= 5(3x + 2y)^4 \frac{\partial}{\partial y}(3x + 2y) \\ &= 10(3x + 2y)^4\end{aligned}$$

5.
$$\begin{aligned}f(s, t) &= \frac{3t}{2s} = \frac{3}{2}s^{-1}t \\ f_s(s, t) &= \frac{3}{2}(-1)s^{-2}t = -\frac{3t}{2s^2} \\ f_t(s, t) &= \frac{3}{2}s^{-1} = \frac{3}{2s}\end{aligned}$$

7.
$$\begin{aligned}z &= xe^{xy} \\ \frac{\partial z}{\partial x} &= x(ye^{xy}) + e^{xy}(1) \\ &= (xy + 1)e^{xy} \\ \frac{\partial z}{\partial y} &= x(e^{xy})(x) = x^2e^{xy}\end{aligned}$$

9.
$$f(x, y) = \frac{e^{2-x}}{y^2} = e^{2-x}y^{-2}$$
$$f_x(x, y) = -e^{2-x}y^{-2} = -\frac{e^{2-x}}{y^2}$$
$$f_y(x, y) = e^{2-x}(-2y^{-3}) = -\frac{2e^{2-x}}{y^3}$$

11.
$$\begin{aligned}f(x, y) &= \frac{2x + 3y}{y - x} \\ f_x(x, y) &= \frac{(y - x)(2) - (2x + 3y)(-1)}{(y - x)^2} \\ &= \frac{5y}{(y - x)^2} \\ f_y(x, y) &= \frac{(y - x)(3) - (2x + 3y)(1)}{(y - x)^2} \\ &= \frac{-5x}{(y - x)^2}\end{aligned}$$

13.
$$z = u \ln v$$
$$\frac{\partial z}{\partial u} = (1)\ln v$$
$$\frac{\partial z}{\partial v} = u\left(\frac{1}{v}\right) = \frac{u}{v}$$

15.
$$f(u,y) = \frac{\ln(x+2y)}{y^2}$$
$$f_x(x,y) = \frac{(y^2)[1/(x+2y)] - \ln(x+2y)(0)}{y^4}$$
$$= \frac{1}{(x+2y)y^2}$$
$$f_y(x,y) = \frac{(y^2)[2/(x+2y)] - \ln(x+2y)(2y)}{y^4}$$
$$= \frac{(y)(2) - (x+2y)\ln(x+2y)(2)}{(x+2y)y^3}$$
$$= \frac{2[y - (x+2y)\ln(x+2y)]}{(x+2y)y^3}.$$

17.
$$f(x,y) = 3x^2 - 7xy + 5y^3 - 3(x+y) - 1$$
$$f_x(x,y) = 6x - 7y - 3$$
$$f_y(x,y) = -7x + 15y^2 - 3$$

$$f_x(-2,1) = -12 - 7 - 3 = -22$$
$$f_y(-2,1) = 14 + 15 - 3 = 26$$

19.
$$f(x,y) = xe^{-2y} + ye^{-x} + xy^2$$
$$f_x(x,y) = e^{-2y} - ye^{-x} + y^2$$
$$f_y(x,y) = -2xe^{-2y} + e^{-x} + 2xy$$

$$f_x(0,0) = 1 \text{ and } f_y(0,0) = 1$$

21.
$$f(x,y) = 5x^4y^3 + 2xy$$
$$f_x = 5(4x^3)y^3 + 2y = 20x^3y^3 + 2y$$
$$f_y = 5x^4(3y^2) + 2x = 15x^4y^2 + 2x$$
$$f_{xx} = \frac{\partial}{\partial x}(f_x)$$
$$= 20(3x^2)y^3 + 0 = 60x^2y^3$$
$$f_{yy} = \frac{\partial}{\partial y}(f_y) = 15x^4(2y) + 0 = 30x^4y$$
$$f_{xy} = \frac{\partial}{\partial y}(f_x)$$
$$= 20x^3(3y^2) + 2(1) = 60x^3y^2 + 2$$
$$f_{yx} = \frac{\partial}{\partial x}(f_y)$$
$$= 15(4x^3)y^2 + 2(1) = 60x^3y^2 + 2$$

23.
$$f(x,y) = e^{x^2 y}$$
$$f_x = 2xye^{x^2 y} \text{ and } f_y = x^2 e^{x^2 y}$$
$$f_{xx} = \frac{\partial}{\partial x}(f_x)$$
$$= 2xy(e^{x^2 y})(2xy) + e^{x^2 y}(2y)$$
$$= 2y(2x^2 y + 1)e^{x^2 y}$$
$$f_{yy} = \frac{\partial}{\partial y}(f_y)$$
$$= x^2(e^{x^2 y})(x^2) = x^4 e^{x^2 y}$$
$$f_{xy} = \frac{\partial}{\partial y}(f_x)$$
$$= 2xy(e^{x^2 y})(x^2) + e^{x^2 y}(2x)$$
$$= 2x(x^2 y + 1)e^{x^2 y}$$
$$f_{yx} = \frac{\partial}{\partial x}(f_y)$$
$$= x^2(e^{x^2 y})(2xy) + e^{x^2 y}(2x)$$
$$= 2x(x^2 y + 1)e^{x^2 y}$$

25.
$$f(s,t) = \sqrt{s^2 + t^2}$$
$$f_s = \frac{1}{2}(s^2 + t^2)^{-1/2}(2s) = s(s^2 + t^2)^{-1/2}$$

Interchanging s and t in the work done from the beginning of the exercise to this point leads to
$$f_t = t(s^2 + t^2)^{-1/2}$$

7.2. PARTIAL DERIVATIVES

Hence,

$$\begin{aligned} f_{ss} &= \frac{\partial}{\partial s}(f_s) \\ &= -\frac{s}{2}(s^2+t^2)^{-3/2}(2s) + (s^2+t^2)^{-1/2} \\ &= -\frac{s^2}{(s^2+t^2)^{3/2}} + \frac{s^2+t^2}{(s^2+t^2)^{3/2}} \\ &= \frac{t^2}{(s^2+t^2)^{3/2}} \\ f_{tt} &= \frac{s^2}{(s^2+t^2)^{3/2}} \end{aligned}$$

(just interchange s and t),

$$\begin{aligned} f_{st} &= \frac{\partial}{\partial t}(f_s) \\ &= s(-\frac{1}{2})(s^2+t^2)^{-3/2}(2t) \\ &= -\frac{st}{(s^2+t^2)^{3/2}} = f_{ts}. \end{aligned}$$

27. Since
$$Q = 60K^{1/2}L^{1/3},$$
the partial derivative
$$\begin{aligned} Q_K &= \frac{\partial Q}{\partial K} = 30K^{-1/2}L^{1/3} \\ &= \frac{30L^{1/3}}{K^{1/2}} \end{aligned}$$

is the rate of change of the output with respect to the capital investment.
For any values of K and L, this is an approximation to the additional number of units that will be produced each week if the capital investment is increased from K to $K+1$ while the size of the labor force is not changed. In particular, if the capital investment K is increased from 900 (thousand) to 901 (thousand) and the size of the labor force is $L = 1,000$, the resulting change in output is

$$\begin{aligned} \Delta Q &= Q_K(900, 1000) \\ &= \frac{30(1,000)^{1/3}}{(900)^{1/2}} \\ &= \frac{30(10)}{30} = 10 \text{ units.} \end{aligned}$$

29. Since $Q = 90K^{1/3}L^{2/3}$

$$Q_K(K,L) = 30K^{-2/3}L^{2/3}$$
$$Q_L(K,L) = 60K^{1/3}L^{-1/3}$$

So with $K = 5,495$ and $L = 4,587$

$$Q_K(5495, 4587) = 30(5495^{-2/3})(4587^{2/3}) \approx 26.60$$
$$Q_L(5495, 4587) = 60(5495^{1/3})(4587^{-1/3}) \approx 63.72$$

(b) An increase of one unit capital ($ 1,000,000) results in an increase in output of 26.60 units, which is less than the 63.72 unit increase in output that results from a unit increase in the labor level. The government should encourage labor employment.

31.
$$F(L,r) = \frac{kL}{r^4}$$

(a)
$$F(3.17, 0.085) = 60,727.24k$$
$$\frac{\partial F}{\partial L} = \frac{k}{r^4} = 19,156.86k$$
$$\frac{\partial F}{\partial r} = -\frac{4kL}{r^5} = -2,857,752.58k$$

(b)
$$F(1.2L, .8r) = \frac{k(1.2L)}{(.8r)^4} = 2.93 F(L,r)$$
$$\frac{\partial F}{\partial r}(1.2L, .8r) = 2.44 \frac{\partial F}{\partial r}(L,r)$$
$$\frac{\partial F}{\partial L}(1.2L, .8r) = 3.66 \frac{\partial F}{\partial L}(L,r)$$

33. The demand for bicycles is
$$F(x,y) = 200 - 24\sqrt{x} + 4(0.1y+4)^{3/2}$$

$x = 324$ is the price of bicycles and $y = 120$ is the price of gasoline in cents. Then
$$\frac{\partial F}{\partial y} = 0.6(0.1y+4)^{1/2}$$

If the change in y is -1 cent, the change in the demand will be

$$\frac{\partial F}{\partial y} = 0.6[(0.1)(120)+4]^{1/2}(-1) = -2.4$$

Thus the demand will decrease by 2.4 bicycles (about 2 bicycles.)

35. $V = \pi R^2 h$, $H = 12$, $R = 3$, $dR = 1$.

$$\frac{\partial V}{\partial R} = 2\pi RH = 226 \text{ cubic cm.}$$

(The actual increase is 264 cubic cm.)

37.
$$D_1(p_1, p_2) = 500 - 6p_1 + 5p_2$$
$$D_2(p_1, p_2) = 200 + 2p_1 - 5p_2$$

$\frac{\partial D_1}{\partial p_2} = 5$ and $\frac{\partial D_2}{\partial p_1} = 2$

Since both partial derivatives are positive for all p_1 and p_2, the commodities are substitute commodities.

39.
$$D_1(p_1, p_2) = 3{,}000 + \frac{400}{p_1 + 3} + 50p_2$$
$$D_2(p_1, p_2) = 2{,}000 - 100p_1 + \frac{500}{p_2 + 4}$$

$\frac{\partial D_1}{\partial p_2} = 50$ and $\frac{\partial D_2}{\partial p_1} = -100$

Since the partial derivatives are opposite in sign for all p_1 and p_2, the commodities are neither substitute nor complementary.

41.
$$D_1(p_1, p_2) = \frac{7p_2}{1 + p_1^2}$$
$$D_2(p_1, p_2) = \frac{p_1}{1 + p_2^2}$$

$\frac{\partial D_1}{\partial p_2} = \frac{7}{1+p_1^2} > 0$ and $\frac{\partial D_2}{\partial p_1} = \frac{1}{1+p_2^2}$

Since both partial derivatives are positive for all p_1 and p_2, the commodities are substitute commodities.

43.
$$z = x^2 - y^2$$
$$\frac{\partial z}{\partial x} = 2x$$
$$\frac{\partial^2 z}{\partial x^2} = 2$$
$$\frac{\partial z}{\partial y} = -2y$$
$$\frac{\partial^2 z}{\partial y^2} = -2$$

Since $\frac{\partial^2 z}{\partial x^2} + \frac{\partial^2 z}{\partial y^2} = 0$ the function satisfies Laplace's equation.

45.
$$z = xe^y - ye^x$$
$$\frac{\partial z}{\partial x} = e^y - ye^x$$
$$\frac{\partial^2 z}{\partial x^2} = -ye^x$$
$$\frac{\partial z}{\partial y} = xe^y - e^x$$
$$\frac{\partial^2 z}{\partial y^2} = xe^y$$

Since $\frac{\partial^2 z}{\partial x^2} + \frac{\partial^2 z}{\partial y^2} = -ye^x + xe^y \neq 0$ the function does not satisfy Laplace's equation.

47. (a) If the price x of the first lawnmower increases, the demand for that same lawnmower should fall. If the price y of the second (competing) lawnmower increases, the demand for the first lawnmower should also increase.

(b) If the price of the first lawnmower increases,

$$D_x < 0, D_y > 0$$

If the price of the second lawnmower increases,

$$D_x > 0, D_y < 0$$

(c) With $D = a + bx + cy$, $D_x = b < 0$ and $D_y = c > 0$ when the price of the first lawnmower increases.

$$d_x = b < 0 \text{ and } D_y = c > 0$$

when the price of the second mower increases.

49.
$$P(x, y, u, v) = \frac{100xy}{xy + uv}$$

$$P_x = \frac{(xy + uv)100y - 100xy^2}{(xy + uv)^2} = \frac{100uvy}{(xy + uv)^2}$$

$$P_y = \frac{(xy + uv)100x - 100x^2y}{(xy + uv)^2} = \frac{100uvx}{(xy + uv)^2}$$

$$P_u = -\frac{100xyv}{(xy + uv)^2}, \quad P_v = -\frac{100xyu}{(xy + uv)^2}$$

All of these partials measure the rate of change of percentage of total blood flow WRT the quantities x, y, u, v.

51.
$$Q(K, L) = 120K^{1/2}L^{1/3}$$

(a)

$$Q_L = 120K^{1/2}\left(\frac{1}{3}L^{-2/3}\right) = 40K^{1/2}L^{-2/3}$$

$$Q_{LL} = -\frac{80}{3}K^{1/2}L^{-5/3}$$

Note that $Q_{LL} = \frac{\partial}{\partial L}\left(\frac{\partial Q}{\partial L}\right)$

or the marginal output (WRT labor). The absolute value of Q_{LL} decreases as L increases. For a fixed level of investment, the effect on output of the addition of one worker- hour of labor is greater when the work force is small.

(b)

$$Q_K = 60K^{-1/2}L^{1/3}$$

$$Q_{KK} = -30K^{-3/2}L^{1/3} < 0$$

Note that $Q_{KK} = \frac{\partial}{\partial K}\left(\frac{\partial Q}{\partial K}\right)$

or the marginal output (WRT capital). The absolute value of Q_{KK} decreases as K increases. For a fixed level of labor, the effect on output of an additional $1,000 of capital investment is greater when the capital investment is small.

53.

$$Q(x, y) = 1,175x + 483y + 3.1x^2y - 1.2x^3 - 2.7y^2$$

For the HP48G, store in $Q(x, y)$ as well as in EXPR.
For the TI-85, Store 71 in Y. 2nd CALC F1 evaluates the function
$(1175x + 483Y + 3.1x^2Y - 1.2x^3 - 2.7Y^2, x, 37)$.

(a)

$$Q(37, 71) = 304,691, \quad Q(38, 71) = 317,310$$

$$Q(37, 72) = 309,031$$

(b) Get and differentiate Q WRT x.
$Q_x(37, 71) = 12,534$.
Now $Q(38, 71) - Q(37, 71)$
$= 317,310 - 304,691 = 12,619$

(c) Get and differentiate Q WRT y.
$Q_y(37, 71) = 4,344$.
Now $Q(37, 72) - Q(37, 71)$
$= 309,031 - 304,691 = 4,340$

7.3 Optimizing Functions of Two Variables

1. If $f(x, y) = 5 - x^2 - y^2$, then $f_x = -2x$ and $f_y = -2y$ which are both equal to 0 only if $x = 0$ and $y = 0$.
Hence, $(0, 0)$ is the only critical point. Since $f_{xx} = -2$, $f_{yy} = -2$, and $f_{xy} = 0$,

$$D = f_{xx}f_{yy} - (f_{xy})^2 = (-2)(-2) - 0 = 4$$

Since $D(0, 0) = 4 > 0$ and $f_{xx}(0, 0) = -2 < 0$, it follows that f has a relative maximum at $(0, 0)$.

3. If $f(x,y) = xy$, then $f_x = y$ and $f_y = x$ which are 0 only when $x = 0$ and $y = 0$.
Hence, $(0,0)$ is the only critical point.
Since $f_{xx} = 0$, $f_{yy} = 0$, and $f_{xy} = 1$,
$D = f_{xx}f_{yy} - (f_{xy})^2 = (0)(0) - 1 = -1$. Since

$$D(0,0) = -1 < 0$$

It follows that f has a saddle point at $(0,0)$.

5.
$$f(x,y) = \frac{1}{x} + \frac{1}{y} - \frac{1}{x+y}$$

$$f_x = -\frac{1}{x^2} + \frac{1}{(x+y)^2} = 0$$

$$f_y = -\frac{1}{y^2} + \frac{1}{(x+y)^2} = 0$$

so $x = y$ (reject $x = -y$).

$$f_{xx} = 2x^{-3} + (x+y)^{-3}$$

$$f_{xy} = (x+y)^{-3}$$

$$f_{yy} = 2y^{-3} + (x+y)^{-3}$$

At $x = y$ $D = \dfrac{3}{x^6} > 0$

and $f_{xx} = \dfrac{7}{4x^6} > 0$

so the surface has relative minima along $y = x$.

7.
$$f(x,y) = 2x^3 + y^3 + 3x^2 - 3y - 12x - 4$$

$$f_x = 6x^2 + 6x - 12 = 6(x+2)(x-1)$$

$$f_y = 3y^2 - 3 = 3(y+1)(y-1)$$

which indicates that $f_x = 0$ when $x = -2$ and $x = 1$, while $f_y = 0$ when $y = -1$ and $y = 1$.
Hence, the critical points of f are $(-2,-1)$, $(-2,1)$, $(1,-1)$, and $(1,1)$.
$f_{xx} = 12x + 6$, $f_{yy} = 6y$, and $f_{xy} = 0$,

$$D = f_{xx}f_{yy} - (f_{xy})^2 = 36y(2x+1)$$

Since $D(-2,-1) = 36(-1)(-3) = 108 > 0$ and $f_{xx}(-2,-1) = -24 + 6 = -18 < 0$, it follows that f has a relative maximum at $(-2,-1)$,
Since $D(-2,1) = 36(1)(-3) = -108 < 0$ it follows that f has a saddle point at $(-2,1)$,

point	D	f_{xx}	type
$(-2,-1)$	>0	<0	max
$(1,1)$	>0	>0	min
$(-2,1)$	<0		saddle
$(1,-1)$	<0		saddle

9.
$$f(x,y) = x^3 + y^2 - 6xy + 9x + 5y + 2$$

$$f_x = 3x^2 - 6y + 9$$

$$f_y = 2y - 6x + 5$$

Setting $f_x = 0$ and $f_y = 0$ gives
$3x^2 - 6y + 9 = 0$ and $2y - 6x + 5 = 0$ or
$x^2 - 2y + 3 = 0$ and $-6x + 2y + 5 = 0$.
Adding these two equations yields
$x^2 - 6x + 8 = 0$, $(x-4)(x-2) = 0$, or $x = 4$ and $x = 2$.
If $x = 4$, the second equation gives
$-24 + 2y + 5 = 0$ or $y = \dfrac{19}{2}$, and if $x = 2$, the second equation gives $-12 + 2y + 5 = 0$ or $y = \dfrac{7}{2}$.
Hence, the critical points are $\left(4, \dfrac{19}{2}\right)$ and $\left(2, \dfrac{7}{2}\right)$.
Since $f_{xx} = 6x$, $f_{yy} = 2$, and $f_{xy} = -6$,

$$D = f_{xx}f_{yy} - (f_{xy})^2 = 12(x-3)$$

Since $D\left(4, \dfrac{19}{2}\right) = 12(4-3) = 12 > 0$ and $f_{xx}(4, \dfrac{19}{2}) = 24 > 0$, it follows that f has a relative minimum at $\left(4, \dfrac{19}{2}\right)$,
since $D\left(2, \dfrac{7}{2}\right) = 12(2-3) = -12 < 0$ it follows that f has a saddle point at $\left(2, \dfrac{7}{2}\right)$.

11.
$$f(x,y) = (x^2 + 2y^2)e^{1-x^2-y^2}$$

$$f_x = -2x(x^2 + 2y^2 - 1)e^{1-x^2-y^2}$$

$$f_y = -2y(x^2 + 2y^2 - 2)e^{1-x^2-y^2}$$

7.3. OPTIMIZING FUNCTIONS OF TWO VARIABLES

Setting $f_x = 0$ and $f_y = 0$ gives
$x^2 + 2y^2 - 1 = 0$ and $x^2 + 2y^2 - 2 = 0$ which produces no solutions.
But $x = 0$ and $y = 0$ is our first solution, so $(0,0)$ is a critical point. Since

$$f_{xx} = e^{1-x^2-y^2}(-6x^2 - 4y^2 + 2 \\ + 4x^4 + 8x^2y^2 - 4x^2)$$

$$f_{yy} = e^{1-x^2-y^2}(-2x^2 - 12y^2 + 4 \\ + 4x^2y^2 + 8y^4 - 8y^2)$$

$$f_{xy} = e^{1-x^2-y^2}(-8xy + 4x^3y + 8xy^3 - 4xy)$$

$f_{xx}(0,0) = 2e$, $f_{yy}(0,0) = 4e$, and $f_{xy}(0,0) = 0$, and since

$$D(0,0) = f_{xx}f_{yy} - (f_{xy})^2 = (4)(2)e^2 > 0$$

and $f_{xx}(0,0) > 0$, it follows that f has a relative minimum at $(0,0)$.

$$f_x = -2x(x^2 + 2y^2 - 1)e^{1-x^2-y^2} = 0$$

and $f_y = -2y(x^2 + 2y^2 - 2)e^{1-x^2-y^2} = 0$

are also satisfied if $x = 0$ and $0^2 + 2y^2 - 2 = 0$, $y = \pm 1$, so at $(0, \pm 1)$.
$f_{xx}(0, \pm 1) = e^{1-0^2-(\pm 1)^2}[-6(0)^2 - 4(\pm 1)^2 + 2 + 4(0)^4 + 8(0)^2(\pm 1)^2 - 4(0)^2] = -2$,
$f_{yy}(0, \pm 1) = -8$, and $f_{xy}(0, \pm 1) = 0$, and since
$D(0, \pm 1) = f_{xx}f_{yy} - (f_{xy})^2 = (-2)(-8) - 0 > 0$
and $f_{xx}(0, \pm 1) < 0$, it follows that f has a relative maximum at $(0, \pm 1)$.
Similarly $y = 0$ and $x^2 + 2(0)^2 - 1 = 0$, $x = \pm 1$, so at $(\pm 1, 0)$ are critical points.
$f_{xx}(\pm 1, 0) = -4$, $f_{yy}(\pm 1, 0) = 2$, and $f_{xy}(\pm 1, 0) = 0$, and since
$D(\pm 1, 0) = f_{xx}f_{yy} - (f_{xy})^2 = (-4)(2) - 0 < 0$,
it follows that f has a saddle point at $(\pm 1, 0)$.

13. If $f(x,y) = x^3 - 4xy + y^3$, then $f_x = 3x^2 - 4y$ and $f_y = -4x + 3y^2$. Setting $f_x = 0$ and $f_y = 0$ gives $y = \dfrac{3x^2}{4}$ and $3y^2 = 4x$.
Solving simultaneously yields $3\left(\dfrac{3x^2}{4}\right)^2 = 4x$, $27x^4 = 64x$, $x(27x^3 - 64) = 0$, or $x = 0$ and $x = \dfrac{4}{3}$. If $x = 0$, then $y = 0$, and if $x = \dfrac{4}{3}$, then $y = \dfrac{4}{3}$.

Hence, the critical points are $(0,0)$ and $\left(\dfrac{4}{3}, \dfrac{4}{3}\right)$.
Since $f_{xx} = 6x$, $f_{yy} = 6y$, and $f_{xy} = -4$,

$$D = f_{xx}f_{yy} - (f_{xy})^2 \\ = (6x)(6y) - (-4)^2 = 4(9xy - 4)$$

Since $D(0,0) = 4(-4) < 0$ it follows that f has a saddle point at at $(0,0)$.
Since $D\left(\dfrac{4}{3}, \dfrac{4}{3}\right) = 4(16 - 4) > 0$ and $f_{xx}\left(\dfrac{4}{3}, \dfrac{4}{3}\right) > 0$, it follows that f has a relative minimum at $\left(\dfrac{4}{3}, \dfrac{4}{3}\right)$.

15.
$$f(x,y) = e^{-(x^2+y^2-6y)}$$
$$f_x = -2xe^{-(x^2+y^2-6y)}$$
$$f_y = -(2y-6)e^{-(x^2+y^2-6y)}$$

which are 0 at $P(0,3)$.

$$f_{xx} = 2(2x^2 - 1)e^{-(x^2+y^2-6y)}$$
$$f_{xy} = 2x(2y-6)e^{-(x^2+y^2-6y)}$$
$$f_{yy} = [-2 + (2y-6)^2]e^{-(x^2+y^2-6y)}$$

At $P(0,3)$, $f_{xx} < 0$, $f_{yy} > 0$, $f_{xy} = 0$ so $D(0,3) < 0$ and f has a saddle poit at $(0,3)$.
Rewrite $f(x,y) = e^{-x^2}e^{9-(y-3)^2}$ and note that the exponent is most positive at $(0,3)$. Thus $(0,3)$ is a maximum.

17.
$$f(x,y) = \dfrac{1}{x^2 + y^2 + 3x - 2y + 1}$$
$$= (x^2 + y^2 + 3x - 2y + 1)^{-1}$$
$$f_x = -(x^2 + y^2 + 3x - 2y + 1)^{-2}(2x + 3)$$
$$f_y = -(x^2 + y^2 + 3x - 2y + 1)^{-2}(2y - 2)$$
$$f_{xy} = 2(2x + 3)(x^2 + y^2 + 3x - 2y + 1)^{-3} \\ (2y - 2)$$
$$f_{xx} = -2(x^2 + y^2 + 3x - 2y + 1)^{-2} \\ + 2(2x + 3)^2(x^2 + y^2 + 3x - 2y + 1)^{-3}$$
$$f_{yy} = -2(x^2 + y^2 + 3x - 2y + 1)^{-2} \\ + 2(x^2 + y^2 + 3x - 2y + 1)^{-3}(2y - 2)^2$$

$f_x = 0$ and $f_y = 0$ at $P\left(-\frac{3}{2}, 1\right)$,
$f_{xx} < 0$, $f_{yy} < 0$, $f_{xy} = 0$, so $D > 0$ and
$P\left(-\frac{3}{2}, 1\right)$ is a maximum.

19.
$$\begin{aligned} f(x,y) &= x\ln\left(\frac{y^2}{x}\right) + 3x - xy^2 \\ &= 2x\ln y - x\ln x + 3x - xy^2 \\ f_x &= 2\ln y - 1 - \ln x + 3 - y^2 \\ f_y &= \frac{2x}{y} - 2xy \\ &= 2x\left(\frac{1}{y} - y\right) \end{aligned}$$

$f_x = 0$ when $\ln x = 1$ or $x = e$,
$f_y = 0$ when $y = \pm 1$.
At $(e, 1)$ $f_{xx} = -\frac{1}{e} < 0$, $f_{xy} = 2 - 2 = 0$, and $f_{yy} = -4e < 0$.
$D > 0$ implies $(e, 1)$ is a maximum.
Since $f(x,y) = f(x,-y)$ $(e, -1)$ is also a maximum.

21. Let x and y denote the sales price of the shirts endorsed by Michael Jordan and Shaq O'Neal, respectively.
The profit $P(x, y)$ is the sum of the number of shirts sold times the difference between the sales price and the cost per shirt. Thus

$$\begin{aligned} P(x,y) &= (x-2)(40 - 50x + 40y) \\ &\quad + (y-2)(20 + 60x - 70y) \\ P_x &= (x-2)(-50) + 40 - 50x + 40y \\ &\quad + (y-2)(60) \\ &= 20(-5x + 5y + 1) \\ P_y &= (x-2)(40) + (y-2)(-70) + 20 \\ &\quad + 60x - 70y \\ &= 20(5x - 7y + 4) \end{aligned}$$

Solving simultaneously yields
$-5x + 5y + (5x - 7y) = -1 + (-4)$, $-2y = -5$,
$y = \frac{5}{2}$, $-5x = -\frac{25 + 2}{2}$, or $x = \frac{27}{10}$.
Hence, the critical point is $(2.7, 2.5)$.

Since $P_{xx} = 20(-5) = -100$, $P_{yy} = -140$, and $P_{xy} = 100$,
$$D = P_{xx}P_{yy} - (P_{xy})^2 = 10^3(14 - 10) > 0$$
and $P_{xx}(2.7, 2.5) < 0$, it follows that P has a relative maximum at $(2.7, 2.5)$.
Thus, to maximize the profit, $x = \$2.7$, the sales price of the Jordan shirts, and $y = \$2.5$, the price of the O'Neal shirts

23. Let x be the length, y the width, and z the height of the box. The surface area of the bottom added to the top is $2xy$. The surface area of the sides is $2xz + 2yz$. Since the volume of the box is 32 ft^3, $xyz = 32$ or $z = \frac{32}{xy}$.
The cost of the bottom is 3 times as expensive as the side and the top is 5 times as expensive. The cost of the surface area is

$$\begin{aligned} C &= 2xz + 2yz + 3xy + 5xy \\ &= \frac{64}{y} + \frac{64}{x} + 8xy \end{aligned}$$

$C_x = 0$ so $y = \frac{8}{x^2}$, $y^2 = \frac{64}{x^4}$
$C_y = 0$ so $x = \frac{8}{y^2}$, $y^2 = \frac{8}{x}$
Equating y^2 leads to $x = 2$ ft., $y = 2$ ft., and $z = 8$ ft.
$S_{xx} = S_{yy} = 16$, $S_{xy} = 8$, so $D > 0$ and the box has minimum cost.

25.
Let $p(x) = 100 - x$
denote the price of one gallon of whole milk and
$$q(y) = 100 - y$$
the price per gallon of skim milk. Further let x be number of gallons of whole milk and y the number of gallons of skim milk.

With cost $C(x,y) = x^2 + xy + y^2$,
the profit is
$$\begin{aligned} P(x,y) &= x(100 - x) + y(100 - y) - x^2 - xy - y^2 \\ &= -2x^2 - 2y^2 + 100x + 100y - xy \end{aligned}$$

7.3. OPTIMIZING FUNCTIONS OF TWO VARIABLES

$P_x = -4x + 100 - y = 0$ if $y = -4x + 100$,
$P_y = -4y + 100 - x = 0$ if $4y = 100 - x$.
Subtracting four times the first equation from the second yields

$$16x - 400 + 4y - x + 100 - 4y = 15x - 300 = 0$$

when $x = 20$. From the first equation $y = 100 - 80 = 20$.
Now $D = (-4)(-4) - 1^2 > 0$ and $P_{xx} < 0$ which signifies that $(20, 20)$ is a maximum.

27.
$$f(x, y) = C + xye^{1-x^2-y^2}$$

$f_x = ye^{1-x^2-y^2}(-2x^2 + 1) = 0$ at $x = \pm\dfrac{\sqrt{2}}{2}$

By symmetry (interchange x and y) $f_y = 0$ when $y = \pm\dfrac{\sqrt{2}}{2}$.

$f_x = f_y = 0$ at $(0,0)$ also.

$$\begin{aligned}
f_{xx} &= y(-4x + 4x^3 - 2x)e^{1-x^2-y^2} \\
f_{xy} &= (-2y^2 + 1)(-2x^2 + 1)e^{1-x^2-y^2} \\
f_y &= x(-2y^2 + 1)e^{1-x^2-y^2} \\
f_{yy} &= -2xy(-2y^2 + 3)e^{1-x^2-y^2}
\end{aligned}$$

At $\left(\dfrac{\sqrt{2}}{2}, \dfrac{\sqrt{2}}{2}\right)$ $f_{xx} < 0$, $f_{yy} < 0$, and $f_{xy} = 0$.

$D > 0$ so $\left(\dfrac{\sqrt{2}}{2}, \dfrac{\sqrt{2}}{2}\right)$ is a maximum.

$(0, 0)$ does not lead to maximum performance.

29. Since the volume of the box is V_0 ft^3,
$xyz = V_0$ or $z = \dfrac{V_0}{xy}$.

$$\begin{aligned}
E(x, y, z) &= \dfrac{k^2}{8m}\left(\dfrac{1}{x^2} + \dfrac{1}{y^2} + \dfrac{x^2 y^2}{V_0^2}\right) \\
E_x &= \dfrac{k^2}{8m}\left(-\dfrac{2}{x^3} + \dfrac{2xy^2}{V_0^2}\right) = 0 \\
E_y &= \dfrac{k^2}{8m}\left(-\dfrac{2}{y^3} + \dfrac{2x^2 y}{V_0^2}\right) = 0
\end{aligned}$$

$E_x = 0$ so $y = \dfrac{V_0}{x^2}$, $y^2 = \dfrac{V_0^2}{x^4}$

$E_y = 0$ so $x^2 = \dfrac{V_0^2}{y^4}$, $y^2 = \dfrac{V_0}{x}$

Equating y^2 leads to $x = y = z = V_0^{1/3}$.

$E_{xx} = E_{yy} = \dfrac{k^2}{mV_0^{4/3}}$, $E_{xy} = \dfrac{k^2}{2mV_0^{4/3}}$, so $D > 0$ and the ground station energy is a minimum.

31. For the domestic market, x machines will sell for

$$60 - \dfrac{x}{5} + \dfrac{y}{20}$$

thousand dollars apiece and the cost of producing each machine is 10 thousand dollars. Thus, the profit from

$$\begin{aligned}
P_1 &= \text{(number of machines)} \\
&\quad \text{(profit per machine)} \\
&= x\left(60 - \dfrac{x}{5} + \dfrac{y}{20} - 10\right) \\
&= x\left(50 - \dfrac{x}{5} + \dfrac{y}{20}\right)
\end{aligned}$$

For the foreign market, y machines will sell for

$$50 - \dfrac{y}{10} + \dfrac{x}{20}$$

thousand dollars apiece and the cost of producing each machine is 10 thousand dollars. Thus, the profit from the foreign market is

$$P_2 = y\left(40 - \dfrac{y}{10} + \dfrac{x}{20}\right)$$

The total profit is

$$\begin{aligned}
P(x, y) &= P_1 + P_2 = x\left(50 - \dfrac{x}{5} + \dfrac{y}{20}\right) \\
&\quad + y\left(40 - \dfrac{y}{10} + \dfrac{x}{20}\right) \\
&= 50x - \dfrac{x^2}{5} + \dfrac{xy}{10} \\
&\quad + 40y - \dfrac{y^2}{10}
\end{aligned}$$

The first-order partial derivatives are

$$P_x = 50 - \dfrac{2x}{5} + \dfrac{y}{10}$$

and $P_y = \dfrac{x}{10} + 40 - \dfrac{y}{5}$

which are equal to 0 when
$50 - \frac{2x}{5} + \frac{y}{10}$ and $\frac{x}{10} + 40 - \frac{y}{5} = 0$,
which are equivalent to $-8x + 2y = -1,000$ and $x - 2y = -400$.
Adding these two equations yields
$-7x = -1,400$ or $x = 200$.
Substituting this into the second equation yields $200 - 2y = -400$ or $y = 300$.
The second-order partial derivatives are
$P_{xx} = -\frac{2}{5}$, $P_{yy} = -\frac{1}{5}$, and $P_{xy} = \frac{1}{10}$.

Hence $D(x,y) = P_{xx}P_{yy} - (P_{xy})^2$
$= \frac{2}{25} - \frac{1}{100} > 0$

Hence, $(200, 300)$ is a relative extremum.
Moreover, since $P_{xx} = -\frac{2}{5} < 0$,
this relative extremum is a relative maximum. Assuming that the relative maximum and absolute maximum are the same, it follows that to generate the largest possible profit, 200 machines should be supplied to the domestic market and 300 to the foreign market.

33. The square of the distance from $S(a,b)$ to $T_1(-5,0)$ is
$d_1^2 = (a+5)^2 + b^2$. Similarly the squares of the distances to the other towns are
$d_2^2 = (a-1)^2 + (b-7)^2$, $d_3^2 = (a-9)^2 + b^2$, and $d_3^2 = a^2 + (b+8)^2$.
The sum of these squares needs to be minimized.

$f = 4a^2 - 10a + 4b^2 + 2b + \text{a constant}$
$f_a = 8a - 10 = 0$
$f_b = 8b + 2 = 0$

Thus $a = \frac{5}{4}$ and $b = -\frac{1}{4}$. Since $f_{aa} = 8$, $f_{bb} = 8$, and $f_{ab} = 0$, $D > 0$ and $S\left(\frac{5}{4}, -\frac{1}{4}\right)$ is the point on the map which minimzes the distance from the television station to the four towns.

35. The sum $S(m,b)$ of the squares of the vertical distances from the three given points is
$S(m,b) = d_1^2 + d_2^2 + d_3^2$

$= (b-1)^2 + (2m+b-3)^2$
$+(4m+b-2)^2$.

To minimize $S(m,b)$, set the partial derivatives

$\frac{\partial S}{\partial m} = 0$ and $\frac{\partial S}{\partial b} = 0$

namely

$\frac{\partial S}{\partial m} = 2(2m+b-3)(2) + 2(4m+b-2)(4)$
$= 40m + 12b - 28 = 0$
$\frac{\partial S}{\partial b} = 2(b-1) + 2(2m+b-3) + 2(4m+b-2)$
$= 12m + 6b - 12 = 0$

Solve the resulting simplified equations
$10m + 3b = 7$ and $6m + 3b = 6$ to get $m = \frac{1}{4}$
and $b = \frac{3}{2}$.
Hence, the equation of the least-squares line is

$y = \frac{x}{4} + \frac{3}{2}$

37. The sum $S(m,b)$ of the squares of the vertical distances from the four given points is

$S(m,b) = (m+b-2)^2 + (2m+b-4)^2$
$+(4m+b-4)^2 + (5m+b-2)^2$

To minimize $S(m,b)$, set the partial derivatives

$\frac{\partial S}{\partial m} = 0$ and $\frac{\partial S}{\partial b} = 0$
$\frac{\partial S}{\partial m} = 2(m+b-2) + 2(2m+b-4)(2)$
$+2(4m+b-4)(4) + 2(5m+b-2)(5)$
$= 92m + 24b - 72 = 0$
$\frac{\partial S}{\partial b} = 2(m+b-2) + 2(2m+b-4)$
$+2(4m+b-4) + 2(5m+b-2)$
$= 24m + 8b - 24 = 0$

Solve the resulting simplified equations
$23m + 6b = 18$ and $3m + b = 3$ to get $m = 0$ and $b = 3$.
Hence, the equation of the least-squares line is $y = 3$.

39. (a)

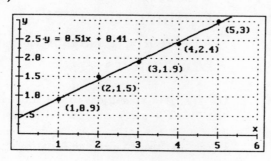

(b) The points are: $(1, 0.9)$, $(2, 1.5)$, $(3, 1.9)$, $(4, 2.4)$, and $(5, 3.0)$.

$$\sum x = 15, \quad \sum y = 9.7, \quad \sum x^2 = 55$$

and $\sum xy = 34.2$

$$m = \frac{n\sum xy - \sum x \sum y}{n\sum x^2 - (\sum x)^2} = 0.51$$

$$b = \frac{\sum x^2 \sum y - \sum x \sum xy}{n\sum x^2 - (\sum x)^2} = 0.41$$

Thus the best fitting line has equation $y = 0.51x + 0.41$.

(c) If $x = 6$, $y = 0.51 \times 6 + 0.41 = 3.47$ (billion dollars.)

41. (a) Let x denote the number of hours after the polls open and y the corresponding percentage of registered voters that have already cast their ballots. Then

x	2	4	6	8	10
y	12	19	24	30	37

x	y	xy	x^2
2	12	24	4
4	19	76	16
6	24	144	36
8	30	240	64
10	37	370	100
$\sum x = 30$	$\sum y = 122$	$\sum xy = 854$	$\sum x^2 = 220$

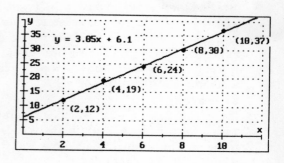

(b) From the formulas

$$m = \frac{n\sum xy - \sum x \sum y}{n\sum x^2 - (\sum x)^2}$$

and $b = \dfrac{\sum x^2 \sum y - \sum x \sum xy}{n\sum x^2 - (\sum x)^2}$

with $n = 5$,

$$m = \frac{5(854) - 30(122)}{5(220) - (30)^2} = \frac{610}{200} = 3.05$$

and $b = \dfrac{220(122) - 30(854)}{5(220) - (30)^2}$

$$= \frac{1,220}{200} = 6.10$$

Hence, the equation of the least-squares line is

$$y = 3.05x + 6.10$$

(c) When the polls close at 8:00 p.m., $x = 12$ and so $y = 3.05(12) + 6.1 = 42.7$, which means that approximately 42.7 % of the registered voters can be expected to vote.

43.
$$f(x, y) = x^2 + y^2 - 4xy,$$

$f_x = 2x - 4y = 0$ when $y = \dfrac{x}{2}$.
$f_y = 2y - 4x = 0$ when $y = 2x$.
Thus $(0,0)$ is a critical point.
$f_{xx} = 2$, $f_{xy} = -4$, and $f_{yy} = 2$, so

$$D(0,0) = 4 - (-4)^2 < 0$$

and $(0,0)$ is a saddle point.
The above is true but not asked for. If $x = 0$, $f(0,y) = y^2$ which is a parabola with a minimum at $(0,0)$ (in the vertical yz–plane).
If $y = 0$, $f(x,0) = x^2$ which is a parabola with a minimum at $(0,0)$ (in the vertical xz–plane).
If $y = x$, $f(x,x) = -2x^2$ which is a parabola with a maximum at $(0,0)$ (in the vertical plane passing through the z–axis and the line $y = x$ in the xy plane).

45.
$$f(x,y) = \frac{x^2 + xy + 7y^2}{x \ln y}$$

$$\begin{aligned}
f_x &= \frac{1}{x^2 \ln^2 y}[x \ln y(2x + y) \\
&\quad -(x^2 + xy + 7y^2) \ln y] \\
&= \frac{x^2 - 7y^2}{x^2 \ln y} \\
f_y &= \frac{1}{x^2 \ln^2 y}[x \ln y(x + 14y) \\
&\quad -(x^2 + xy + 7y^2)(x/y)] \\
&= \frac{1}{xy \ln^2 y}[y \ln y(x + 14y) - x^2 - xy - 7y^2)]
\end{aligned}$$

$f_x = 0$ when $x = \pm\sqrt{7}y$. Substitute into $f_y = 0$ to get $y^2 \ln y(\sqrt{7} + 14) = (14 + \sqrt{7})y^2$ or $y = e$. Thus the critical points are $P(\pm\sqrt{7}e, e)$.

47.
$$\begin{aligned}
f(x,y) &= 2x^4 + y^4 - x^2(11y - 18) \\
f_x &= 8x^3 - 2x(11y - 18) \\
f_y &= 4y^3 - 11x^2
\end{aligned}$$

$f_x = 0$ and $f_y = 0$ at $P(0,0)$, so $P(0,0)$ is a critical point.
$x^2 = \frac{4}{11}y^3$, so

$$8x^3 - 2x(11y - 18) = 0$$

when $8\left(\frac{4}{11}y^3\right) - 2(11y - 18) = 0$
for which $y = -3.3544$. This is impossible since x^2 can't be negative.

7.4 Constrained Optimization: The Method of Lagrange Multipliers

1. For $f(x,y) = xy$ subject to the constraint that $g(x,y) = x + y = 1$, the partial derivatives are $f_x = y$, $f_y = x$, $g_x = 1$, and $g_y = 1$.
Hence, the three Lagrange equations are

$$y = \lambda, \; x = \lambda, \; x + y = 1$$

From the first two equations, $x = y$ which, when substituted into the third equation gives $2x = 1$ or $x = \frac{1}{2}$.
Since $x = y$, the corresponding value for y is $y = \frac{1}{2}$. Thus, the constrained maximum is
$$f\left(\frac{1}{2}, \frac{1}{2}\right) = \frac{1}{4}.$$

3. For $f(x,y) = x^2 + y^2$ subject to the constraint that $g(x,y) = xy = 1$, the partial derivatives are $f_x = 2x$, $f_y = 2y$, $g_x = y$, and $g_y = x$.
Hence, the three Lagrange equations are

$$2x = \lambda y, \; 2y = \lambda x, \; xy = 1$$

Multiply the first equation by y and the second by x to get $2xy = \lambda y^2$ and $2xy = \lambda x^2$. Set the two expressions for $2xy$ equal to each other to get $\lambda y^2 = \lambda x^2$, $y^2 = x^2$, or $x = \pm y$. (Note that another solution of the equation $\lambda y^2 = \lambda x^2$ is $\lambda = 0$, which implies that $x = 0$ and $y = 0$, which is not consistent with the third equation.)
If $y = x$, the third equation becomes $x^2 = 1$, which implies that $x = \pm 1$ and $y = \pm 1$.
If $y = -x$, the third equation becomes $-x^2 = 1$, which has no solutions. Thus, the two points at which the constrained extrema can occur are $(1,1)$ and $(-1,-1)$.
Since $f(1,1) = 2$ and $f(-1,-1) = 2$, it follows that the minimum value is 2 and it is attained at the two points $(1,1)$ and $(-1,-1)$.

5. For $f(x,y) = x^2 - y^2$ subject to the constraint that $g(x,y) = x^2 + y^2 = 4$, the partial

derivatives are $f_x = 2x$, $f_y = -2y$, $g_x = 2x$, and $g_y = 2y$.
Hence, the three Lagrange equations are
$$2x = 2\lambda x, \quad -2y = 2\lambda y, \quad x^2 + y^2 = 4$$

From the first equation, either $\lambda = 1$ or $x = 0$. If $x = 0$, the third equation becomes $y^2 = 4$ or $y = \pm 2$.
From the second equation, either $\lambda = -1$ or $y = 0$. If $y = 0$, the third equation becomes $x^2 = 4$ or $x = \pm 2$.
If neither $x = 0$ nor $y = 0$, the first equation implies $\lambda = 1$ while the second equation implies $\lambda = -1$, which is impossible.
Hence, the only points at which the constrained extrema can occur are $(0, -2)$, $(0, 2)$, $(-2, 0)$, and $(2, 0)$. Since $f(0, -2) = -4$, $f(0, 2) = -4$, $f(-2, 0) = 4$, and $f(2, 0) = 4$, it follows that the constrained minimum is -4 and is attained at the two points $(0, -2)$ and $(0, 2)$.

7. For $f(x, y) = x^2 - y^2 - 2y$ subject to the constraint that $g(x, y) = x^2 + y^2 = 1$, the partial derivatives are $f_x = 2x$, $f_y = -2y - 2$, $g_x = 2x$, and $g_y = 2y$.
Hence, the three Lagrange equations are
$$2x = 2\lambda x, \quad -2y - 2 = 2\lambda y$$
$$x^2 + y^2 = 1$$

From the first equation, either $\lambda = 1$ or $x = 0$. If $\lambda = 1$, the second equation becomes $-2y - 2 = 2y$, $4y = -2$, or $y = -\frac{1}{2}$. From the third equation, $x^2 + \left(\frac{-1}{2}\right)^2 = 1$, or $x = \pm \frac{\sqrt{3}}{2}$.
If $x = 0$, the third equation becomes $0^2 + y^2 = 1$ or $y = \pm 1$ (λ, obtained from the second equation, is immaterial).
Hence, the only points at which the constrained extrema can occur are $\left(-\frac{\sqrt{3}}{2}, -\frac{1}{2}\right)$, $\left(\frac{\sqrt{3}}{2}, -\frac{1}{2}\right)$, $(0, -1)$, and $(0, 1)$. Since
$$f\left(\frac{\sqrt{3}}{2}, \frac{1}{2}\right) = f\left(-\frac{\sqrt{3}}{2}, \frac{1}{2}\right)$$

$$= \frac{3}{4} - \frac{1}{4} - 2\left(\frac{-1}{2}\right) = \frac{1}{2} + 1 = \frac{3}{2}$$
$f(0, -1) = 0^2 - (-1)^2 - 2(-1) = -1 + 2 = 1$, and $f(0, 1) = -1 - 2 = -3$, it follows that the constrained maximum is $\frac{3}{2}$ and the constrained minimum is -3.

9.
$$\begin{aligned} f(x, y) &= 2x^2 + 4y^2 - 3xy - 2x - 23y + 3 \\ g(x, y) &= x + y - 15 = 0 \\ f_x &= 4x - 3y - 2 \\ f_y &= 8y - 3x - 23 \\ g_x &= g_y = 1 \end{aligned}$$

The three Lagrange equations are:
$$4x - 3y - 2 = \lambda$$
$$-3x + 8y - 23 = \lambda$$
$$x + y = 15$$

The first two lead to $7x - 11y = -21$.
Substitute $y = 15 - x$ to obtain $18x = 144$ or $x = 8$ and $y = 7$.
The minimum is $f(8, 7) = -18$.

11.
$$\begin{aligned} f(x, y) &= e^{xy} \\ g(x, y) &= x^2 + y^2 - 4 = 0 \\ f_x &= ye^{xy}, \text{ and } f_y = xe^{xy} \\ g_x &= 2x \text{ and } g_y = 2y \end{aligned}$$

The three Lagrange equations are:
$$ye^{xy} = 2\lambda x$$
$$xe^{xy} = 2\lambda y$$
$$x^2 + y^2 - 4 = 0$$

Dividing the first two leads to $\frac{y}{x} = \frac{x}{y}$, or $x^2 = y^2$.
Substitute in $x^2 + y^2 = 4$ to obtain $x = \pm\sqrt{2}$ and $y = \pm\sqrt{2}$.
The minimum is $f(-\sqrt{2}, -\sqrt{2}) = e^{-2}$, the maximum is $f(\sqrt{2}, \sqrt{2}) = e^2$.

13.
$$f(x,y) = xyz$$
$$g(x,y) = x + 2y + 3z - 24 = 0$$
$$f_x = yz,\ f_y = xz,\ \text{and}\ f_z = xy$$
$$g_x = 1,\ g_y = 2,\ \text{and}\ g_z = 3$$

The three Lagrange equations are:
$$yz = \lambda$$
$$xz = 2\lambda$$
$$xy = 3\lambda$$

Dividing the first two leads to $y = \frac{x}{2}$, dividing the first by the third leads to $z = \frac{x}{3}$. Substitute in $x + 2y + 3z = 24$ to obtain $x = 8$, $y = 4$, and $z = \frac{8}{3}$.
The maximum is $f(8, 4, 8/3) = \frac{256}{3}$.

15.
$$f(x,y) = x + 2y + 3z$$
$$g(x,y) = x^2 + y^2 + z^2 - 16 = 0$$
$$f_x = 1,\ f_y = 2,\ \text{and}\ f_z = 3$$
$$g_x = 2x,\ g_y = 2y,\ \text{and}\ g_z = 2z$$

The three Lagrange equations are:
$$1 = 2\lambda x$$
$$2 = 2\lambda y$$
$$3 = 2\lambda z$$

Dividing the first two leads to $y = 2x$, dividing the first by the third leads to $z = 3x$. Substitute in $x^2 + y^2 + z^2 = 16$ to obtain $x = \pm 2\sqrt{\frac{2}{7}}$, $y = \pm 4\sqrt{\frac{2}{7}}$, and $z = \pm 6\sqrt{\frac{2}{7}}$.
The maximum is $28\sqrt{\frac{2}{7}}$ and the minimum is $-28\sqrt{\frac{2}{7}}$.

17. Let f denote the amount of fencing needed to enclose the pasture, x the side parallel to the river and y the sides perpendicular to the river. Then,
$$f(x,y) = x + 2y$$

The goal is to minimize this function subject to the constraint that the area $g(x,y) = xy = 3,200$. The partial derivatives are $f_x = 1$, $f_y = 2$, $g_x = y$, and $g_y = x$. Hence, the three Lagrange equations are
$$1 = \lambda y,\ 2 = \lambda x,\ xy = 3,200$$

From the first equation, $\lambda = \frac{1}{y}$. From the second equation $\lambda = \frac{2}{x}$. Setting the two expressions for λ equal to each other gives $\frac{1}{y} = \frac{2}{x}$ or $x = 2y$, and substituting this into the third equation yields $2y^2 = 3,200$, $y^2 = 1,600$, or $y = \pm 40$.
Only the positive value is meaningful in the context of this problem. Hence, $y = 40$, and (since $x = 2y$), $x = 80$. That is, to minimize the amount of fencing, the dimensions of the field should be 40 meters by 80 meters.

19. Let f denote the volume of the parcel. Then,
$$f(x,y) = x^2 y$$

The girth $4x$ plus the length y can be at most 108 inches. The goal is to maximize this function $f(x,y)$ subject to the constraint $g(x,y) = 4x + y = 108$.
The partial derivatives are $f_x = 2xy$, $f_y = x^2$, $g_x = 4$, and $g_y = 1$.
Hence, the three Lagrange equations are
$$2xy = 4\lambda,\ x^2 = \lambda,\ 4x + y = 108$$

From the first equation, $\lambda = \frac{xy}{2}$, which, combined with the second equation, gives $\frac{xy}{2} = x^2$ or $y = 2x$.
(Another solution is $x = 0$, which is impossible in the context of this problem.)
Substituting $y = 2x$ into the third equation gives $6x = 108$ or $x = 18$, and since $y = 2x$, the corresponding value of y is $y = 36$.
Hence, the largest volume is
$f(18, 36) = (18)^2(36) = 11,664$ cubic inches.

7.4. CONSTRAINED OPTIMIZATION: THE METHOD OF LAGRANGE MULTIPLIERS

21. The volume is $V = \pi r^2 h$, and the amount of material
$$M = 2\pi r^2 + 2\pi r h$$
$$M_r = 4\pi r + 2\pi h$$
$$M_h = 2\pi r$$
$g = \pi r^2 h - 6.89\pi = 0$, $g_r = 2\pi rh$, $g_h = \pi r^2$.
Using Lagrange multipliers we get
$$4\pi r + 2\pi h = 2\pi\lambda rh$$
$$2\pi r = \pi\lambda r^2$$
This last equation gives $\lambda = \dfrac{2}{r}$ which leads to $2r = h$ when used in the previous equation. The constraint shows that $\pi r^2 h = 6.89\pi$, or $h = \dfrac{4 \times 6.89}{h^2}$ and $h = \sqrt[3]{27.56} = 3.02$ inches, $r = 1.51$ inches.

23.
$$f(x,y) = 50x^{1/2}y^{3/2}, \; g = x + y - 8 = 0$$
$$f_x = 25x^{-1/2}y^{3/2}, \; f_y = 75x^{1/2}y^{1/2}$$
$$g_x = 1, \; g_y = 1,$$
$$25x^{-1/2}y^{3/2} = \lambda, \; 75x^{1/2}y^{1/2} = \lambda$$
Equating λ leads to $25x^{-1/2}y^{3/2} = 75x^{1/2}y^{1/2}$ or $y = 3x$.
Substituting in the constraint equation leads to $4x = 8$ or $x = 2$ thousand dollars for development and $y = 6$ thousand dollars for promotion.

25. From problem 24, the three Lagrange equations were
$$20x^{-2/3}y^{2/3} = \lambda, \; 40x^{1/3}y^{-1/3} = \lambda$$
$$x + y = 120$$
from which it was determined that the maximal output occurs when $x = 40$ and $y = 80$.
Substituting these values in the first equation gives
$$\lambda = 20(40)^{-2/3}(80)^{2/3} = 31.75$$
which implies that the maximal output will increase by approximately 31.75 units if the available money is increased by one thousand dollars and allocated optimally.

27.
$$S(R,H) = 2\pi R^2 + 2\pi RH$$
$$g(R,H) = \pi R^2 H = V$$
$$S_R = 4\pi R + 2\pi H, \; S_H = 2\pi R$$
$$g_R = 2\pi RH, \; g_H = \pi R^2$$
We have $4\pi R + 2\pi H = 2\lambda\pi RH$
or $\lambda = \dfrac{2R + H}{RH}$
and $2\pi R = \lambda\pi R^2$ or $\lambda = \dfrac{2}{R}$
Equating these yields $2RH = 2R^2 + RH$
or $H = 2R$ ($R \neq 0$).

29.
We seek to maximize $s(d_0, d_i) = d_0 + d_i$
subject to $g(d_0, d_i) = \dfrac{1}{d_0} + \dfrac{1}{d_i} = \dfrac{1}{L}$
$s_{d_0} = 1$, $s_{d_i} = 1$, $g_{d_0} = -\dfrac{1}{d_0^2}$, $g_{d_i} = -\dfrac{1}{d_i^2}$.
This leads to $\lambda = -d_0^2$ and $\lambda = -d_i^2$, from which $d_0 = d_i$.
Substituting into g yields $d_0 = d_i = 2L$.

Therefore $s_{max} = 4L$

31. Let k be the cost per cm^2 of the bottom and sides. Then the cost of the top is $2k$ per cm^2 and the cost of the interior partitions is $\dfrac{2k}{3}$ per cm^2.
We seek to minimize the cost of the box, i.e.
$$C(x,y) = k(x^2 + 4xy) + 2kx^2 + \dfrac{2k}{3}(2xy)$$
subject to $g(x,y) = x^2 y = 800$.
We have $C_x = 6kx + \dfrac{16}{3}ky$.
$$C_y = \dfrac{16}{3}kx, \; g_x = 2xy, \; g_y = x^2$$
Then $6kx + \dfrac{16}{3}ky = 2\lambda xy$ and $\dfrac{16}{3}kx = \lambda x^2$
Solving each of these for λ and equating yields $x = \dfrac{8}{9}y$. Substituting into g reveals $\dfrac{9}{8}x^3 = 800$ from which
$$x = 8.9 \text{ and } y = 10.04$$

33.
$$E(x,y,z) = \frac{k^2}{8m}\left(\frac{1}{x^2} + \frac{1}{y^2} + \frac{1}{z^2}\right)$$
$$g(x,y) = xyz - V_0 = 0$$
$$E_x = \frac{k^2}{8m}\left(-\frac{2}{x^3}\right)$$
$$E_y = \frac{k^2}{8m}\left(-\frac{2}{y^3}\right)$$
$$E_z = \frac{k^2}{8m}\left(-\frac{2}{z^3}\right)$$
$$g_x = yz,\ g_y = xz,\ \text{and}\ g_z = xy$$

The three Lagrange equations are:
$$\frac{k^2}{8m}\left(-\frac{2}{x^3}\right) = \lambda yz$$
$$\frac{k^2}{8m}\left(-\frac{2}{y^3}\right) = \lambda xz$$
$$\frac{k^2}{8m}\left(-\frac{2}{z^3}\right) = \lambda xy$$

Dividing the first two leads to $y^2 = x^2$ or $y = x > 0$, dividing the first by the third leads to $z^2 = x^2$ or $z = x > 0$.
Substitute in $xyz = V_0$ to obtain $x = y = z = V_0^{1/3}$.

35.
$$f(x,y,z) = xyz$$
$$g(x,y) = 15xy + 12(2yz + xz) + 20xz - 8{,}000 = 0$$
$$f_x = yz,\ f_y = xz,\ \text{and}\ f_z = xy$$
$$g_x = 15y + 32z$$
$$g_y = 15x + 24z$$
$$g_z = 24y + 32x$$

The three Lagrange equations are:
$$yz = \lambda(15y + 32z)$$
$$xz = \lambda(15x + 24z)$$
$$xy = \lambda(24y + 32x)$$

Dividing the first two leads to $y = \frac{4x}{3}$,
dividing the first by the third leads to $z = \frac{5x}{8}$.
Substitute in $15xy + 24yz + 32xz = 8{,}000$ to obtain
$$x = \frac{20\sqrt{3}}{3},\ y = \frac{80\sqrt{3}}{9},\ z = \frac{25\sqrt{3}}{6}.$$

37. $\lambda = P_y = \dfrac{64{,}000}{49} - 1{,}000 = 306.122$
(for each \$1,000).
Since the change in this promotion/development is \$100, the corresponding change in profit is \$30.61. (Remember that the Lagrange multiplier is the change in maximum profit for a 1 (thousand) dollar change in the constraint.)

39.
$$U = 100x^{0.25}y^{0.75}$$
$$U_x = 25x^{-0.75}y^{0.75}$$
$$U_y = 75x^{0.25}y^{-0.25}$$
$$g = 2x + 5y - 280 = 0$$
$$g_x = 2\ \text{and}\ g_y = 5$$

(a) Solving
$$25x^{-0.75}y^{0.75} = 2\lambda$$
$$75x^{0.25}y^{-0.25} = 5\lambda$$
simultaneously leads to $\lambda = 15x^{0.25}y^{-0.25}$ and substituting back gives
$$30x^{0.25}y^{-0.25} = 25x^{-0.75}y^{0.75}$$
$6x = 5y$, and using the constraint equation $2x + 6x = 280$. Thus $x = 35$ and $y = 42$.

(b)
$$\lambda = 15 \times 35^{0.25} \times 42^{-0.25} = 14.33$$
which approximates the change in maximum utility due to an additional \$1.00 in available funds.

41. $\lambda \approx \Delta u$ if $\Delta k = \$1$. Since
$$U(x,y) = x^\alpha y^\beta,\ \alpha x^{\alpha-1}y^\beta = \lambda a$$

7.4. CONSTRAINED OPTIMIZATION: THE METHOD OF LAGRANGE MULTIPLIERS

and $k = ax + by$, it follows that

$$\begin{aligned}
\lambda &= \frac{\alpha x^{\alpha-1} y^\beta}{a} = \frac{\alpha y^\beta}{a x^{1-\alpha}} \\
&= \left(\frac{\alpha}{a}\right)\left(\frac{k\beta}{b}\right)^\beta \left(\frac{a}{k\alpha}\right)^{1-\alpha} \\
&= \left(\frac{\alpha}{a}\right)\left(\frac{k\beta}{b}\right)^\beta \left(\frac{a}{k\alpha}\right)^\beta \\
&= \left(\frac{\alpha}{a}\right)\left(\frac{k\beta a}{bk\alpha}\right)^\beta = \frac{\alpha \beta^\beta a^{\beta-1}}{a^\beta b^\beta} \\
&= \frac{a^{\beta-1}\beta^\beta}{a^{\beta-1}b^\beta} = \left(\frac{\alpha}{a}\right)^\alpha \left(\frac{\beta}{b}\right)^\beta
\end{aligned}$$

43. Let $Q(x,y)$ be the production level curve subject to $px + qy = k$. The three Lagrange equations then are $Q_x = \lambda p$, $Q_y = \lambda q$, and $px + qy - k = 0$. From the first two equations $\frac{Q_x}{p} = \frac{Q_y}{q}$.

45. We want to find extrema of $f(x,y) = x - y$ subject to the constraint
$g(x,y) = x^5 + x - 2 - y = 0$. $f_x = 1$, $f_y = -1$, $g_x = 5x^4 + 1$, $g_y = -1$. This leads to the three Lagrange equations

$$1 = \lambda(5x^4 + 1), \quad -1 = \lambda(-1)$$
$$x^5 + x - 2 - y = 0$$

From the second one we get $\lambda = 1$. From the first $x = 0$ and from the third it follows that $y = -2$.
Thus a possible extremum occurs at the point $(0, -2)$.
But $f(1,0) = 1$, $f(0,-2) = 2$, and $f(-1,-4) = 3$, which shows that $f(x,y) = x - y$ is not maximized nor minimized for $(0,-2)$. The function $y = x^5 + x - 2$ and several level curves are sketched.

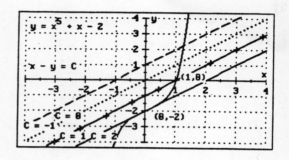

47. Find

$$\frac{\partial P}{\partial K} \text{ as well as } \frac{\partial P}{\partial L}$$

and

$$\frac{\partial C}{\partial K} \text{ as well as } \frac{\partial C}{\partial L}$$

The three Lagrange equations are

$$\frac{\partial P}{\partial K} = \lambda \frac{\partial C}{\partial K}, \quad \frac{\partial P}{\partial L} = \lambda \frac{\partial C}{\partial L}, \quad C(K,L) = A$$

Divide the first two equations to eliminate λ and presto!

$$\frac{\frac{\partial P}{\partial K}}{\frac{\partial P}{\partial L}} = \frac{\frac{\partial C}{\partial K}}{\frac{\partial C}{\partial L}} \text{ or } \frac{\frac{\partial P}{\partial K}}{\frac{\partial C}{\partial K}} = \frac{\frac{\partial P}{\partial L}}{\frac{\partial C}{\partial L}}$$

49.

$$\begin{aligned}
F(x,y) &= xe^{xy^2} + \frac{y}{x} + x\ln(x+y) \\
F_x &= xy^2 e^{xy^2} + e^{xy^2} - \frac{y}{x^2} + \frac{x}{x+y} + \ln(x+y) \\
F_y &= 2x^2 y e^{xy^2} + \frac{1}{x} + \frac{x}{x+y} \\
\frac{dy}{dx} &= -\frac{F_x}{F_y} \\
&= -\frac{xy^2 e^{xy^2} + e^{xy^2} - \frac{y}{x^2} + \frac{x}{x+y} + \ln(x+y)}{xe^{xy^2} + \frac{y}{x} + x\ln(x+y)}
\end{aligned}$$

51.

$$f(x,y) = \ln(x + 2y)$$
$$g(x,y) = xy + y - 5 = 0$$
$$f_x = \frac{1}{x+2y}, \; f_y = \frac{2}{x+2y}$$
$$g_x = y, \; g_y = x+1$$

The three Lagrange equations are

$$\frac{1}{x+2y} = \lambda y$$

$$\frac{2}{x+2y} = \lambda(x+1)$$

$$xy + y = 5$$

Dividing the first two yields $y = \frac{1}{2}(x+1)$, which is substituted in the third one to give $(x+1)^2 = 10$ or $x = 2.1623$, $y = 1.5812$. The graphing utility reveals $f(2.1623, 1.5812) = 1.6724$.

53.

$$f(x,y) = xe^{x^2-y}$$
$$g(x,y) = x^2 + 2y^2 - 1 = 0$$
$$f_x = (2x^2+1)e^{x^2-y}$$
$$f_y = -xe^{x^2-y}, \; g_x = 2x, \; g_y = 4y$$

The three Lagrange equations are

$$(2x^2+1)e^{x^2-y} = 2\lambda x$$

$$-xe^{x^2-y} = 4\lambda y, \; x^2 + 2y^2 = 1$$

Dividing and simplifying the first two equations leads to $x^2 = -\frac{2y}{4y+1}$. Now substitute in the constraint to get

$$8y^3 + 2y^2 - 6y - 1 = 0$$

From the graphing utility we get $y = 0.832$ and $y = -0.9184$ which produce $x^2 < 0$ and are rejected.
If $y = -0.1636$ then $x = 0.9468$ and $f(0.9468, -0.1636) = 2.7329$.

7.5 Double Integrals Over Rectangular Regions

1.

$$\int_0^1 \int_1^2 x^2 y \, dx \, dy$$
$$= \int_0^1 \left[\int_1^2 x^2 y \, dx\right] dy$$
$$= \int_0^1 \left[\frac{x^3}{3}y\Big|_1^2\right] dy$$
$$= \int_0^1 \left[\frac{8}{3}y - \frac{1}{3}y\right] dy$$
$$= \frac{7}{6}y^2\Big|_0^1 y^2 = \frac{7}{6}$$

3.

$$\int_0^{\ln 2} \int_{-1}^0 2xe^y \, dx \, dy$$
$$= \int_0^{\ln 2} \left[\int_{-1}^0 2xe^y \, dx\right] dy$$
$$= \int_0^{\ln 2} \left[x^2 e^y\Big|_{-1}^0\right] dy$$
$$= \int_0^{\ln 2} [-e^y] dy = -e^y\Big|_0^{\ln 2} = -1.$$

5.

$$\int_1^3 \int_0^1 \frac{2xy}{x^2+1} \, dx \, dy$$
$$= \int_1^3 \left[\int_0^1 \frac{2xy}{x^2+1} \, dx\right] dy$$
$$= \int_1^3 \left[y \ln(x^2+1)\Big|_0^1\right] dy$$
$$= \int_1^3 y \ln 2 \, dy = \ln 2 \left(\frac{1}{2}\right) y^2\Big|_1^3 = 4\ln 2$$

7.

$$\int_0^4 \int_{-1}^1 x^2 y \, dy \, dx$$
$$= \int_0^4 \left[\int_{-1}^1 x^2 y \, dy\right] dx$$
$$= \int_0^4 \left[\frac{y^2}{2}x^2\Big|_{-1}^1\right] dx = 0$$

7.5. DOUBLE INTEGRALS OVER RECTANGULAR REGIONS

9.
$$\int_2^3 \int_1^2 \frac{x+y}{xy} dy dx$$
$$= \int_2^3 \int_1^2 \left[\frac{1}{y} + \frac{1}{x}\right] dy dx$$
$$= \int_2^3 \left[\ln(y) + \frac{y}{x}\right]\Big|_1^2 dx$$
$$= (x \ln 2 + \ln x)\Big|_2^3 = \ln 2 + \ln \frac{3}{2} = \ln 3$$

11.
$$\int_0^1 \int_0^2 (6 - 2x - 2y) dy dx$$
$$= \int_0^1 [2(6-2x) - 4] dx$$
$$= [2(6x - x^2) - 4x]\Big|_0^1 = 6$$

13.
$$\int_1^2 \int_1^3 \frac{1}{xy} dy dx$$
$$= \int_1^2 \frac{\ln y}{x}\Big|_1^3 dx$$
$$= (\ln 3)(\ln 2)$$

15.
$$\int_0^1 \int_0^2 x e^{-y} dy dx$$
$$= \int_0^1 (-xe^{-y})\Big|_0^2 dx$$
$$= \int_0^1 x(1 - e^{-2}) dx$$
$$= \frac{e^2 - 1}{2e^2}$$

17. The area of the rectangular region is 15.

$$f_{ave} = \frac{1}{15} \int_{-2}^3 \int_{-1}^2 xy(x - 2y) dy dx$$
$$= \frac{1}{15} \int_{-2}^3 \left(\frac{x^2 y^2}{2} - \frac{2xy^3}{3}\right)\Big|_{-1}^2 dx$$
$$= \frac{1}{15} \int_{-2}^3 \left(2x^2 - \frac{16x}{3} - \frac{x^2}{2} - \frac{2x}{3}\right) dx$$
$$= \frac{1}{15} \int_{-2}^3 (1.5 x^2 - 6x) dx$$
$$= \frac{1}{15} \left(\frac{x^3}{2} - 3x^2\right)\Big|_{-2}^3$$
$$= 0.1667$$

19. The area of the rectangular region is 2.

$$f_{ave} = \frac{1}{2} \int_0^1 \int_0^2 x e^{x^2 y} dx dy$$
$$= \frac{1}{4} \int_0^1 e^{x^2 y}\Big|_0^2 dy$$
$$= \frac{1}{4} \int_0^1 (e^{4y} - 1) dy$$
$$= \frac{1}{4} \left(\frac{e^{4y}}{4} - y\right)\Big|_0^1$$
$$= \frac{e^4 - 5}{16} = 3.1$$

21.
$$\int_2^5 \int_1^3 \frac{1}{y^2} \ln(xy) \frac{dx}{x} dy$$
$$= \int_2^5 \frac{1}{y^2} \frac{\ln^2(xy)}{2}\Big|_1^3 dy$$
$$= \int_2^5 \frac{1}{2y^2} (\ln^2 3y - \ln^2 y) dy$$
$$= \frac{1}{2} \int_2^5 \frac{1}{y^2} (\ln 3y^2)(\ln 3) dy$$
$$= \frac{\ln 3}{2} \int_2^5 \left[\frac{\ln 3}{y^2} + 2 \frac{\ln y}{y^2}\right] dy$$
$$= \frac{3 \ln^2 3}{20} + \ln 3 \left[-\frac{\ln y}{y}\Big|_2^5 + \int_2^5 y^{-2} dy\right] = 0.5377$$

Note:

$$\ln^2 3y - \ln^2 y = (\ln 3y + \ln y)(\ln 3y - \ln y)$$
$$= (\ln 3y^2)(\ln 3)$$

For $I_1 = \int \ln y \frac{dy}{y^2}$, by parts

$$f = \ln y, \quad f' = \frac{dy}{y}, \quad g = y^{-2} dy, \quad G = -\frac{1}{y}$$

$$I_1 = -\frac{\ln y}{y} + \int y^{-2}dy$$
$$= -\frac{\ln y}{y} - \frac{1}{y} + C$$

23.

$$2\int_1^2 \int_0^1 e^{-2x}e^{-y}dxdy$$
$$= \int_1^2 (-e^{-2x})\Big|_0^1 e^{-y}dy$$
$$= (1-e^{-2})(e^{-1}-e^{-2}) = 0.201$$

25.

$$\int_0^1 \int_0^1 x^2 e^{x^2 y}dydx$$
$$= \int_0^1 x^{x^2 y}\Big|_0^1 dx$$
$$= \int_0^1 (e^{x^2}-1)dx = 1.463 - 1 = 0.463$$

The last integral above was evaluated with a graphing utility.

27.

$$Q_{av} = \frac{1}{35}\int_0^7 \int_0^5 (2x^3 + 3x^2y + y^3)dxdy$$
$$= \frac{1}{35}\int_0^7 (0.5x^4 + x^3y + xy^3)\Big|_0^5 dy$$
$$= \frac{1}{7}(0.5\times 125y + 0.5\times 25y^2 + 0.25y^4)\Big|_0^7$$
$$= \frac{943}{4} = 235.75$$

29.

$$P(x,y) = \int_{70}^{89}\int_{100}^{125} [(x-30)(70+5x-4y) + (y-40)(80-6x+7y)]\,dx\,dy$$
$$= \int_{70}^{89}\int_{100}^{125} [5x^2 + 7y^2 + 160x - 10xy - 80y - 5,300]\,dx\,dy$$
$$= \int_{70}^{89} [1.6667x^3 + 7xy^2 + 80x^2 - 5x^2y - 80xy - 5,300x]\Big|_{100}^{125}\,dy$$
$$= \int_{70}^{89} [1,909,218.75 + 175y^2 - 30,125y]\,dy$$
$$= [1,909,218.75y + 58.33y^3 - 15,062.5y^2]\Big|_{70}^{89}$$
$$= 1.1826(10^7)$$

The area is $(125-100)(89-70) = 475$.
The average profit is $\frac{1.1826(10^7)}{475} = 24,896.5$ or \$24,896,500.

31.

$$E(x,y) = \frac{90}{5,280}(2x+y^2) \text{ miles}$$
$$E_{av} = \frac{0.01705}{12}\int_0^3\int_0^4 (2x+y^2)dxdy$$
$$= 0.00142\int_0^3 (16+4y^2)dy$$
$$= 0.00142(16y + 1.333y^3)\Big|_0^3 = 630 \text{ ft.}$$

33. The probability is

$$P = \frac{1}{12}\int_0^3\int_0^3 e^{-x/4}e^{-y/3}dydx$$
$$= -\frac{1}{4}\int_0^3 e^{-x/4}e^{-y/3}\Big|_0^3 dx$$
$$= (e^{-1}-1)\int_0^3 \left(-\frac{1}{4}e^{-x/4}dx\right) = 0.3335.$$

35.

$$V(x,y) = 90e^{-x^2-y} \text{ thousand dollars}$$
$$V_{av} = \frac{90}{200}\int_{-5}^5\int_{-10}^{10} e^{-x^2}e^{-y}dxdy$$
$$= \frac{9}{20}\int_{-10}^{10} e^{-x^2}(-e^{-y})\Big|_{-5}^5 dx$$
$$= \frac{9}{20}(1.772)(e^5 - e^{-5}) \approx 118.37 \text{ thousand dollar}$$

The average property in this congressional district is \$118,370.

REVIEW PROBLEMS   217

37.
$$V = -\int_0^2 \int_0^3 x(-xe^{-xy}dy)dx$$
$$= \int_0^2 x(1-e^{-3x})dx$$
$$= x\left(x+\frac{1}{3}e^{-3x}\right)\Big|_0^2 - \int_0^2 \left(x+\frac{1}{3}e^{-3x}\right)dx$$
$$= \left[2\left(2+\frac{e^{-6}}{3}\right) - \left(\frac{x^2}{2} - \frac{e^{-3x}}{9}\right)\right]\Big|_0^2$$
$$= 1.8908 \text{ cubic units}$$

39.
$$\int_0^1 \int_0^2 xe^{-x-y}dydx$$
$$= \int_0^1 \int_0^2 xe^{-x}e^{-y}dydx$$
$$= \int_0^1 (-xe^{-x}e^{-y})\Big|_0^2 dx$$
$$= \int_0^1 xe^{-x}(-e^{-2}+1)dx$$
$$= (1-e^{-2})\int_0^1 xe^{-x}dx$$
$$= (1-e^{-2})\left(-xe^{-x}\Big|_0^1 + \int_0^1 e^{-x}dx\right)$$
$$= (1-e^{-2}) - [e^{-1}+(1-e^{-1})]$$
$$= (1-e^{-2})(1-2e^{-1}) = 0.2285$$

Review Problems

1. (a)
$$f(x,y) = 2x^3y + 3xy^2 + \frac{y}{x}$$
$$f_x = 6x^2y + 3y^2 - \frac{y}{x^2}$$
$$f_y = 2x^3 + 6xy + \frac{1}{x}$$

(b)
$$f(x,y) = (xy^2+1)^5$$
$$f_x = 5(xy^2+1)^4(y^2)$$
$$f_y = 5(xy^2+1)^4(2xy) = 10xy(xy^2+1)^4$$

(c)
$$f(x,y) = xye^{xy}$$
$$f_x = xye^{xy}(y) + e^{xy}(y) = y(xy+1)e^{xy}$$
$$f_y = xye^{xy}(x) + e^{xy}(x) = x(xy+1)e^{xy}$$

(d)
$$f(x,y) = \frac{x^2-y^2}{2x+y}$$
$$f_x = \frac{(2x+y)(2x) - (x^2-y^2)(2)}{(2x+y)^2}$$
$$= \frac{2(x^2+xy+y^2)}{(2x+y)^2}$$
$$f_y = \frac{(2x+y)(-2y) - (x^2-y^2)(1)}{(2x+y)^2}$$
$$= \frac{-4xy-y^2-x^2}{(2x+y)^2}$$

(e)
$$f(x,y) = \ln\frac{xy}{x+3y} = \ln x + \ln y - \ln(x+3y)$$
$$f_x = \frac{1}{x} - \frac{1}{x+3y}$$
$$f_y = \frac{1}{y} - \frac{3}{x+3y}$$

2. (a)
$$f(x,y) = x^2 + y^3 - 2xy^2$$
$$f_x = 2x - 2y^2, \ f_y = 3y^2 - 4xy.$$
Now $f_{xx} = 2$, $f_{yy} = 6y - 4x$,
and $f_{xy} = f_{yx} = -4y$

(b)
$$f(x,y) = e^{x^2+y^2}$$

$$\begin{aligned}
f_x &= 2xe^{x^2+y^2}, \; f_y = 2ye^{x^2+y^2} \\
f_{xx} &= 2(2x^2+1)e^{x^2+y^2} \\
f_{yy} &= 2(2y^2+1)e^{x^2+y^2} \\
f_{xy} &= f_{yx} = 2xe^{x^2+y^2}(2y) \\
&= 4xye^{x^2+y^2}
\end{aligned}$$

(c)
$$f(x,y) = x \ln y$$

$$f_x = \ln y \text{ and } f_y = \frac{x}{y}$$

$$f_{xx} = 0, \; f_{yy} = -\frac{x}{y^2}$$

$$f_{xy} = f_{yx} = \frac{1}{y}$$

3.
$$Q = 40K^{1/3}L^{1/2}$$

denotes the total output, where K denotes the capital investment and L the size of the labor force.
The marginal product of capital is

$$\frac{\partial Q}{\partial K} = \frac{40}{3}K^{-2/3}L^{1/2} = \frac{40L^{1/2}}{3K^{2/3}}$$

which is approximately the change ΔQ in output due to one (thousand dollar) unit increase in capital. When $K = 125$ (thousand) and $L = 900$,

$$\Delta Q \approx \frac{\partial Q}{\partial K} = \frac{40(900)^{1/2}}{3(125)^{2/3}} = 16 \text{ units.}$$

4. The marginal product of labor is the partial derivative $\frac{\partial Q}{\partial L}$. To say that this partial derivative increases as K increases is to say that its derivative with respect to K is positive, that is

$$\frac{\partial^2 Q}{\partial K \partial L} > 0$$

5. (a) $f(x,y) = x^2 - y$ when $f = 2$, so $x^2 - y = 2$ or $y = x^2 - 2$ which is a parabola opening upward with vertex at $(0, -2)$, and when $f = -2$, which is a parabola opening upward with vertex at $(0, 2)$.

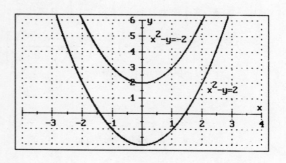

(b) If $f(x,y) = 6x + 2y$, then when $f = 0$, $6x + 2y = 0$ or $y = -3x$ which is a straight line through the origin with slope -3; when $f = 1$, $6x + 2y = 1$ or $y = -3x + \frac{1}{2}$ which is a straight line with y-intercept $\frac{1}{2}$ and slope -3; and when $f = 2$, $6x + 2y = 2$ or $y = -3x + 1$ which is a straight line with y-intercept 1 and slope -3

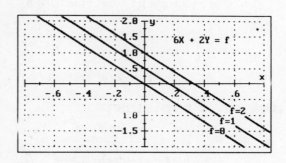

6. (a) $f(x,y) = x^2 - y^3$, $f_x = 2x$, $f_y = -3y^2$, and so

$$\frac{dy}{dx} = -\frac{f_x}{f_y} = \frac{2x}{3y^2}$$

REVIEW PROBLEMS

which is the slope of the tangent line at any point on the level curve.
When $x = 1$ and $f = 2$, $2 = 1 - y^3$ or $y = -1$. At the point $(x,y) = (1,-1)$, the slope is

$$m = \frac{dy}{dx} = \frac{2(1)}{3(-1)^2} = \frac{2}{3}.$$

(b) $f(x,y) = xe^y$, $f_x = e^y$, $f_y = xe^y$, and so

$$\frac{dy}{dx} = -\frac{f_x}{f_y} = -\frac{e^y}{xe^y} = -\frac{1}{x}$$

which is the slope of the tangent line at any point on the level curve.
When $x = 2$ the slope is

$$m = \frac{dy}{dx} = -\frac{1}{2}$$

7.
$$Q(x,y) = 60x^{1/3}y^{2/3}$$

where x denotes the number of skilled workers. The combination of x and y for which output will remain at the current level are the coordinates of the points (x, y) that lie on the constant-production curve $Q = k$, where k is the current level of output.
For any value of x, the slope of this constant-production curve is an approximation to the change in unskilled labor y that should be made to offset a one-unit increase in skilled labor x so that the level of output will remain constant. Thus,

$$\Delta Q = \text{change in unskilled labor}$$
$$\approx \frac{dQ}{dx} = -\frac{Q_x}{Q_y}$$
$$= -\frac{20x^{-2/3}y^{2/3}}{40x^{1/3}y^{-1/3}} = -\frac{y}{2x}$$

When $x = 10$ and $y = 40$,

$$\Delta Q \approx \frac{dQ}{dx} = -\frac{40}{2(10)} = -2$$

that is, the level of unskilled labor should be decreased by approximately 2 workers.

8. (a)

$$f(x,y) = x^3 + y^3 + 3x^2 - 18y^2 + 81y + 5$$
$$f_x = 3x^2 + 6x$$
$$f_y = 3y^2 - 36y + 81$$

To find the critical points, set $f_x = 0$ and $f_y = 0$.
Thus $3x^2 + 6x = 3x(x + 2) = 0$ or $x = 0$ and $x = -2$. Similarly,

$$3y^2 - 36y + 81 = 3(y - 3)(y - 9)$$

or $y = 3$ and $y = 9$. Hence, the critical points of f are $(0,3)$, $(0,9)$, $(-2,3)$, and $(-2,9)$.
Since $f_{xx} = 6x + 6$, $f_{yy} = 6y - 36$, and $f_{xy} = 0$,

$$D = f_{xx}f_{yy} - (f_{xy})^2 = 36(x+1)(y-6)$$

Since

$$D(0,3) = 36(1)(-3) = -108 < 0$$

f has a saddle point at $(0,3)$.
Since

$$D(0,9) = 36(1)(3) = 108 > 0$$

and $f_{xx}(0,9) = 6 > 0$, f has a relative minimum at $(0,9)$. Since

$$D(-2,3) = 36(-1)(-3) = 108 > 0$$

and $f_{xx}(-2,3) = -6 < 0$, f has a relative maximum at $(-2,3)$. Since

$$D(-2,9) = 36(-1)(3) = -108 < 0$$

f has a saddle point at $(-2,9)$.

(b)
$$f(x,y) = x^2 + y^3 + 6xy - 7x - 6y$$
$$f_x = 2x + 6y - 7$$
$$f_y = 3y^2 + 6x - 6$$

To find the critical points, set $f_x = 0$ and $f_y = 0$.

Thus $2x + 6y - 7 = 0$ and $3y^2 + 6x - 6 = 0$
or $2x + 6y - 7 = 0$ and $2x + y^2 - 2 = 0$.
Subtracting the two equations gives
$y^2 - 6y + 5 = 0$, $(y - 1)(y - 5) = 0$, or
$y = 1$ and $y = 5$.
When $y = 1$, the first equation gives
$2x + 6 - 7 = 0$ or $x = \dfrac{1}{2}$ and when $y = 5$,
the first equation gives $2x + 30 - 7 = 0$ or
$x = -\dfrac{23}{2}$.
Hence, the critical points of f are
$\left(\dfrac{1}{2}, 1\right)$, $\left(-\dfrac{23}{2}, 5\right)$.
Since $f_{xx} = 2$, $f_{yy} = 6y$, and $f_{xy} = 6$,

$$D = f_{xx}f_{yy} - (f_{xy})^2 = (2)(6y) - 36 = 12(y-3)$$

Since $D\left(\dfrac{1}{2}, 1\right) = 12(-2) = -24 < 0$

f has a saddle point at $\left(\dfrac{1}{2}, 1\right)$; and
since

$$D\left(-\dfrac{23}{2}, 5\right) = 12(2) = 24 > 0$$

and $f_{xx}\left(-\dfrac{23}{2}, 5\right) = 2 > 0$, f has a
relative minimum at $\left(-\dfrac{23}{2}, 5\right)$.

(c)
$$\begin{aligned} f(x,y) &= 3x^2y + 2xy^2 - 10xy - 8y^2 \\ f_x &= 6xy + 2y^2 - 10y = 0 \\ f_y &= 3x^2 + 4xy - 10x - 16y \\ &= 2y(3x + y - 5) = 0 \end{aligned}$$

at $O(0,0)$. With $y = 0$, $f_y = 0$ when
$x = \dfrac{10}{3}$.
From f_x, when $y = -3x + 5$, substitute
into $f_y = 0$ to get

$$9x^2 - 58x + 80 = 0$$

which has roots $x = \dfrac{40}{9}$ and $x = 2$.
The critical points are $O(0, 0)$,
$P\left(\dfrac{40}{9}, -\dfrac{25}{3}\right)$, $Q(2, -1)$, and $R\left(\dfrac{10}{3}, 0\right)$.

$f_{xx} = 6y$, $f_{xy} = 6x + 4y - 10$,
$f_{yy} = 4x - 16$

	(0,0)	(2,−1)	(4.44,−8.33)	(3.33,0)
f_{xx}	0	−6	50	0
f_{xy}	−10	−2	−16.6666	10
f_{yy}	−16	−8	1.7777	−2.6667
D	< 0	> 0	< 0	< 0

$P\left(\dfrac{40}{9}, -\dfrac{25}{3}\right)$, $R\left(\dfrac{10}{3}, 0\right)$, and $O(0,0)$
are saddle points while $Q(2, -1)$ is a
relative maximum.

(d)
$$f(x,y) = xe^{2x^2 + 5xy + 2y^2}$$

$$f_x = e^{2x^2 + 5xy + 2y^2}(4x^2 + 5xy + 1)$$

$$f_y = xe^{2x^2 + 5xy + 2y^2}(5x + 4y)$$

$f_y = 0$ when $x = 0$ but $f_x \neq 0$ then.
$f_y = 0$ when $y = -\dfrac{5x}{4}$ which is
substituted into $f_x = 0$ to give $x = \pm\dfrac{2}{3}$.

$$\begin{aligned} f_{xx} &= e^{2x^2+5xy+2y^2}[8x + 5y \\ &\quad + (4x^2 + 5xy + 1)(4x + 5y)] \\ f_{xy} &= e^{2x^2+5xy+2y^2}[5x \\ &\quad + (4x^2 + 5xy + 1)(5x + 4y)] \\ f_{yy} &= xe^{2x^2+5xy+2y^2}[4 + (5x + 4y)^2] \end{aligned}$$

	(0.666,−0.833)	(−0.666, 0.8333)
f_{xx}	0.7076	−0.7076
f_{xy}	2.0218	−2.0218
f_{yy}	1.6174	−1.6174
D	< 0	< 0

$P\left(-\dfrac{2}{3}, \dfrac{5}{6}\right)$ and $Q\left(\dfrac{2}{3}, -\dfrac{5}{6}\right)$ are saddle
points.

9. For
$$f(x,y) = x^2 + 2y^2 + 2x + 3$$
subject to the constraint that
$$g(x,y) = x^2 + y^2 = 4$$

the partial derivatives are $f_x = 2x + 2$, $f_y = 4y$, $g_x = 2x$, and $g_y = 2y$.
Hence, the three Lagrange equations are

$$2x + 2 = 2\lambda x, \quad 4y = 2\lambda y, \quad x^2 + y^2 = 4$$

From the first equation $\lambda = 1 + \dfrac{1}{x}$.
From the second equation, $\lambda = 2$ or $y = 0$.
If $y = 0$, the third equation gives $x = \pm 2$.
If $y \neq 0$, setting the two expressions for λ equal to each other yields $1 + \dfrac{1}{x} = 2$ or $x = 1$.
From the third equation, $y = \pm\sqrt{3}$.
Hence, the points at which the constrained extrema can occur are $(-2, 0)$, $(2, 0)$, $(1, \sqrt{3})$, and $(1, -\sqrt{3})$.
Since $f(-2, 0) = 3$, $f(2, 0) = 11$, and $f(1, -\sqrt{3}) = f(1, \sqrt{3}) = 12$, it follows that the constrained maximum is 12 which is attained at $(1, -\sqrt{3})$ as well as $(1, \sqrt{3})$, and the constrained minimum is 3 which is attained at $(-2, 0)$.

10. Let x denote the length of the rectangle, y the width, and $f(x, y)$ the corresponding area. Then

$$f(x, y) = xy$$

Since the rectangle is to have a fixed perimeter, the goal is to maximize $f(x, y)$ subject to the constraint that

$$g(x, y) = x + y = k$$

for some constant k.
The partial derivatives are $f_x = y$, $f_y = x$, $g_x = 1$, and $g_y = 1$.
The three Lagrange equations are

$$y = \lambda, \quad x = \lambda, \quad x + y = k$$

From the first two equations, $x = y$, which implies that the rectangle of greatest area is a square.

11. Let x denote the amount spent on development and y the amount spent on promotion in thousand dollars. The profit

$P(x, y) = $ (number of units sold)
(price per unit − cost per unit)
−total amount spent on
development and promotion

Number of units sold: $\dfrac{250y}{y+2} + \dfrac{100x}{x+5}$ The selling price is \$350 per unit and the cost \$150 per unit.
Hence, the price per unit minus the cost per unit is \$200 or $\dfrac{1}{5}$ of a thousand dollars.
Putting it all together,

$$P(x, y) = \frac{50y}{y+2} + \frac{20x}{x+5} - x - y$$

$$P_x = \frac{100}{(x+5)^2} - 1$$

$$P_y = \frac{100}{(y+2)^2} - 1$$

To find the critical points, set $P_x = 0$ and $P_y = 0$.
Thus $\dfrac{100}{(x+5)^2} - 1 = 0$
$(x+5)^2 = 100$, $x + 5 = 10$, or $x = 5$,

and $\dfrac{100}{(y+2)^2} - 1 = 0$

$(y+2)^2 = 100$, $y + 2 = 10$, or $y = 8$.
Hence, the critical point is $(5, 8)$.
Since $P_{xx} = -\dfrac{200}{(x+5)^3}$,

$$P_{yy} = -\frac{200}{(y+2)^3} \text{ and } P_{xy} = 0$$

$$D(5, 8) = P_{xx} P_{yy} - (P_{xy})^2 = \frac{40,000}{(10)^3 (10)^3}$$

$$P_{xx}(5, 8) = -\frac{200}{1,000} < 0$$

it follows that $P(x, y)$ has a relative maximum at $(5, 8)$. Assuming that the absolute maximum and the relative maximum are the same, it follows that to maximize the profit, \$5,000 should be spent on development and \$8,000 should be spent on promotion.

12. From problem 11, the profit function is

$$P(x, y) = \frac{50y}{y+2} + \frac{20x}{x+5} - x - y$$

The constraint is $g(x, y) = x + y = 11$ thousand dollars. Hence, the partial derivatives are

$$P_x = \frac{100}{(x+5)^2} - 1, \ P_y = \frac{100}{(y+2)^2} - 1$$

$g_x = 1$, and $g_y = 1$.
The three Lagrange equations are

$$\frac{100}{(x+5)^2} - 1 = \lambda, \ \frac{100}{(y+2)^2} - 1 = \lambda$$

$$x + y = 11$$

From the first two equations,

$$(x+5)^2 = (y+2)^2$$

or $y = x + 3$. From the third equation,
$x + (x + 3) = 11$ or $x = 4$.
Since $y = x + 3$, the corresponding value of y is $y = 7$.
Hence, to maximize profit, $4,000 should be spent on development and $7,000 should be spent on promotion.
Note that if $y + 2 = -(x + 5)$, $y = -x - 7$, $x + (-x - 7) = 11$, which is impossible.

13. The increase maximal in profit M resulting from an increase in available money by one thousand dollars is

$$\Delta M \approx \frac{dM}{dk} = \lambda$$

where, from problem 12,

$$\lambda = \frac{100}{(x+5)^2} - 1$$

Since the optional allocation of 11 thousand dollars is $x = 4$ and $y = 7$, the increase in maximal profit resulting from the decision to spend 12 thousand dollars is

$$\Delta M \approx \lambda = \frac{100}{(x+5)^2} - 1 = \frac{100}{9^2} - 1$$
$$= \frac{100}{81} - 1 = 0.235$$

thousand or $235.

14.
$$f(x,y) = \frac{12}{x} + \frac{18}{y} + xy$$

Suppose y is fixed, (say at $y = 1$), then f is very large when x is quite small.
f is also large when x is large, with a dip in the values of f between these extremes.
The same reasoning applies to y when x is fixed.

$$f_x = -\frac{12}{x^2} + y, \ f_y = -\frac{18}{y^2} + x$$

To find the critical points, set $f_x = 0$ and $f_y = 0$. Thus $y = \frac{12}{x^2} > 0$ and $x = \frac{18}{y^2} > 0$.
Substituting leads to

$$y = \frac{12}{x^2} = \frac{12}{\left(\frac{18}{y^2}\right)^2} = \frac{12y^4}{18^2}$$

or $y = 0$ (which is not in the domain of the function) and $12y^3 = 18^2$, $y^3 = 27$, $y = 3$.
The corresponding value for x is $x = \frac{18}{3^2} = 2$.
Hence, the critical points of f is $(2, 3)$.
Since

$$f_{xx} = \frac{24}{x^3}, \ f_{yy} = \frac{36}{y^3}$$

$f_{xy} = 1$,

$$D = f_{xx}f_{yy} - (f_{xy})^2 = \left(\frac{24}{x^3}\right)\left(\frac{36}{y^3}\right) - 1 > 0$$

for some (x, y).
Since $D(2, 3) = \frac{(24)(36)}{(2^3)(3^3)} - 1 > 0$
and $f_{xx}(2, 3) > 0$,
f has a relative minimum at $(2, 3)$.

15. (a)

$$\int_0^1 \int_{-2}^0 (2x + 3y)dy\,dx$$
$$= \int_0^1 \left[2xy + \frac{3y^2}{2}\right]\Big|_{-2}^0 dx$$
$$= \int_0^1 [0 - (-4x + 6)]dx = -4$$

(b)

$$\int_0^1 \int_0^2 e^{-x-y}dy\,dx$$

REVIEW PROBLEMS

$$= \int_0^1 \int_0^2 e^{-x} e^{-y} dy dx$$
$$= \int_0^1 (-e^{-x} e^{-y})\Big|_0^2 dx$$
$$= \int_0^1 (-e^{-x} e^{-2} + e^{-x}) dx$$
$$= (1 - e^{-2}) \int_0^1 e^{-x} dx$$
$$= (1 - e^{-2})(-e^{-x})\Big|_0^1$$
$$= (1 - e^{-2})(-e^{-1} + 1) = 0.5466$$

(c)
$$\int_0^1 \int_0^2 x\sqrt{1-y}\, dx\, dy$$
$$= \int_0^1 \frac{x^2}{2}\sqrt{1-y}\Big|_0^2 dy$$
$$= 2\int_0^1 (1-y)^{1/2} dy$$
$$= -\frac{4}{3}(1-y)^{3/2}\Big|_0^1 = \frac{4}{3}$$

(d)
$$\int_0^1 \int_{-1}^1 xe^{2y} dy dx$$
$$= \int_0^1 \left(\frac{1}{2}\right) xe^{2y}\Big|_{-1}^1 dx$$
$$= \int_0^1 \left(\frac{xe^2}{2} - \frac{xe^{-2}}{2}\right) dx$$
$$= \frac{e^2 - e^{-2}}{2} \int_0^1 x\, dx$$
$$= \frac{e^2 - e^{-2}}{4}$$

(e)
$$\int_0^2 \int_{-1}^1 \frac{6xy^2}{x^2+1} dy dx$$
$$= \int_0^2 \frac{2xy^3}{x^2+1}\Big|_{-1}^1 dx$$

$$= 2\int_0^2 \frac{2x}{x^2+1} dx$$
$$= 2\ln(x^2+1)\Big|_0^2 = 2\ln 5$$

(f)
$$I = \int_1^e \int_1^e (\ln x + \ln y) dy dx$$
$$= \int_1^e [y(\ln x) + (y\ln y - y)]\Big|_1^e dx$$
$$= \int_1^e [(e-1)\ln x + 1] dx$$
$$= [(e-1)(x\ln x - x) + x]\Big|_1^e = 3.4366$$

16. Describe R by $-1 \leq x \leq 2$ and $0 \leq y \leq 3$. Then, the area of R is

$$\int\!\!\int_R (6x^2 y) dA = \int_{-1}^2 \int_0^3 6x^2 y\, dy\, dx$$
$$= \int_{-1}^2 3x^2 y^2\Big|_0^3 dx$$
$$= \int_{-1}^2 27x^2 dx$$
$$= 9x^3\Big|_{-1}^2 = 9(8+1) = 81$$

17.
$$\int\!\!\int_R (x+2y) dA$$
$$= \int_0^1 \int_{-2}^2 (x+2y) dy dx$$
$$= \int_0^1 (xy + y^2)\Big|_{-2}^2 dx$$
$$= \int_0^1 4x\, dx = 2x^2\Big|_0^1 = 2$$

18.
$$V = \int_0^2 \int_0^3 2xy\, dy\, dx$$
$$= \int_0^2 xy^2\Big|_0^3 dx$$
$$= 9\int_0^2 x\, dx = 9\frac{x^2}{2}\Big|_0^2 = 18$$

19.
$$V = \int_1^2 \int_2^3 xe^{-y}\,dy\,dx$$
$$= \int_1^2 (-xe^{-y})\Big|_2^3 dx$$
$$= \int_1^2 (x)(e^{-2} - e^{-3})\,dx$$
$$= (e^{-2} - e^{-3})\frac{3}{2} = 0.1283$$

20.
$$\int_3^5 \int_{-1}^2 xy^2\,dx\,dy$$
$$= \int_3^5 \frac{x^2 y^2}{2}\Big|_{-1}^2 dy$$
$$= \int_3^5 \frac{3y^2}{2}\,dy$$
$$= \frac{y^3}{2}\Big|_3^5 = 0.5(125 - 27) = 49$$

The average value is $\dfrac{49}{6}$

21.
$$f(x,y) = 6e^{-2x}e^{-3y}$$

$P(0 \leq x \leq 1 \text{ and } 0 \leq y \leq 2)$
$$= \int_0^2 \int_0^1 6e^{-2x}e^{-3y}\,dx\,dy$$
$$= \int_0^2 (-3e^{-2x}e^{-3y})\Big|_0^1 dy$$
$$= \int_0^2 (-3e^{-2}e^{-3y} + 3e^{-3y})\,dy$$
$$= 3(1 - e^{-2}) \int_0^2 e^{-3y}\,dy$$
$$= -(1 - e^{-2})e^{-3y}\Big|_0^2 = 0.8625.$$

22. The probability is
$$P = \frac{1}{8} \int_0^5 \int_0^5 e^{-x/4}e^{-y/2}\,dy\,dx$$
$$= -\frac{1}{4} \int_0^5 e^{-x/4}e^{-y/2}\Big|_0^5 dx$$
$$= (1 - e^{-5/2})\frac{1}{4} \int_0^5 e^{-x/4}\,dx$$
$$= -(1 - e^{-5/2})e^{-x/4}\Big|_0^5$$
$$= (1 - e^{-5/2})(1 - e^{-5/4}) = 0.6549$$

23. Let x, y, and z denote the three numbers. Their sum is $x + y + z = 20$, so $z = 20 - x - y$ and the product is
$$P = xy(20 - x - y) = 20xy - x^2 y - xy^2$$
$$P_x = 20y - 2xy - y^2 = y(20 - 2x - y)$$
$$P_y = 20x - x^2 - 2xy = x(20 - x - 2y)$$

Both vanish when $x = 0$, $y = 0$, leading to a minimum, namely $P = 0$. Thus we'll consider $xy \neq 0$.
$2x + y - 20 = 0$ and $x + 2y - 20 = 0$.
$2x + y - 20 - 2(x + 2y - 20) =$
$2x + y - 20 - 2x - 4y + 40 = -3y + 20 = 0$,
or $y = \dfrac{20}{3}$.

Now $x = 20 - 2\dfrac{20}{3} = \dfrac{20}{3}$. $P_{xx} = -2y$, $P_{yy} = -2x$, $P_{xy} = 20 - 2x - 2y$, and so

$$D\left(\frac{20}{3}, \frac{20}{3}\right)$$
$$= (-2)\frac{20}{3}(-2)\frac{20}{3} - \left(20 - 2\frac{20}{3} - \frac{40}{3}\right)^2 > 0$$

Since $P_{xx}\left(\dfrac{20}{3}, \dfrac{20}{3}\right) < 0$, $\left(\dfrac{20}{3}, \dfrac{20}{3}, \dfrac{20}{3}\right)$ is a relative maximum.

24. With $z = 60 - 2x - 3y$, the sum of the squares of the three numbers is
$$S = x^2 + y^2 + (60 - 2x - 3y)^2$$

Thus $S_x = 2x + 2(60 - 2x - 3y)(-2)$
$= 2(5x + 6y - 120) = 0$
$S_y = 2y + 2(60 - 2x - 3y)(-3)$
$= 4(5y + 3x - 90) = 0$

These equations lead to
$15x + 18y - 15x - 25y = 360 - 450$, $7y = 90$,

or $y = \dfrac{90}{7}$. Thus $3x = 90 - \dfrac{450}{7}$,
$x = 30 - \dfrac{150}{7} = \dfrac{60}{7}$, and
$z = 60 - \dfrac{120}{7} - \dfrac{270}{7} = \dfrac{30}{7}$. With $S_{xx} = 10$, $S_{yy} = 20$, and $S_{xy} = 12$,

$$D = (10)(20) - (12)^2 = 200 - 144 > 0$$

which signifies that $P\left(\dfrac{60}{7}, \dfrac{90}{7}, \dfrac{30}{7}\right)$ is a minimum since $S_{xx} > 0$.

25. $y^2 - z^2 = 10$ and

$$D = d^2 = (\sqrt{x^2 + y^2 + z^2})^2 = x^2 + y^2 + z^2$$

Substituting for $y^2 = z^2 + 10$ in the equation for the square of the distance yields

$$D = x^2 + 10 + z^2 + z^2 = x^2 + 2z^2 + 10$$

$D_x = 2x = 0$ when $x = 0$ and $D_z = 0$ when $z = 0$. It follows that $y = \pm\sqrt{10}$.
The critical points are $(0, \pm\sqrt{10})$.
Since $D_{xx} = 2$, $D_{xz} = 0$, and $D_{zz} = 4$,
$D = 8 - 0 > 0$ which indicates a minimum at $(0, \pm\sqrt{10})$ since $D_{xx} > 0$.

26. The sum $S(m, b)$ of the squares of the vertical distances from the four given points is

$$\begin{aligned}S(m,b) &= (m+b-1)^2 + (m+b-2)^2 \\ &\quad + (3m+b-2)^2 + (4m+b-3)^2\end{aligned}$$

To minimize $S(m, b)$, set the partial derivatives

$$\dfrac{\partial S}{\partial m} = 0 \text{ and } \dfrac{\partial S}{\partial b} = 0$$

namely

$$\begin{aligned}\dfrac{\partial S}{\partial m} &= 2(m+b-1) + 2(m+b-2) \\ &\quad + 2(3m+b-2)(3) + 2(4m+b-3)(4) \\ &= 2(27m + 9b - 21) = 0\end{aligned}$$

$$\begin{aligned}\dfrac{\partial S}{\partial b} &= 2(m+b-1) + 2(m+b-2) \\ &\quad + 2(3m+b-2) + 2(4m+b-3) \\ &= 18m + 8b - 16 = 0\end{aligned}$$

Solve the resulting simplified equations $9m + 3b = 7$ and $9m + 4b = 8$ to get $m = \dfrac{4}{9}$ and $b = 1$.
Hence, the equation of the least-squares line is

$$y = \dfrac{4x}{9} + 1$$

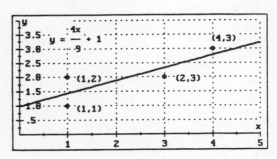

27. (a) Let x denote the monthly advertising expenditure and y the corresponding sales (both measured in units of $1,000). Then

x	3	4	7	9	10
y	78	86	138	145	156

x	y	xy	x^2
3	78	234	9
4	86	344	16
7	138	966	49
9	145	1305	81
10	156	1560	100
$\sum x = 33$	$\sum y = 603$	$\sum xy = 4,409$	$\sum x^2 = 255$

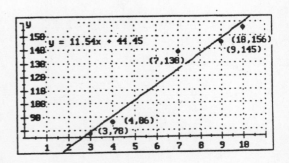

(b) From the formulas

$$m = \frac{n\sum xy - \sum x \sum y}{n\sum x^2 - (\sum x)^2}$$

$$b = \frac{\sum x^2 \sum y - \sum x \sum xy}{n\sum x^2 - (\sum x)^2}$$

with $n = 5$,

$$m = \frac{5(4,409) - 33(603)}{5(255) - (33)^2} = 11.54$$

$$b = \frac{255(603) - 33(4,409)}{5(255) - (33)^2} = 44.45$$

Hence, the equation of the least-squares line is

$$y = 11.54x + 44.45$$

(c) If the monthly advertising expenditure is $5,000, then $x = 5$ and $y = 11.54(5) + 44.45 = 102.15$. Thus the monthly sales will be approximately $102,150.

28.

$$T(x,y) = 2x^2 - xy + y^2 - 2y + 1$$

The area is $A = 4$, thus the average temperature is

T_{av}
$$= \frac{1}{4}\int_{-1}^{1}\int_{-1}^{1}(2x^2 - xy + y^2 - 2y + 1)dydx$$
$$= \frac{1}{4}\int_{-1}^{1}\left(2x^2y - \frac{xy^2}{2} + \frac{y^3}{3} - y^2 + y\right)\Big|_{-1}^{1}dx$$
$$= \frac{1}{4}\int_{-1}^{1}\left(4x^2 + \frac{2}{3} + 2\right)dx$$
$$= \frac{1}{12}(4x^3 + 8x)\Big|_{-1}^{1} = 2$$

So long, Arnold, and "Bon Voyage, mon ami".

29. (a) The problem is to minimize the total time $T(x,y)$, where

$$T = \frac{\sqrt{(1.2)^2 + x^2}}{2} + \frac{\sqrt{(2.5)^2 + y^2}}{4} + \frac{4.3 - (x+y)}{6}$$

$$\frac{\partial T}{\partial x} = \frac{1}{2}\left[\frac{1}{2}\frac{2x}{\sqrt{(1.2)^2 + x^2}}\right] - \frac{1}{6}$$

$$\frac{\partial T}{\partial y} = \frac{1}{4}\left[\frac{1}{2}\frac{2y}{\sqrt{(2.5)^2 + y^2}}\right] - \frac{1}{6}$$

$$\frac{\partial T}{\partial x} = \frac{\partial T}{\partial y} = 0 \text{ when}$$
$$\frac{1}{2}\frac{x}{\sqrt{(1.2)^2 + x^2}} = \frac{1}{6} \text{ and}$$
$$\frac{1}{4}\frac{y}{\sqrt{(2.5)^2 + y^2}} = \frac{1}{6} \text{ which leads to}$$
$x = 0.424$ and $y = 2.236$.
We must also check the "boundary" cases where the team travels straight "up" from A to B to C
and the one where they take the diagonal from A to D.
The possibilities are:
Case 1: A to B to C to D
Case 2: "Broken diagonal" A to M to N to D
Case 3: Diagonal A to D

Case	x	y	Time
1	0	0	$\frac{1.2}{2} + \frac{2.5}{4}$
			$+\frac{4.3}{6} = 1.942$
2	.424	2.236	$\frac{1.273}{2} + \frac{3.354}{4}$
			$+\frac{1.64}{6} = 1.748$
3	1.395	2.905	$\frac{1.84}{2} + \frac{3.833}{4}$
			$+\frac{0}{6} = 1.878$

The minimum time is the broken diagonal.

(b) For the second team, the time is

$$T = \frac{\sqrt{(1.2)^2 + x^2}}{1.7} + \frac{\sqrt{(2.5)^2 + y^2}}{3.5} + \frac{4.3 - (x+y)}{6}$$

$$\frac{\partial T}{\partial x} = \frac{1}{1.7}\left[\frac{x}{\sqrt{(1.2)^2 + x^2}}\right] - \frac{1}{6}$$

$$\frac{\partial T}{\partial y} = \frac{1}{3.5}\left[\frac{y}{\sqrt{(2.5)^2 + y^2}}\right] - \frac{1}{6}$$

$\frac{\partial T}{\partial x} = \frac{\partial T}{\partial y} = 0$ when $x = 0.3545$ and $y = 1.7955$.

Case	x	y	Time
1	0	0	$\frac{1.2}{1.7}+\frac{2.5}{3.5}$
			$+\frac{4.3}{6.3}=2.103$
2	.3545	1.7955	$\frac{1.251}{1.7}+\frac{3.078}{3.5}$
			$+\frac{2.15}{6.3}=1.957$
3	1.395	2.905	$\frac{1.84}{1.7}+\frac{3.833}{3.5}$
			$+\frac{0}{6.3}=2.177$

The first team wins by
$1.957 - 1.748 = 0.209$ hours (about 14 minutes).

(c) Writing Exercise —
Answers will vary.

30. With

$$h = \frac{kV^{1/3}}{D^{2/3}} = kV^{1/3}D^{-2/3}$$

$$\frac{\partial h}{\partial V} = \frac{1}{3}kV^{-2/3}D^{-2/3}$$

$$\frac{\partial h}{\partial D} = \frac{-2}{3}kV^{1/3}D^{-5/3}$$

The desired ratio is

$$\frac{\frac{\partial h}{\partial V}}{\frac{\partial h}{\partial D}} = \frac{\frac{1}{3}kV^{-2/3}D^{-2/3}}{\frac{-2}{3}kV^{1/3}D^{-5/3}} = -\frac{D}{2V}.$$

31. With $Q = x^a y^b$, $Q_x = ax^{a-1}y^b$
and $Q_y = bx^a y^{b-1}$.

$$xQ_x + yQ_y = x(ax^{a-1}y^b) + y(bx^a y^{b-1})$$
$$= (a+b)x^a y^b = (a+b)Q$$

If $a + b = 1$ then $xQ_x + yQ_y = (a+b)Q = Q$.

32.

$$S(W,t) = \left(\frac{W-B}{k}\right)t + \frac{W(W-B)}{k^2 g}(e^{-kgt/W}-1)$$

(a)

$$\frac{\partial S}{\partial W} = \frac{t}{k} + \frac{1}{k^2 g}(2W-B)(e^{-kgt/W}-1) + \frac{t(W-B)}{kW}e^{-kgt/W}$$

(change in depth WRT weight)

$$\frac{\partial S}{\partial t} = \frac{W-B}{k} - \frac{W-B}{k}e^{-kgt/W}$$

(change in depth WRT time)

If $\frac{\partial S}{\partial t}$ were 0 then $e^{-kgt/W} = 1$
but $-\frac{kgt}{W} \neq 0$. Therefore $\frac{\partial S}{\partial t} \neq 0$.

If $\frac{\partial S}{\partial W} = 0$ then a small change in W
would produce no change in S. This
seems unlikely.

(b) Using a graphing utility we find $\frac{\partial S}{\partial t} = 10$
at $t = 5.7$ sec. At this time the container
will be $S(2,417,5.7) = 28.46$ meters
deep.

(c) Writing exercise —
Answers will vary.